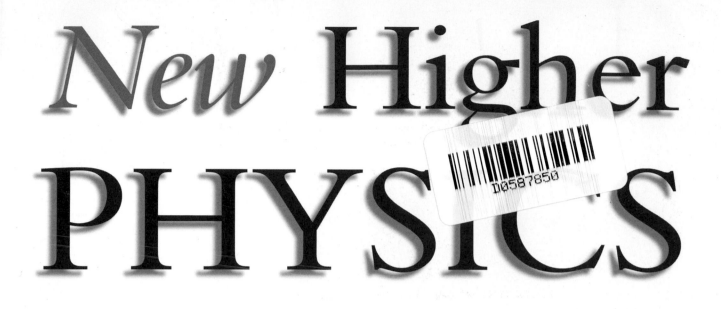

New Higher PHYSICS

Adrian Watt

Editor: Jim Page

HODDER
GIBSON
PART OF HACHETTE LIVRE UK

Although every effort has been made to ensure that website addresses are correct at time of going to press, Hodder Gibson cannot be held responsible for the content of any website mentioned in this book. It is sometimes possible to find a relocated web page by typing in the address of the home page for a website in the URL window of your browser.

Orders: please contact Bookpoint Ltd, 130 Milton Park, Abingdon, Oxon OX14 4SB. Telephone: (44) 01235 827720. Fax: (44) 01235 400454. Lines are open 9.00–5.00, Monday to Saturday, with a 24-hour message answering service. Visit our website at www.hoddereducation.co.uk. Hodder Gibson can be contacted direct on: Tel: 0141 848 1609; Fax: 0141 889 6315; email: hoddergibson@hodder.co.uk

© Adrian Watt 1999, 2002
First edition published in 1999
This edition published in 2002 by
Hodder Gibson, an imprint of Hodder Education, part of Hachette Livre UK.
2a Christie Street,
Paisley PA1 1NB

Impression number   10 9 8 7 6 5
Year   2010 2009 2008

Typeset in Times by J&L Composition, Filey, North Yorkshire
Printed in Spain for Hodder Gibson, 2A Christie Street, Paisley, PA1 1NB, Scotland

A catalogue record for this title is available from the British Library

ISBN-13: 978-0-340-84776-3

Acknowledgments

The publishers would like to thank the following individuals, institutions and companies for permission to reproduce photographs in this book. Every effort has been made to trace ownership of copyright. The publishers would be happy to make arrangements with any copyright holder whom it has not been possible to contact:

Action-Plus/Alain Aubard (44 right), DPPI (45), /R. Francis (164), /Tony Henshaw (60); /Mike Hewitt (65), /Glyn Kirk (53), /Peter Tarry (77), /Neil Tingle (38), Andrew Lambert (30, 196, 194 both, 185); Life File/Joan Blencowe (44 left), /Lionel Moss (71), /Jan Suttle (65); Meteosat (206 bottom); PA New (76, 128, 258 left); Science Photo Library (166, 167, 209, 210, 124, 168), /Chris Butler (168 left), /CERN (237), /Ray Ellis (223 right), /Fundamental Photos (90), /John Greim (183 top), /John Heseltine (170 top), /Laguna Design (168 right), /Maximilian Stock Ltd (257), /Peter Menzel (89), /Hank Morgan (170 bottom), /Novosti (259), /David Nunuk (206 top), /David Parker (223 left), /Alfred Pasieka (130 top), /Photo Researchers (258 right), /Andrew Syred (124), /Takeshi Takahara (41), /Charles D. Winters (231); Dave Stranock (105, 126 both, 130 bottom)

The publisher would also like to thank the Scottish Qualifications Authority for permission to reproduce copyright material from the Higher arrangements documents and from past examination papers. Any answers or comments are the sole responsibility of the author and have not been provided or approved by the Authority.

Contents

Waves and radiation

Revision guide

Index

Answers

Preface

Preface

'wisdom resteth in the heart of him that hath understanding'

Proverbs 14:33

The writer of this proverb undoubtedly had his mind on eternal things, nevertheless, the truth of his saying outlines what this book tries to achieve.

This book aims to improve the reader's knowledge and understanding of physics by covering the content required for the Scottish Qualification Authority's Higher Still courses in Physics.

To the pupil

New Higher Physics has been written with you, the student, in mind. The course content is divided up into small sections which are highly illustrated with appropriate diagrams. There are sections of **Worked examples** and **Consolidation questions** to help you test your knowledge as you progress through the book. The **Exam style questions** are based on content from a number of sections and are set at the same standard as those you will find in SQA course examinations. Chapter 10 provides **Revision guides** as well as a **Formula index** and some advice on **Avoiding common mistakes**.

To the teacher

New Higher Physics has been written to cover the contents statements detailed in the Higher Still arrangements documents for Higher Physics. We are indebted to those who have compiled these courses for the trouble that they have taken to build in a natural progression between the Intermediate 2 and Higher levels. In nearly all areas, the Intermediate 2 content statements detail material which would be met naturally along the way by those advancing from Standard Grade to Higher Physics.

The material in the original edition of *New Higher Physics* has been reconfigured *2nd edition* to cover only those contents statements detailed in the SQA's arrangements for Higher Physics. Material from the first edition, which was required exclusively at Intermediate 2 has been omitted from the second edition. (The Intermediate 2 material is now covered in some detail by 'Intermediate 2 Physics' by McCormick and Baillie.) Ideas and concepts taught in earlier courses, which are required as background knowledge for Higher Physics have been included in this 2nd edition as they are regarded as essential preparation for higher level. A quick guide as to how content is developed in each chapter can be seen in the revision outlines in Chapter 10. Some chapters also include ideas extending beyond the scope of what would normally be expected at higher level. These sections are clearly labelled as being '... *a little further*' and it is my hope that some students will appreciate this extension material.

The second major change is this second edition is the provision of answers to numerical problems. While every effort has been made to guarantee their accuracy, I apologise in advance for mistakes encountered and will welcome notification of errors, via the publisher.

Once again I am indebted to Sandra and our children who have tolerated much during the compilation of this second edition. Jim Page's depth of knowledge; his sustained encouragement and continual good humour has again been invaluable. Jim and I have both been encouraged by the way that the first edition of New Higher Physics has been received and trust that the second edition addresses those matters brought to our attention and will meet with the approval of teaching colleagues.

Once again it is my earnest desire that this book will simplify, clarify and extend the knowledge and understanding of physics and that in so doing, some may come to share my own enjoyment of this wonderful subject.

Adrian Watt
Ratho, Edinburgh

Mechanics & matter

PART 1

1.1 Average speed

Calculating average speeds

To compare the performance of different modes of transport, scientists have defined average speed as;

$$\text{Average speed} = \frac{\text{Total distance travelled}}{\text{Total journey time}}$$

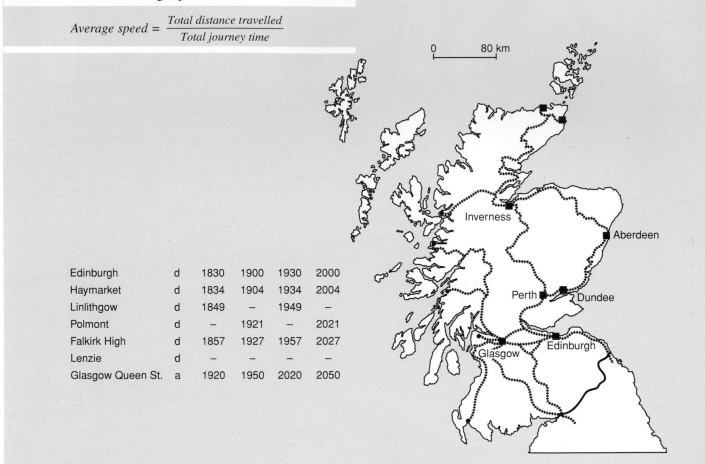

Edinburgh	d	1830	1900	1930	2000
Haymarket	d	1834	1904	1934	2004
Linlithgow	d	1849	–	1949	–
Polmont	d	–	1921	–	2021
Falkirk High	d	1857	1927	1957	2027
Lenzie	d	–	–	–	–
Glasgow Queen St.	a	1920	1950	2020	2050

Figure 1.1.1

Figure 1.1.1 shows part of a train timetable. The train travels the 75km between Edinburgh Waverley and Glasgow Queen St in 50 minutes (0.83 hours). Its average speed is;

$$\text{Average speed} = \frac{75\ km}{0.83\ h}$$

$$\text{Average speed} = 90\ km\,h^{-1}$$

For some parts of the journey the *actual* speed of the train would have been above 90 km h^{-1} while along other sections of track it would have been going slower or could have stopped. The value of 90 km h^{-1} is the *average* speed between the beginning and end of the journey.

In Glasgow the train has a wait of 30 minutes before making the return journey, also lasting 50 minutes. Therefore the *average* speed for the *entire* journey is now given by;

$$\text{Average speed} = \frac{(75 + 75)\ km}{(0.83 + 0.5 + 0.83)\ h}$$

$$\text{Average speed} = 69\ km\,h^{-1}$$

This also shows us that a train travelling at a constant speed of 69 km h^{-1} would take the same time to complete the entire journey.

Measuring an average speed

EXPERIMENT This involves accurately measuring the time taken to travel a known distance. Measuring distances with a ruler is sufficiently accurate if the distance is more than a few centimetres, but measuring short time intervals accurately by manually starting and stopping a stopwatch is more difficult. To overcome this problem you can use a system of light gates like that shown in figure 1.1.4. When the card on the trolley interrupts the light beam at A the electronic timing system starts. Timing continues until the card interrupts the light beam at B. The clock within the computer measures the time between start and stop signals very accurately. You could measure the distance between A and B and calculate the average speed.

Displaying speeds on a graph

Calculating the average speed is one way of summarising the train's journey. A more detailed way to follow the train's progress is to record its speed while it is travelling.

2

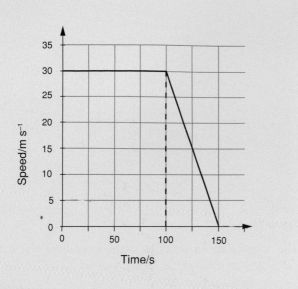

Figure 1.1.2

The graph in figure 1.1.2 shows the speed of the train as it travels along one section of the track. **We can calculate the length of this section of track by finding the area under this speed versus time graph.**

$$\text{Total distance travelled} = \text{Area of } \square + \text{Area of } \triangleright$$

$$\text{Total distance travelled} = 100 \times 30 + \frac{1}{2} \times (50 \times 30)$$

$$\text{Total distance travelled} = 3750 \text{ m}$$

$$\text{Average speed} = \frac{\text{Total distance travelled}}{\text{Total journey time}}$$

$$\text{Average speed} = \frac{3750 \text{ m}}{150 \text{ s}}$$

$$\text{Average speed} = 25 \text{ m s}^{-1}$$

Distance versus time graphs

The graph in figure 1.1.2 shows the speed of a train at any time during the 150 seconds it takes to cover a 3750 m section of track. The graph in figure 1.1.3 shows the distance-time graph for the *same* motion.

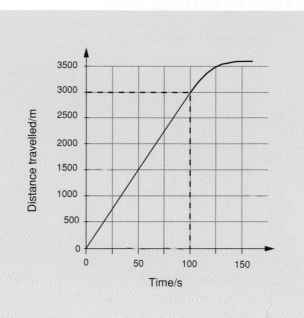

Figure 1.1.3

During each of the first 100 seconds the train gets 30 m further away from its starting point. This gives the steady slope on the distance-time graph. Between the 100th and 150th second the train continues to move away from the starting point, but as it is slowing down the distance covered in each successive second reduces. Therefore the gradient of the line decreases.

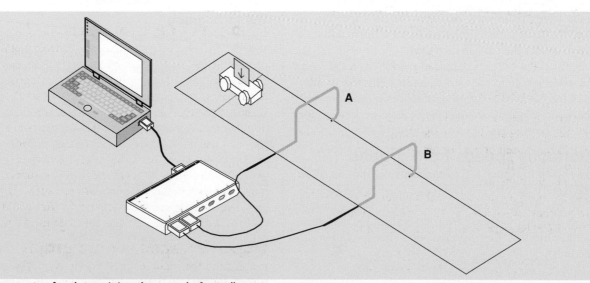

Figure 1.1.4 Apparatus for determining the speed of a trolley

1.2 Instantaneous speed

Introduction

Figure 1.2.1 shows a speed-time graph for a cyclist's journey from central Edinburgh to the Forth Road Bridge. The total distance of 16 km was covered in 45 minutes (2700 seconds), so the average speed for the entire journey was 5.9 m s^{-1}.

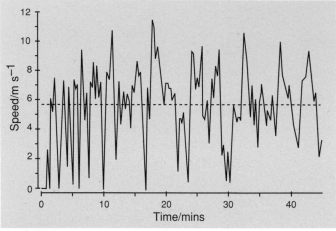

Figure 1.2.1

During the downhill sections the cyclist had a faster than average speed and on uphill sections the speed was less.
Calculating the average speed over a very short time interval gives a value for the speed at one particular instant. This is called the **instantaneous speed**. The graph is a record of the bicycle's instantaneous speed. We can see that for most of the journey the average speed and instantaneous speeds differ but there are a few occasions when the cyclist is actually travelling at his average speed of 5.9 m s^{-1}.

Measuring instantaneous speed

Over the past few years speed cameras have been introduced on many roads. As a car passes a speed camera a beam of electro-magnetic radiation is emitted from the camera and reflected back from the moving car. The changes to the beam caused by its reflection from the moving car, allow the car's speed to be determined. If the car is travelling too fast the camera will take a photograph so that the car can be traced.
In some areas, as the car passes special markings on the road, the camera takes a second photograph a few milliseconds after the first. As the distance between road markings and the time interval between the photographs being taken are known, the car's speed can be confirmed.

Instantaneous speeds from graphs

The graph shown in figure 1.2.2 represents the motion of an object which starts at rest, gets faster and reaches a maximum speed of 40 m s^{-1} after 8 seconds. The speed at any instant during the 8 seconds is found from the values on the graph axes. For this motion, the instantaneous speed is 20 m s^{-1} after 4 seconds.

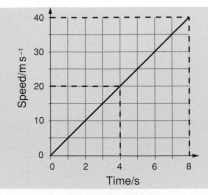

Figure 1.2.2

The average speed over the eight seconds of the motion is calculated by using the equation;

$$Average\ speed = \frac{Total\ distance\ travelled}{Total\ journey\ time}$$

$$Total\ distance\ travelled = Area\ of\ \triangle$$

$$Total\ distance\ travelled = \frac{1}{2} \times (8 \times 40) = 160\ m$$

$$Average\ speed = \frac{160\ m}{8\ s}$$

$$Average\ speed = 20\ m\,s^{-1}$$

It is not purely by chance that the average speed for the entire motion and the instantaneous speed after 4 seconds are identical. As you will see in later chapters, the graph in figure 1.2.2 showing a steady increase in speed, represents a **uniform acceleration. For a uniform acceleration the average speed equals the instantaneous speed at a time half way through the journey.** For this type of motion the average speed is given by;

$$Average\ speed = \frac{Initial\ speed + Final\ speed}{2}$$

The tachograph

Heavy goods vehicles are required by law to keep a record of each journey. The tachograph produces a special type of speed-time graph. The speed at any instant, the length of a journey and the duration of breaks taken by the driver can be found from the tachograph chart.

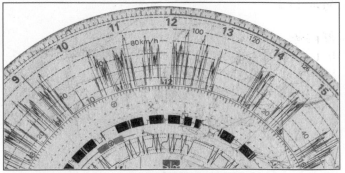

Figure 1.2.3 Section of a tachograph chart showing the vehicle's speed between 9 a.m. and 3 p.m.

Instantaneous speeds from distance-time graphs

Plotting a distance-time graph is the simplest way of recording the distance travelled by any moving object.

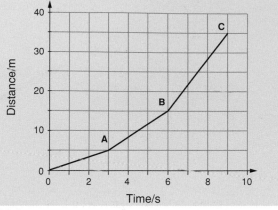

Figure 1.2.4

In the distance-time graph of figure 1.2.4 the constant gradient section for the first 3 seconds represents a constant speed. From A to B the slope of the graph is again constant. The section from B to C also shows a constant speed but since the slope is greater the object must have been moving faster for the final 3 seconds. The speeds from O to A, A to B and from B to C are 1.7 m s^{-1}, 3.3 m s^{-1} and 6.7 m s^{-1} respectively.

The slope of the line changes suddenly at points A and B so the speed must also have increased rapidly at these points.

The curve in figure 1.2.5 shows a distance-time graph where the speed is increasing steadily throughout the motion. In figure 1.2.6 the slope of the line reduces during the motion so the moving object is gradually slowing down.

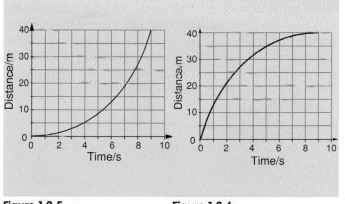

Figure 1.2.5   **Figure 1.2.6**

Measuring instantaneous speed

Instantaneous speed has already been defined as the average speed over a very short time interval. In laboratory experiments we use apparatus similar to that shown in figure 1.2.7 to measure the speed of a rolling trolley. As the front of the card placed on top of the trolley passes into the light beam a timer starts. Timing continues while the beam is interrupted but stops when the end of the card emerges from the light gate.

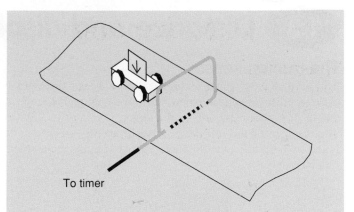

Figure 1.2.7 Measuring instantaneous speed

Knowing the length of the card and the time it takes to pass through the light beam, we can calculate the average speed during this short time interval. When the card is narrow we can assume that the instantaneous speed of the trolley as the midpoint of the interrupt card passes through the light beam is given by;

$$\text{Speed at light gate} = \frac{\text{Length of card}}{\text{Time that light beam was interrupted}}$$

EXPERIMENT It is possible to find the instantaneous speed of a trolley after it has travelled different distances along an inclined runway. (see figure 1.2.8). The computer measures the time during which the light beam is interrupted. Telling the computer the length of the card will allow it to calculate the speed automatically. Measure the distance from the start line to each of the lines marked A to E on the runway. Position the light gate above line A, release the trolley from the start position and measure the speed at A. Repeat this measurement a few times and calculate the mean value for the speed at A. Measure the speeds at each of the other marked positions (remembering to go back to the original start position each time) and plot a graph showing the trolley's speed at each point. Is the trolley accelerating?

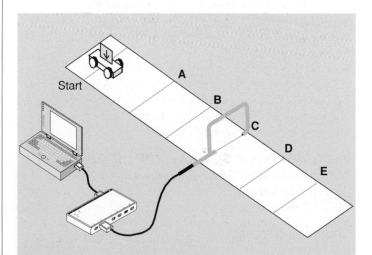

Figure 1.2.8 Measuring the variation of speed with distance

1.3 Distance and displacement

The motion sensor

This is a device which automatically measures distance. It emits a pulse of ultrasound which travels through the air at 330 m s^{-1} before rebounding off a target.

The distance of the reflecting object from the sensor can be found by measuring the time interval between the emitted and reflected pulses.

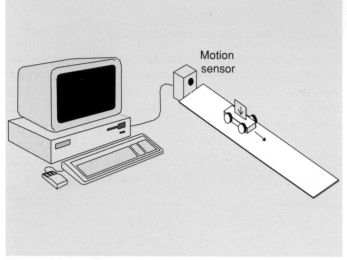

Figure 1.3.1 A motion sensor device

When a trolley is moving away from the sensor the distance can be measured at regular time intervals and a graph plotted. This type of measurement is best done by a computer where the rapid timings and calculations can be done quickly and results displayed on the monitor.

If the trolley gets faster as it moves away from the motion sensor the graph would look like that shown in figure 1.3.2.

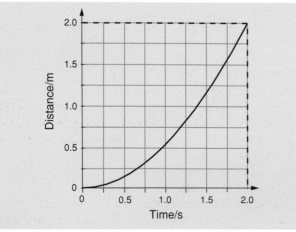

Figure 1.3.2

Changing direction

If the trolley used in figure 1.3.1 is placed 2.5 m away from the motion sensor and pushed up the inclined runway towards the motion sensor it slows down. The distance between the trolley and the motion sensor changes as shown in figure 1.3.3.

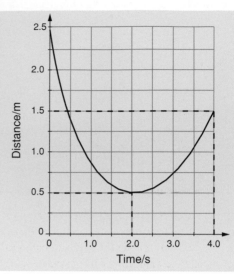

Figure 1.3.3

The trolley comes to rest momentarily after 2 seconds when it is 0.5 m away from the motion sensor. It then starts to roll back down the runway and after 4 seconds it is 1.5 m away from the motion sensor.

How far has the trolley travelled during the 4 seconds? The distance travelled must include the uphill and downhill sections.

Total distance travelled = 2 m + 1 m
Total distance travelled = 3 m

Stating that the trolley has travelled 3 m in the first 4 seconds of the motion allows its average speed to be calculated but gives little information about the position of the trolley at the end of the experiment.

More information is given by stating that, after 4 seconds, the trolley is 1 m further up the runway than at the beginning and is travelling back towards the start. This description **specifying the position of the trolley** is called a **displacement**. Like other quantities where the size and direction need to be stated, displacement is a **vector** quantity. (You will learn more about vectors in chapter 2.1).

Displacement

In an orienteering event the competitors begin at the start line and end at the finish line having visited certain intermediate markers. The competitors decide on their own route between the markers depending on the terrain and the conditions.

The displacement of each athlete at the finishing line is the same even though the distances they have run might differ.

Calculating displacements

In an orienteering event a competitor runs north east (bearing 045) for 3 km and then north west (bearing 315) for 1.25 km. In order to calculate her displacement after she has run the total distance of 4.25 km we can draw a **scale diagram** with the second vector starting where the first finishes, see figure 1.3.4. The **resultant** is then drawn from the start of the first to the end of the final vector. You can find the overall magnitude of the displacement from the scale diagram by measuring the length of the resultant and then using the scale to calculate the actual distance involved.

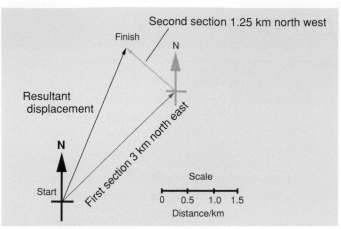

Figure 1.3.4 Scale diagram

In this example the turn from north east to north west results in a right angled triangle so the size of the displacement of the finish from the start can be found using Pythagoras' theorem. You must also state the direction of the displacement, which can be measured with a protractor. In this case the displacement should be stated as;

> *Displacement – 3.25 km at 67° north of east*
> *(or bearing 023)*

The mathematical sine and cosine rules can be used to calculate the size and directions of vectors. The rules can also be used if the triangle is a right angled one.

Displacement from graphs

When considering motion in a straight line the direction part of the displacement requirements can be stated using + or – signs.

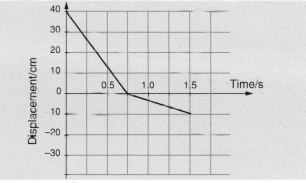

Figure 1.3.5

Figure 1.3.5 shows an object starting 40 cm away from the observation point, moving *towards* it and reaching the point after 0.75 seconds. Its motion continues on past the observation point and after a further 0.75 seconds it is 10 cm beyond the observation point. In this graph we have used the + sign to represent distances on one side of the observation point while the – sign shows distances on the other side. The object's displacement after 1.5 seconds could be stated as –10 cm. Merely stating the displacement as –10 cm does not tell us the direction in which the object is moving. However, the trend from the graph allows us to say that the object is moving away from the chosen observation point.

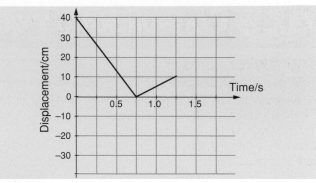

Figure 1.3.6

In the graph in figure 1.3.6 the displacement is always positive. This means that the moving object reaches, but never passes, the observation point. After the 1.25 seconds of journey the displacement is +10 cm from the observation point even though the distance travelled is 50 cm.

Q To take best advantage of the wind and tides a yacht follows a course of 60° east of north (bearing 060) at a speed of 5 km h^{-1} for 3 hours, and then for a further 2 hours it travels at 3 km h^{-1} on a course due north (000) to the finish line. What is the yacht's displacement from the start?

A

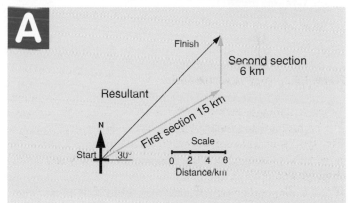

Figure 1.3.7

From the scale drawing, the finish is 19 km from the start in a direction 45° east of north, (bearing 045).

Motion sensor experiments

EXPERIMENT In this experiment you will use apparatus similar to figure 1.3.1 to 'follow' the motion of a trolley rolling towards and rebounding from a block clamped at the bottom of a runway. On the graph you should be able to identify the region where the trolley is:

1 travelling down the runway
2 in contact with the block
3 travelling back up the runway after the rebound.

You might also like to try giving the trolley a push up the runway and recording the subsequent motion of the trolley as it slows down on its way up the runway, stops momentarily at the highest point and begins to roll back down. Again you should be able to identify regions on the graph where the trolley is:

1 travelling up the runway
2 stopped momentarily at the top
3 travelling back down the runway.

Worked example 1

A student uses the apparatus shown in figure 1.4.1 to investigate the motion of a trolley as it travels down a slope.

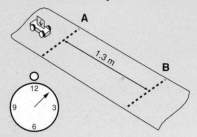

fig 1.4.1 Investigating the motion of a trolley

The results taken by the student are recorded in the table shown below.

Attempt	1	2	3	4	5	6
Time to travel from line A to line B (s)	1.4	1.5	1.4	1.5	1.3	1.6

Table 1.4.1

a) What is the distance travelled by the trolley when it moves from line A to line B?

b) Find:

(i) The average time the trolley takes to travel from A to B

(ii) The average speed of the trolley as it travels from A to B.

c) The student notices that the trolley is gradually getting faster as it moves along the runway. Explain why you would expect the instantaneous speed at B to be greater than $0.9 \, \text{m s}^{-1}$.

Answers and comments

In this question the time to travel from the line marked A to line B is measured by a student using a stopwatch.

a) By looking at the diagram we can see that lines A and B are marked as being 1.3 m apart.

b) (i) When the trolley moves from A to B it travels 1.3 m. The average time for the trolley to move from A to B is found by calculating the mean of the results.

$$Average \ travel \ time = \frac{1.4 + 1.5 + 1.4 + 1.5 + 1.3 + 1.6}{6}$$

$$Average \ travel \ time = 1.45 \ s$$

(ii) From the answers above:

$$Average \ speed = \frac{Total \ distance \ travelled}{Total \ journey \ time}$$

$$Average \ speed = \frac{1.3}{1.45}$$

$$Average \ speed = 0.9 \ m \, s^{-1}$$

c) The trolley is getting faster so its instantaneous speed at the end of the journey will be greater than the average speed for the entire journey. Since the average speed for the entire journey is $0.9 \, \text{m s}^{-1}$ the instantaneous speed at B should be greater.

Worked example 2

A runner in an orienteering competition starts by running 500 m due south (bearing 180) and then runs 1200 m west (bearing 270) to the finishing line.

a) Calculate the displacement of the runner when she reaches the finishing line, relative to her position at the start.

b) The course takes her a total time of 8 minutes. Calculate her average speed.

Answers and comments

a) Like all vector questions it is best to start by drawing a sketch.

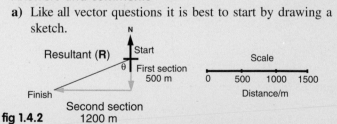

fig 1.4.2

The resultant can be found by Pythagoras' theorem.

$$R^2 = 1200^2 + 500^2$$
$$R = 1300 \ m$$

$$tan \ \theta = \frac{1200}{500} \quad \theta = 67°$$

The finishing line is 1300 m from the start on a bearing of 247 (180+67).

b) Average speed is found from;

$$Average \ speed = \frac{Total \ distance \ travelled}{Total \ journey \ time} = \frac{500 + 1200}{8 \times 60}$$

$$Average \ speed = 3.54 \ m \, s^{-1}$$

Questions

1 Which of the following statements is true?

  A Average speed is defined as the total distance travelled during any time interval.

  B A car travelling at an average speed of $80\ km\,h^{-1}$ will travel 160 km in 30 minutes

  C The actual speed at every instant during a train journey will be equal to the average speed.

  D A car travelling at an average speed of $60\ km\,h^{-1}$ will travel 300 km in 6 hours.

  E Average speed is defined as the total distance travelled divided by the total journey time.

2 Which of the following situations involves the highest average speed?

  A A runner who travels 800 m in 2½ minutes

  B A junior sprinter who runs 100 m in 12.5 seconds

  C A cyclist travelling 2 km in 5 minutes

  D A scooter travelling for 10 minutes at $30\ km\,h^{-1}$

  E A steam train covering 800 m in 2 minutes

3 Figure 1.4.3 shows the speed of a train as it travels along a certain section of track.

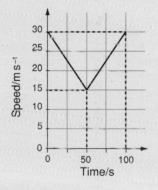

fig 1.4.3

  a) What time does the train take to travel along this section of track?

  b) What is the total length of this section of track?

  c) What is the average speed of the train during its journey along this section of track?

4 The glider on the air track shown in figure 1.4.4 is released from the start position and passes through light gates placed at different points along the track.

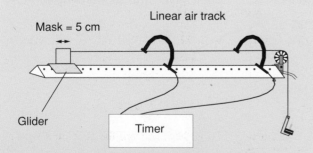

fig 1.4.4 Measuring the speed of a glider

The 5 cm long card attached to the glider takes 0.1 s to pass through the first light gate and 0.085 s to pass through the second light gate.

  a) Calculate the instantaneous speeds of the glider at each of the light gates.

  b) What is the increase in speed as the glider travels from one light gate to the other?

5 Figure 1.4.5 shows the speed-time graph for an object starting from rest and increasing its speed as it travels.

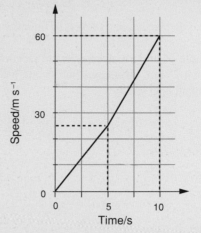

fig 1.4.5

  a) What is the speed of the object 5 seconds after starting?

  b) What is the average speed over the first 5 seconds?

  c) What distance does the object travel during the first 5 seconds?

  d) What is its average speed between times of 5 and 10 seconds?

6 Figure 1.4.6 shows a distance-time graph for a toy car being pushed uphill towards an observer.

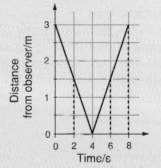

fig 1.4.6

  a) How far is the observer from the point where the car starts?

  b) How far does the car travel between 2 seconds and 6 seconds after release?

7 A robot in the production line at a car factory moves forward for 1.2 m before turning right and travelling 1.6 m before turning right again and travelling another 1.2 m

  a) How far has the robot travelled during this operation?

  b) What is the robot's final displacement from its starting position?

8 An unpowered glider is towed at constant altitude by a light plane which travels due north (000) for a distance of 500 m. After being released from the plane the glider flies, at the same height in a north easterly direction (045) for a further 3 km. What is the glider's displacement from its release position?

1.5 Velocity

Introduction

We use displacement to define average velocity;

$$Average\ velocity = \frac{Displacement}{Total\ time\ for\ journey}$$

Velocity, like displacement, is a **vector** quantity requiring both size and direction to be stated.

In the example of the orienteering event (section 1.3), the girl ran a total of 4.25 km. If this took her 1 hour, her average speed over the ground is given by;

$$Average\ speed = \frac{Total\ distance\ travelled}{Total\ time\ for\ journey}$$

$$Average\ speed = \frac{4.25}{1}$$

$$Average\ speed = 4.25\ km\,h^{-1}$$

However, her *average velocity* is given by;

$$Average\ velocity = \frac{Displacement}{Total\ time\ for\ journey}$$

$$Average\ velocity = \frac{3.25\ km\ at\ 67°\ north\ of\ east}{1\ h}$$

$$Average\ velocity = 3.25\ km\,h^{-1}\ at\ 67°\ north\ of\ east\ (023)$$

In this example the size of the speed and velocity are different.

Combining velocities

In situations where more than one velocity acts on an object the resultant can be found by drawing a scale diagram.

For example, a ball rolling across the deck of a ship sailing due east will be moving in an easterly direction with the ship. In addition, it will also be moving northwards due to its own motion.

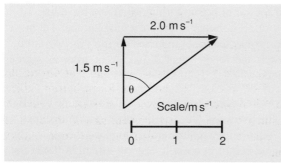

Figure 1.5.1 A ball rolling on the deck of a ship

The **resultant velocity** can be found by calculation or by drawing a scale diagram.

Figure 1.5.2 Scale diagram

In this case, we can state the combined velocity as 2.5 m s⁻¹ at an angle of 53° east of north (053).

Velocities from graphs

Mostly, we only have to consider motion in a straight line so there are only ever two possible directions of travel: 'to the left' and 'to the right', or in other situations it might be more helpful to use 'up' and 'down' or north and south. Alternatively, we can use the signs + and – to distinguish opposite directions of travel. The choice of which to make positive is purely arbitrary but needs to be made at the start of an example or question and *must be applied consistently throughout*.

The graph in figure 1.5.3 was obtained using a motion sensor to follow the motion of a trolley and calculate the velocity as it travelled down a slope.

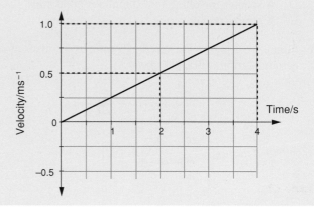

Figure 1.5.3

From this graph we are able to find the velocity at any instant during the 4 seconds of the journey. The velocity is always positive, showing that the trolley has continued to move in the same direction. In this example, movement down the runway away from the motion sensor has been chosen as the positive direction.

Changing direction

A trolley pushed up the slope will slow down as it approaches the top, stop momentarily and then move off back down the runway getting faster as it nears the bottom. The velocity-time graph for such a motion might be as shown in figure 1.5.4.

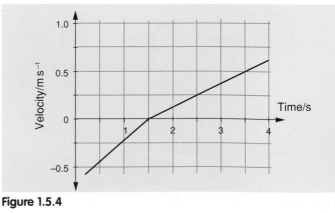

Figure 1.5.4

From the graph in figure 1.5.4 we can deduce the following:

1 At $t = 1.5$ s the velocity is 0 ms^{-1} so the trolley is momentarily stationary
2 At $t = 1.5$ s the velocity is about to change from a negative to a positive value
3 After $t = 1.5$ s the trolley moves off in the opposite direction, returning towards its starting point
4 From $t = 0$ to $t = 1.5$ s the trolley is slowing down and after $t = 1.5$ s it is speeding up.

Figure 1.5.5, like figure 1.5.4 represents a situation where a moving object slows down and then speeds up again. A *very* important difference, however, is that in figure 1.5.5 the sign of the velocity does not change. *There is no change in the direction of travel.*

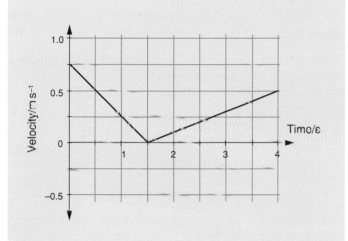

Figure 1.5.5

Figure 1.5.5 represents a situation where the moving object slows down and stops momentarily before moving off again *in the same direction*, getting faster as its goes.

Distance and displacement from velocity-time graphs

Calculating the area under a speed-time graph gives the distance travelled by a moving object. **Similarly, calculating the area under a velocity-time graph gives the displacement.**

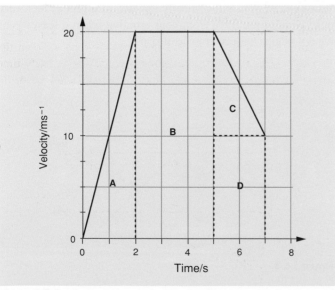

Figure 1.5.6

Figure 1.5.6 shows a velocity-time graph for a journey which has three distinctive parts;

1 an acceleration
2 a constant velocity
3 a deceleration.

There are no negative velocities on the graph therefore the motion is all in the same direction. By calculating the area under the graph you can show that the displacement after 7 seconds is 110 m in the direction of the initial motion.

The velocity-time graph shown in figure 1.5.7 would be what you would expect for the motion of a ball thrown vertically upwards.

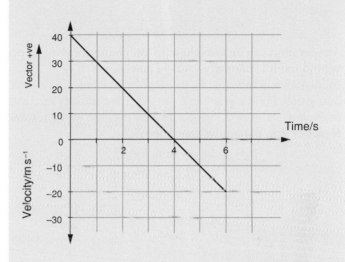

Figure 1.5.7

On its upward journey, *after leaving the hand*, the ball slows down, stops momentarily at the highest point before speeding up on the way back down.

After 4 seconds the ball is stationary for a brief instant at the top of its path. We can find this height by calculating the area under the line from time $t = 0$ s to $t = 4$ s.

$$Displacement\ at\ highest\ point = \frac{1}{2} \times 4 \times 40$$

$$Displacement\ at\ highest\ point = 80\ m\ (upwards)$$

Between time $t = 4$ s and $t = 6$ s the velocity is negative. This means that the ball is moving back towards the starting position. You can work out the distance fallen during this time from:

$$Distance\ fallen = \frac{1}{2} \times (6 - 4) \times 20$$

$$Distance\ fallen = 20\ m$$

By combining these two results we can say that during the 6 seconds of its motion, the ball rose 80 m before falling back 20 m. Therefore, after 6 seconds it is 60 m above its starting point. We state its displacement as +60 m, the sign of the displacement being consistent with our chosen convention.

1.6 Acceleration

Introduction

Motor car manufacturers, keen to show how quickly their latest sports car can speed up, publish the time the car takes to go from 0–60 mph (miles per hour). These figures enable the performance of different cars to be compared.

A more scientific way of defining how quickly a moving object can change its velocity is to define the acceleration.

$$\text{Acceleration} = \frac{\text{Change in velocity}}{\text{Time for the change}}$$

$$\text{Acceleration} = \frac{\text{Final velocity} - \text{initial velocity}}{\text{Time for the change}}$$

Acceleration can be thought of as the change in velocity per second and it is normally given in units of $m\,s^{-2}$.

Acceleration is indicated by the slope or gradient of a velocity-time graph. For the graph in figure 1.6.1 the velocity increases throughout. However, the acceleration over the first 3 seconds is greater than during the final 5 seconds.

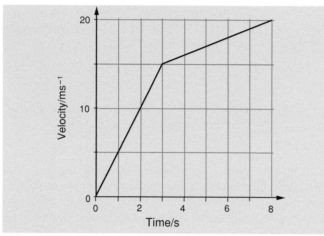

Figure 1.6.1

The velocity changes from 0 to 15 $m\,s^{-1}$ during the first 3 seconds therefore.

$$\text{Acceleration} = \frac{\text{Final velocity} - \text{initial velocity}}{\text{Time for the change}}$$

$$\text{Acceleration} = \frac{15 - 0}{3 - 0}$$

$$\text{Acceleration} = 5\ m\,s^{-2}$$

Similarly, the acceleration during the final 5 seconds of this motion has reduced to 1 $m\,s^{-2}$.

Uniform accelerations

If the velocity-time graph for a motion is a straight line we call this a **uniform (or constant) acceleration.**

The velocity-time graph of figure 1.6.2 shows two straight lines so it represents two different *constant* accelerations. The velocity values are always positive so the entire motion is in the same direction.

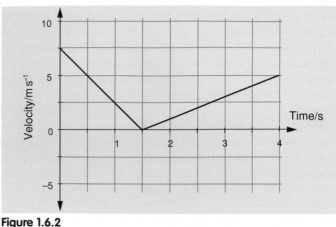

Figure 1.6.2

In the first part of figure 1.6.2 the magnitude of the velocity is decreasing. The moving object is slowing down. The acceleration can be found using;

$$\text{Acceleration} = \frac{\text{Change in velocity}}{\text{Time for the change}}$$

$$\text{Acceleration} = \frac{0 - 7.5}{1.5 - 0}$$

$$\text{Acceleration} = -5\ m\,s^{-2}$$

This acceleration value of -5 $m\,s^{-2}$ shows that the object's velocity is *decreasing* by 5 $m\,s^{-1}$ every second.

During the second part of the graph in figure 1.6.2 the magnitude of the velocity increases so the moving object is speeding up. The acceleration is $+2$ $m\,s^{-2}$.

Acceleration, like velocity and displacement is a vector quantity, but the symbols + and – can be used *either* to signify direction or to show speeding up and slowing down! It is good practice to use + and – to signify direction, and the words acceleration for getting faster and deceleration for slowing down.

When you throw a ball vertically upwards you accelerate it very quickly using the force from your arm, but when it leaves your hand it begins to decelerate as it rises.

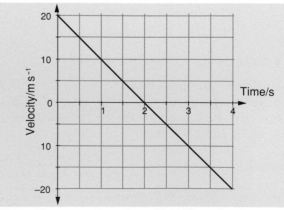

Figure 1.6.3

In the graph of figure 1.6.3 the ball leaves the hand at time $t = 0$ with an initial velocity of $+20$ m s^{-1} (20 m s^{-1} in the upwards direction). This reduces to zero as the ball decelerates on its upwards journey. The ball then speeds up as it falls, reaching a velocity of -20 m s^{-1} (20 m s^{-1} downwards) in just over 4 seconds. The velocity of the ball has both positive and negative values indicating a change in direction. However the straight line has a *single gradient* showing that the acceleration during both the upwards and downwards motions is the same.

Measuring accelerations

To measure the acceleration of an object, you need to determine a change in velocity and the time for this change to occur. To find the change in velocity an initial velocity and a final velocity are needed. In mathematical notation the final velocity, initial velocity and the time for the change to occur are given the symbols v, u, and t. So the acceleration can be defined as;

$$a = \frac{v - u}{t}$$

Final velocity = v
Initial velocity = u
Time for change to occur = t

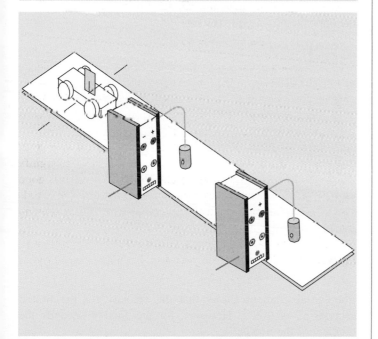

Figure 1.6.4 Measuring initial and final velocity

It is possible to measure acceleration using the apparatus shown in Figure 1.6.4. As the card on top of the rolling trolley interrupts each of the light beams, the velocity of the trolley at that point can be calculated from;

$$Velocity\ at\ light\ gate = \frac{Length\ of\ card}{Time\ that\ light\ beam\ was\ interrupted}$$

The timing system connected to the light gates must be capable of measuring the times for which each light gate is interrupted (t_1 and t_2) along with the time taken to travel between the light gates (t_3).

$$Initial\ velocity,\ u = \frac{Length\ of\ card}{t_1}$$

$$Final\ velocity,\ v = \frac{Length\ of\ card}{t_2}$$

$$Acceleration,\ a = \frac{v - u}{t_3}$$

A simpler way of measuring the acceleration uses a single light gate and double interrupt card as shown in figure 1.6.5. Careful thought will show that when this card passes through the light gate it produces the three time measurements needed to calculate the initial velocity, the final velocity and acceleration.
EXPERIMENT Using the apparatus shown in figure 1.6.5 it is possible to determine the acceleration of a trolley rolling down an inclined slope.

You could use the apparatus to investigate which of the following factors affect the acceleration;
1 the angle of the inclined slope
2 the mass of the accelerating trolley
3 the position from which the trolley is released.

Investigate each of these factors separately. You should measure the acceleration for a number of identical attempts and calculate a mean. You should also work out the range of the readings and the **uncertainty** in each reading, using:

$$Uncertainty = \frac{Range\ of\ readings}{Number\ of\ attempts}$$

By looking at the mean values and the uncertainties you can identify trends or confirm mathematical relationships.

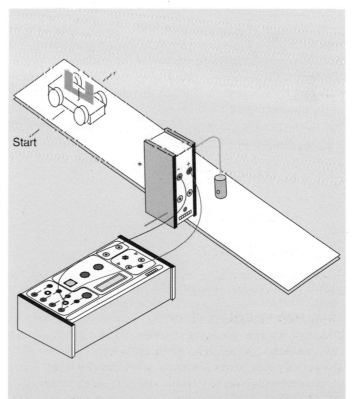

Start

Figure 1.6.5 Acceleration measurement with one light gate

1.7 Displacement-time graphs

Introduction

A graph is often used to display a sequence of information presented in a table. By considering the line drawn through the points on the graph we can see trends. We can also use the line to estimate values not given in the table, so a graph contains more information than is shown in a table.

Time (s)	0	1	2	3	4	5	6
Displacement (cm)	0	10	20	30	40	50	60

Table 1.7.1

Figure 1.7.1 shows this data presented as a graph.

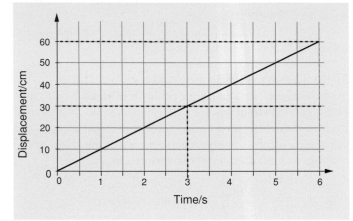

Figure 1.7.1

Any variation where the quantity on the 'y-axis' changes by equal amounts as the quantity on the 'x-axis' increases by equal amounts gives rise to a straight line graph with a steady slope. This is sometimes referred to as a graph with a **constant gradient**.

In displacement-time graphs the size of the gradient shows how quickly the displacement is changing as time passes. **The gradient of a displacement-time graph therefore gives the velocity of the moving object**. Since the graph of figure 1.7.1 is a straight line it has only one gradient and therefore represents a constant velocity. The size of the gradient can be calculated using:

$$Gradient = \frac{Change\ in\ 'y\text{-}axis'\ variable}{Change\ in\ 'x\text{-}axis'\ variable}$$

$$Gradient = \frac{60 - 30\ (cm)}{6 - 3\ (s)}$$

$$Gradient = 10\ cm\,s^{-1}$$
$$Velocity = 10\ cm\,s^{-1}$$

Figure 1.7.2 shows two straight lines. Each represents a constant velocity during a certain time interval.

The gradient of the line for the first 2 seconds shows a velocity of 2 m s^{-1} whereas the gradient for the next 4 seconds shows a constant velocity of 0.5 m s^{-1}. Once again these trends could be recognised if the information were presented in a table;

Time (s)	0	1	2	3	4	5	6
Displacement (m)	0	2	4	4.5	5	5.5	6

Table 1.7.2

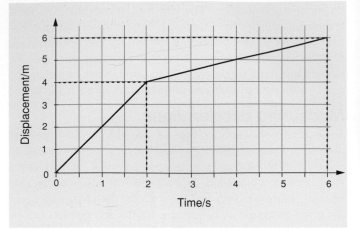

Figure 1.7.2

In each of the first 2 seconds the displacement increases by 2 metres whereas between times $t = 2$ s and time $t = 6$ s the displacement increases by only 0.5 metres every second.

In all the examples shown so far the objects move in a single direction. Figure 1.7.3 shows the displacement initially increasing but then reducing. The final part of the motion is in the opposite direction to the first.

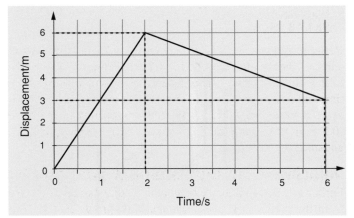

Figure 1.7.3

The initial gradient shows that the velocity for the first 2 seconds is +3 m s^{-1}. However, for the final 4 seconds the gradient of the line is;

$$Gradient = \frac{Change\ in\ 'y\text{-}axis'\ variable}{Change\ in\ 'x\text{-}axis'\ variable}$$

$$Gradient = \frac{3 - 6}{6 - 2}$$

$$Gradient = -0.75\ m\,s^{-1}$$
$$Velocity = -0.75\ m\,s^{-1}$$

For the final 4 seconds of its motion this object has a velocity of –0.75 m s^{-1}. This means that it is returning along its initial path with a speed of 0.75 m s^{-1}.

A change in the sign of the gradient indicates a change in direction of travel.

Acceleration from displacement-time graphs

In each of the examples so far we have considered only constant velocities. When a trolley speeds up as it moves away from a reference point the displacement-time graph would be as shown in figure 1.7.4.

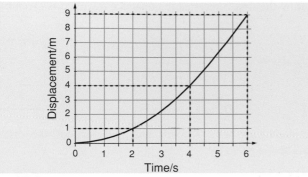

Figure 1.7.4

In the first 2 seconds the trolley moves 1 m. In the next 2 seconds the trolley moves a further 3 m while during the final 2 seconds the trolley moves 5 m.

Table 1.7.3 shows displacement-time values for the motion graphed in Figure 1.7.4.

Time (s)	0	1	2	3	4	5	6
Displacement (m)	0	0.25	1	2.25	4	6.25	9

Table 1.7.3

Between $t=1$ and $t=2$ the trolley moves 0.75 m, between $t=2$ and $t=3$ the distance travelled is 1.25 m. From $t=3$ to $t=4$ the trolley travels 1.75 m. The distance travelled increases with time but not by the same amount each second. However, the *extra* distance travelled in consecutive seconds is the same (0.5 m in this example). This type of variation is typical of a **constant acceleration**.

Velocities from displacement-time graphs

Earlier you saw that the velocity at any point can be calculated from the gradient of a displacement-time graph. For motions involving speeding up or slowing down the displacement-time graph is curved. Therefore to find the velocity at any time we need to consider the slope of tangents touching the curve at particular times.

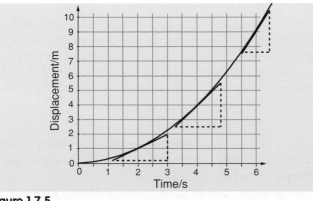

Figure 1.7.5

Figure 1.7.5 shows that the slope of the tangents is increasing with time indicating a speeding up. Figure 1.7.5 shows that the trolley is moving faster as the displacement increases. This motion is described as an acceleration away from a fixed observation point.

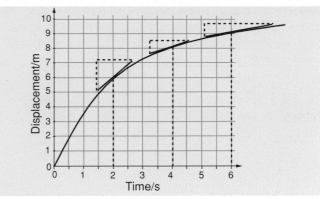

Figure 1.7.6

Figure 1.7.6 again shows an increasing displacement but the gradient of the displacement-time curve reduces showing that the object is slowing down.

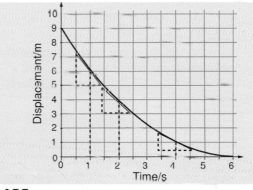

Figure 1.7.7

On the graph of figure 1.7.7 both the displacement and gradient decrease with time. Again this object is slowing down. The decreasing displacement shows that it is moving towards the observation point.

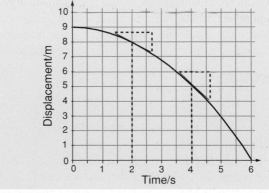

Figure 1.7.8

In the graph of figure 1.7.8 the displacement decreases but the gradient increases with time showing that the object is getting faster as it moves towards the observation point.

1.8 Velocity-time graphs

Introduction

The velocity-time graph of figure 1.8.1 represents a constant velocity of 7.5 m s^{-1} due north for 4 seconds.

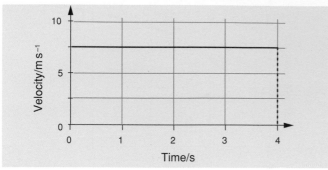

Figure 1.8.1

Since the motion is in only *one direction* the average speed and average velocity are numerically the same so we can say;

> Distance travelled = Average speed × time
> Distance travelled = 7.5 × 4
> Distance travelled = 30 m

If the graph shows the motion of a cyclist riding due north, his displacement after 4 seconds is 30 m due north of the start. The size of the displacement is equal to the distance travelled because the motion is in only one direction.

The area under the line in the graph of figure 1.8.1 can be found from:

> Area under line = Length × Height

But since the dimensions of the rectangle represent physical quantities, we can say;

> Area under line = Time × Velocity
> The displacement = Area under line.

For this simple example, where the velocity is constant throughout the motion, it is easy to see that the distance travelled is also given by the area under the velocity-time graph. More generally, we would say that distance travelled is found from the area under the line in a speed-time graph while displacement is calculated from the area under the line in a velocity-time graph. Even in situations where the magnitudes of a distance and displacement are equal, the displacement must be properly stated by including a direction term to fully specify the position.

Acceleration

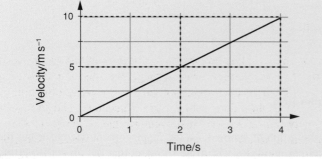

Figure 1.8.2

When a cyclist riding away from a fixed point speeds up, the graph for the motion is as shown in figure 1.8.2. We can find the displacement after certain times by calculating the area under the line.

During the first 2 seconds the distance travelled is given by;

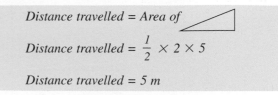

> Distance travelled = Area of
> Distance travelled = $\frac{1}{2}$ × 2 × 5
> Distance travelled = 5 m

Between times $t=2$ seconds and $t=4$ seconds the distance travelled is given by;

> Distance travelled = Area of   + Area of
> Distance travelled = $\frac{1}{2}$ × (4 − 2) × (10 − 5) + 2 × 5
> Distance travelled = 15 m

In these two examples the motion has only been in one direction, so the distance travelled is the same as the size of the displacement.

The graph of figure 1.8.3 shows a situation where an object, initially moving at 5 m s^{-1} to the left, slows down steadily and stops momentarily before moving off to the right.

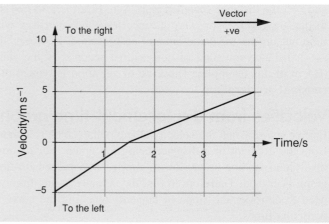

Figure 1.8.3

After 1.5 seconds the velocity is 0 m s^{-1} indicating that the object has stopped momentarily. For the second part of its journey, the object is moving to the right back along the initial path. We can calculate the distance travelled and find the displacement after 4 seconds by considering motion in each direction separately.

> Distance moved before stopping =
> Area under line from = 0 to t = 1.5 second;
>
> Distance moved before stopping = $\frac{1}{2}$ × 1.5 × 5
>
> Distance moved before stopping = 3.75 m

Similarly;

> Distance moved on the return journey =
>   Area under line from t = 1.5 to t = 4 seconds
>
> Distance moved on the return journey = $\frac{1}{2} \times (4 - 1.5) \times 5$
>
> Distance moved on the return journey = 6.25 m

From these calculations we can see that the object travelled 3.75 m to the left before coming to rest. It then travelled 6.25 m to the right. Therefore;

> The total distance travelled = 3.75 + 6.25 m
> Total distance travelled = 10 m.

However, to calculate the displacement after 4 seconds we need to take account of the change in direction. Since movement to the left is regarded as in the negative direction while movement to the right is in the positive direction;

> Displacement after 4 seconds = –3.75 + 6.25 m
> Displacement = + 2.5 m

Therefore after 4 seconds the object is 2.5 m to the right of its starting position.

The gradients of velocity-time graphs

In figure 1.8.2 the velocity changes by 5 m s^{-1} in the first 2 seconds and by a further 5 m s^{-1} in the next 2 seconds. Therefore, for the entire duration of this motion the change in velocity per second could be stated as 2.5 m s^{-1} every second or 2.5 m s^{-2}. As you know, the change in velocity per second is called the acceleration. Figure 1.8.2 represents a velocity-time graph characteristic of a constant (or uniform) acceleration.

Figure 1.8.4 shows the graph obtained for a cyclist accelerating away from traffic lights.

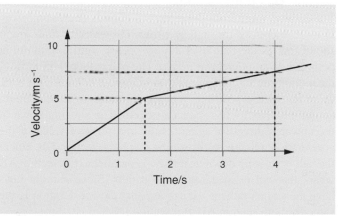

Figure 1.8.4

This velocity-time graph has two straight sections showing that there were two different accelerations, perhaps caused by a gear change. For the first 1.5 seconds there was one constant acceleration and for the next 2.5 seconds there was a different, smaller but still constant acceleration. You can calculate the values of the acceleration using;

$$Acceleration = \frac{Change\ in\ velocity}{Time\ for\ the\ change}$$

For the acceleration from $t=0$ to $t=1.5$ seconds;

$$a = \frac{v-u}{t}$$

$$a = \frac{5-0}{1.5}$$

$$a = 3.3\ m\,s^{-2}$$

For the acceleration from $t=1.5$ to $t=4$ seconds;

$$a = \frac{v-u}{t}$$

$$a = \frac{7.5-5}{2.5}$$

$$a = 1\ m\,s^{-2}$$

The acceleration-time graph for the motion of this cyclist would be as shown in figure 1.8.5.

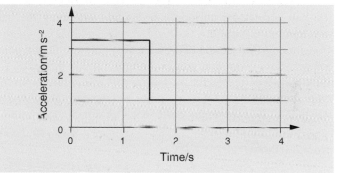

Figure 1.8.5

Changing direction

The graph in figure 1.8.3 shows a slowing down for 1.5 seconds followed by an acceleration for the next 2.5 seconds. If the acceleration values are calculated for this motion, the acceleration-time graph is as shown in figure 1.8.6.

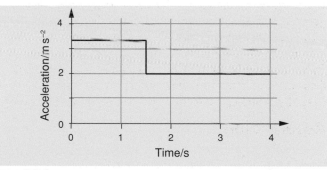

Figure 1.8.6

You might be surprised that the first section of this graph shows a positive acceleration even through the object was slowing down! The object is initially moving in a negative direction (to the left) so the initial velocity (u) is negative. The change in velocity (v-u) gives a positive value for the acceleration. If you are careful to use the correct + and – signs for the velocities when determining accelerations the calculations will give the correct sign for the acceleration.

1.9 Graphs for a bouncing ball

Introduction

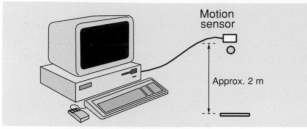

Figure 1.9.1 Studying a bouncing ball

When a volleyball, held 2 m above a level floor is released, it speeds up as it falls. The ball is then in contact with the floor for a short time before starting to move back towards the release point. During the contact with the ground some of the ball's kinetic energy is transferred into other forms of energy so it never quite returns to the original height. During the descent the ball accelerates uniformly and while ascending the ball decelerates uniformly.

Considering this motion provides a very interesting end point for our study of motion graphs. Figures 1.9.2, 1.9.3 and 1.9.4 show the set of graphs for the falling ball. For these graphs the motion sensor is the reference point and, somewhat unusually, *downwards* is taken as the positive vector direction

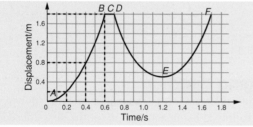

Figure 1.9.2

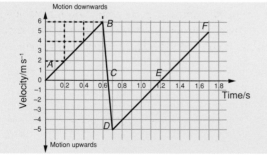

Figure 1.9.3

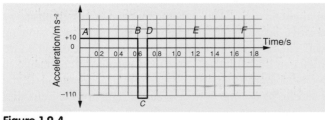

Figure 1.9.4

Interpreting the graphs

Section AB on the graphs

As the ball descends it gets further away from the motion sensor, so the displacement increases. During the time it is falling, its velocity increases but the acceleration remains constant. The ball falls a distance of 1.8 m in 0.6 seconds, reaching a velocity of +6 m s^{-1} just before hitting the ground.

$$Acceleration = \frac{Final\ velocity - initial\ velocity}{Time\ for\ the\ change} = \frac{6-0}{0.6} = 10\ m\,s^{-2}$$

Section BCD on the graphs

At point B on the graphs the ball makes contact with the ground, so the displacement stops increasing. The velocity decreases as the ball begins to squash. Eventually at C the ball has compressed as much as it can and begins to decompress. This makes it begin to move upwards, but by only a small amount so the section CD on the displacement-time graph is still almost horizontal. As the ball leaves its contact with the ground it has the greatest upwards velocity. The ball has been in contact with the ground for 0.1 seconds and its velocity has changed from 6 m s^{-1} downwards (+6 m s^{-1}) to 5 m s^{-1} upwards (−5 m s^{-1}). The change in velocity can be calculated.

$$Change\ in\ velocity = v - u = -5 - 6$$
$$Change\ in\ velocity = -11\ m\,s^{-1}$$

This change occurs in 0.1 seconds so the acceleration is both large and negative (−110 m s^{-2}). The negative gradient of the section BCD on the velocity-time graph confirms the negative or upwards acceleration.

Section DE on the graphs

The displacement from the reference point is decreasing from D to E, therefore the ball is rising. The velocity-time graph shows that as the ball rises it slows down, until eventually at point E it is at its greatest height. The ball is then 0.55 m from the motion sensor confirming that it does not bounce right back up to its release position. When it has risen to its maximum height the velocity has reduced to zero. The slowing down in the upwards direction again results in a positive value of the acceleration. On the velocity-time graph the section DE, where the ball is slowing down as it rises, is parallel to the section AB where the ball is speeding up as it falls. Since parallel lines have the same gradient we can say that the acceleration is the same for both section AB and section DE and is 10 m s^{-2}.

Section EF on the graphs

After the ball has reached its highest point in the first bounce it again starts to fall. The displacement increases and the velocity again increases in the positive direction. The downwards acceleration is the same as for the section AB. The time for the downward motion (section EF) is the same as that for the upwards motion (section DE) and the speed with which the ball hits the ground at point F is the same as the speed with which it left the ground at point D.

1.10 Consolidation questions

Worked example 1

A train travelling due north along a section of track increases its velocity steadily from 10 m s^{-1} to 20 m s^{-1} in 1 minute.

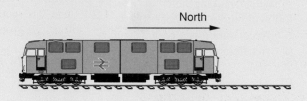

fig 1.10.1 Train travelling due north

a) Draw a graph of velocity against time to show how the velocity of the train increases. Make the scales on the axes go up as far as 25 m s^{-1} and 100 seconds.

b) What is the time taken for the train to increase its velocity from 10 m s^{-1} north to 15 m s^{-1} north?

c) The motion of the train continues but at the instant that the train reaches a velocity of 25 m s^{-1} the brakes are applied. The train then slows down until it stops 100 seconds after the start.
Complete the graph to show the velocity of the train throughout the 100 s of the journey.

Answers and comments

a) Values given in the question must be selected and a set of axes drawn with a suitable scale. The word 'steadily' in the question means that we can draw a straight line graph since the acceleration is constant.

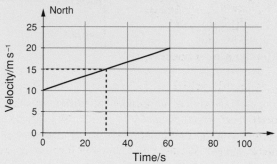

fig 1.10.2

b) The time when the velocity is 15 m s^{-1} north can be found from the x-axis of the graph.
The velocity is 15 m s^{-1} due north 30 seconds after the start.

c)

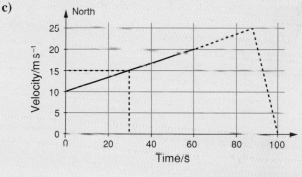

fig 1.10.3

Worked example 2

A computer model of the velocity-time graph for a cyclist changing up through some gears is as shown in figure 1.10.4.

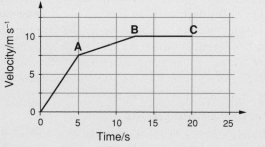

fig1.10.4

a) Calculate the acceleration during the section labelled AB.

b) Is the acceleration during the section labelled OA bigger or smaller than during the section labelled AB?

c) Describe the motion of the cycle during the section labelled BC. What is the value of the acceleration during this time?

Answers and comments

a) The velocity values and the time for the velocity to change from its value at A to the value at B can be found from the graph and substituted into the equation.

$$a = \frac{v - u}{t}$$

$$a = \frac{10 - 7.5}{12.5 - 5}$$

$$a = 0.33 \ m \, s^{-2}$$

b) Acceleration is given by the slope of the line on a velocity-time graph, so since the slope during OA is steeper than during AB, the acceleration is bigger during OA than during AB.

c) The part of the line on the velocity-time graph labelled BC shows a velocity which is not changing so the acceleration is 0 m s^{-2}.

Questions

1 A student making notes about velocity writes down the following three statements;

  I Velocity is a vector quantity.

  II To state a velocity value properly you must indicate the direction.

  III A velocity can be regarded as the speed in a given direction.

  Which of these statements is/are correct
  A I, II and III
  B I and II only
  C II and III only
  D I only
  E I and III only

2 Figure 1.10.5 shows a number of different shapes of velocity-time graph.

  During which of the motions does the direction of the velocity change?

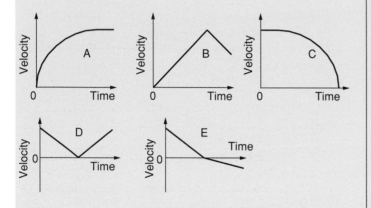

fig 1.10.5

3 A light aircraft flying at 50 m s^{-1} due north (bearing 000) is blown off course by a wind with a velocity of 30 m s^{-1} which is blowing from the south west (bearing 225).

fig 1.10.6 An aircraft flying due north

  a) What is the velocity of the plane as seen by an observer on the ground?

  b) At a particular instant the plane is directly over a marker. How far away from this position is the plane 5 minutes later?

4 To relieve the boredom during a long flight a golfer takes his putter and putts a golf ball 10 m along the airplane's aisle towards the front of the plane and into a practice hole.

  a) The ball takes 10 seconds to travel along the aisle and the plane is flying at 500 m s^{-1}. What is the velocity of the ball relative to the ground?

  b) How far, relative to the ground, does the ball travel?

5 A student making notes on acceleration writes down the following three statements;

  I Acceleration is a vector quantity.

  II Acceleration is defined as the change in velocity in unit time.

  III The gradient of a velocity-time graph gives acceleration.

  Which of these statements about acceleration is/are true?
  A I, II and III only
  B I and II only
  C II and III only
  D II only
  E I and III only

6 Car A can accelerate from 3 m s^{-1} to 24 m s^{-1} in 6 seconds. Car B is quoted as being able to accelerate from rest to 30 m s^{-1} in 9 seconds. Which car has the greater acceleration?

7 A cyclist, travelling north along a straight level road, reaches a maximum speed of 12 m s^{-1} by accelerating at 2.5 m s^{-2} for 3 seconds. What was the velocity before the acceleration started?

8 In the summer, when a well is dry, a stone dropped from rest at the top takes 2.2 seconds to hit the bottom. During its fall the acceleration of the stone is 9.8 m s^{-2}.

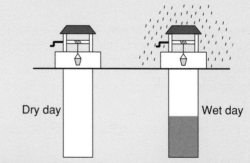

fig 1.10.7 Wet and dry wells

  a) What is the speed of the stone immediately before hitting the bottom of the well?

  b) By calculation, show that the average velocity of the stone during its fall is 10.8 m s^{-1} downwards.

  c) What is the depth of the well?

  d) On a day when there is some water in the well a dropped stone takes 1.5 seconds to fall, from rest, from the top into the water. How deep is the water in the well?

9 A student making notes on displacement-time graphs writes down the following statements;

I   The gradient of a displacement-time graph gives the velocity.

II   A straight line on a displacement-time graph indicates a steady velocity.

III When the gradient of a displacement-time graph changes from a positive to a negative value the direction of movement changes.

Which of these statements is/are correct?

A   I and III only

B   I and II only

C   II and III only

D   I only

E   I, II and III

10 The displacement of an object relative to a reference point is shown in the graph shown in figure 1.10.8.

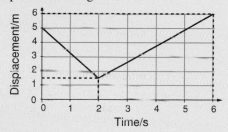

fig 1.10.8

a)   How far away from the reference point is the object at the start of its motion?

b)   After 6 seconds how far is the object from the reference point?

c)   What is the total distance travelled during the 6 seconds of the motion?

d)   Is the displacement after 6 seconds the same as the distance travelled during the 6 seconds? Explain your answer.

11 The displacement of a moving object is recorded and displayed at different times in the following table.

Time (s)	0	1	2	3	4	5
Displacement (m)	0	5	20	45	80	125

Table 1.10.1

The motion during this time could be described as which of the following?

A   Stationary

B   A uniform velocity

C   A constant acceleration

D   A gradually reducing speed

E   A constant speed in a negative direction

12 Figure 1.10.9 shows the velocity-time graph for an object initially moving at 10 m s⁻¹ to the right, away from a fixed point.

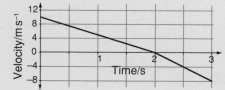

fig1.10.9

a)   Describe the motion of the object during the first 2 seconds.

b)   What is the acceleration of the object during the first 2 seconds?

c)   How far does the object travel during the first 2 seconds?

d)   How far does the object travel between times of 2 and 3 seconds?

e)   What is the object's displacement after 3 seconds?

13 Figures 1.10.10 and 1.10.11 show the velocity-time graphs for two 'Hot-Rods' in a dragster race.

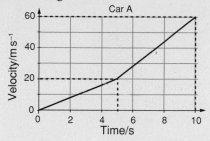

fig 1.10.10

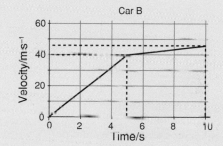

fig 1.10.11

a)   How far does each car travel in the first 5 seconds of its motion?

b)   Which car travels the greater distance during the 10 seconds?

c)   Both cars start at the same time and car B crosses the finish line after 8 seconds. How far behind is car A?

14 In an experiment a ball, held directly underneath a motion sensor, is released.

A pupil makes the following statements about the graph;

I   At point P the ball is held directly underneath the motion sensor.

II   Between Q and R the ball is in contact with the ground.

III Between S and T the ball is rising.

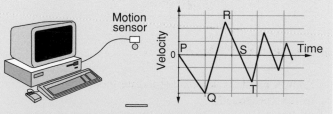

fig 1.10.12 An experiment with a ball

Which of these statements is/are correct?

A   I, II and III

B   I and II only

C   I and III only

D   II and III only

E   III only

Past paper question

The velocity of a trolley on a slope can be investigated using a computer and a sensor as shown below.

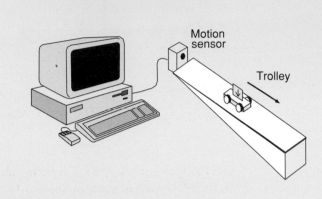

fig 1.11.1 Motion sensor

The sensor emits ultrasound pulses which are reflected from the trolley. The computer measures the time between emitted and reflected pulses and uses this information to calculate the velocity at regular intervals.

In an investigation, the trolley is given a sharp push up the slope and then released. The diagram below shows the resulting velocity-time graph as displayed on the screen.

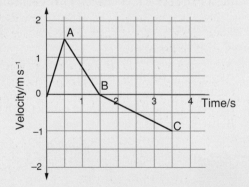

fig 1.11.2

Point A on the graph corresponds to the instant at which the trolley is released.

a) At what time is the trolley at its maximum distance from the sensor? You must justify your answer.

b) On graph paper draw the corresponding acceleration-time graph of the motion. **SQA 1992**

Answers and comments

After the trolley is released there are no forces pushing it up the runway. The trolley slows down after it is released and its velocity reduces to zero. It then moves back towards its starting position.

a) When it is furthest away from the motion sensor, the trolley will stop momentarily as it changes direction. Therefore, when it is furthest away from the motion sensor its velocity is $0 \, \text{m s}^{-1}$.

From the graph we can see that its velocity is zero at the point labelled 'B', but the question asks for the **time** so we must state that, its maximum displacement is after a time of 1.5 seconds.

b) Acceleration is calculated from the formula;

$$Acceleration = \frac{Change\ in\ velocity}{Time\ for\ the\ change}$$

$$Acceleration = \frac{Final\ velocity - initial\ velocity}{Time\ for\ the\ change}$$

$$a = \frac{v-u}{t}$$

For the section OA

$$a = \frac{v-u}{t} = \frac{1.5-0}{0.5}$$

$$a = 3 \, m\,s^{-2}$$

For the section AB

$$a = \frac{v-u}{t} = \frac{0-1.5}{1}$$

$$a = -1.5 \, m\,s^{-2}$$

For the section BC

$$a = \frac{v-u}{t} = \frac{-1-0}{2}$$

$$a = -0.5 \, m\,s^{-2}$$

All that remains now is to plot these values on a graph remembering to make each acceleration last for the correct time.

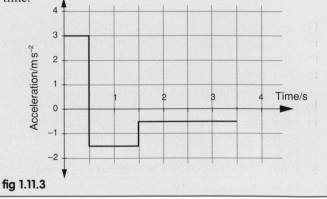

fig 1.11.3

Past paper question

A barge is travelling with a velocity of 2 m s⁻¹ due west, along a canal. A girl runs with a speed of 4.8 m s⁻¹ from X to Y across the deck of the barge as shown.

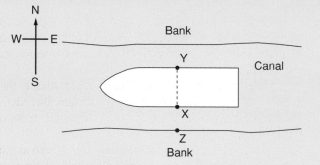

fig 1.11.4 Diagram of a barge

By drawing a scale diagram or otherwise, find the resultant velocity of the girl relative to someone standing at point Z on the bank of the canal. **SQA 1992**

Answers and comments

To the observer on the bank the girl will have two components to her velocity; one caused by the motion of the barge westwards, the other caused by the motion of the girl across the deck. The resultant velocity is found by drawing a scale diagram.

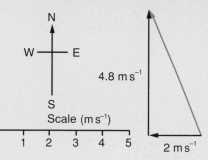

fig 1.11.5 Scale diagram

Therefore, the velocity of the girl = 5.2 m s⁻¹ at 23° west of north (337).

Questions

1 A car which has a constant acceleration of 2 m s⁻² has a velocity of 10 m s⁻¹ at a certain time.
   a) What is the velocity of the car 10 seconds later?
   b) What is the average speed of the car during this acceleration?
   c) What distance does the car travel whilst accelerating?

2 Car drivers must take care with the distance they leave between their car and the one in front. The time interval between seeing a problem on the road ahead and reacting by pressing the brake is assumed to be 0.2 seconds. This time interval is commonly called 'the thinking time' or 'reaction time'.
   a) If the car is travelling at 40 m s⁻¹ what distance does it travel during the driver's thinking time?
   b) Draw a graph to show the car coming to a halt 3 seconds after the driver presses the brake pedal.
   c) What is the minimum safe distance that a driver should leave between her car and the car in front if she is travelling at 40 m s⁻¹ in a car which comes to rest 3 seconds after the brakes are applied?
   d) Draw on the graph a plot to show another example in which she stops the car in the same time, but from an initial speed of 20 m s⁻¹.

3 During a time trial, a cyclist rides due north at 18 km h⁻¹ into a head wind of 3 km h⁻¹. He turns round after 7.5 km and rides back to his starting point.
   a) What is the resultant velocity of the cyclist during his journey north to the turning point?
   b) How long does he take to reach the turning point?
   c) What is his resultant velocity on the journey back to his starting point if he continues to ride at a speed of 18 km h⁻¹?
   d) What is the total time that he takes to complete the 15 km time trial course?

4 A student investigating the acceleration of a trolley along different slopes uses the apparatus shown in figure 1.11.6.

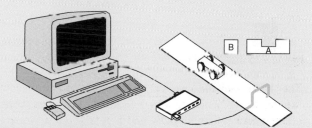

fig 1.11.6 Measuring acceleration

   a) Which of the cards labelled A or B should she attach to the trolley? Explain your answer.
   b) She measures the acceleration four times for each of the different heights of the end of the slope. Her results are shown table 1.11.1. Calculate the average acceleration for each of the heights.
   c) Draw a graph of average acceleration against height of the end of the runway for this experiment.
   d) Calculate the uncertainty in each of the average acceleration values.

| Height (m) | Acceleration (m s⁻²) | | | | Average acceleration |
	1st	2nd	3rd	4th	
0.2	1.10	1.15	1.13	1.08	
0.3	1.60	1.62	1.65	1.68	
0.4	2.20	2.22	2.18	2.20	
0.5	2.73	2.77	2.72	2.75	
0.6	3.35	3.35	3.28	3.31	

Table 1.11.1

5 The velocity-time graphs for two objects, A and B moving in opposite directions away from the same point are shown in figure 1.11.7 and figure 1.11.8.

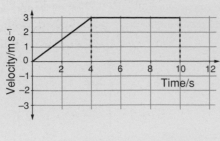

fig 1.11.7

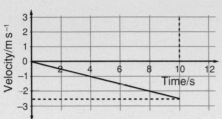

fig 1.11.8

a) The objects leave the starting point at the same time. Which is furthest away from the starting point after 10 seconds?

b) How far apart are the objects after 10 seconds?

6 Displacement, distance, speed, velocity and acceleration are all quantities commonly used in describing motion. Which of the following are all vector quantities?
A   Distance, speed and acceleration
B   Displacement, speed and velocity
C   Displacement and velocity only
D   Displacement, velocity and acceleration
E   Distance and velocity only

7 The displacement of an object moving in a straight line in a single direction is;
A   The same throughout its motion.
B   Fully described by stating only the distance travelled.
C   Of a different magnitude to the distance travelled.
D   Constantly increasing.
E   The same magnitude as the distance travelled in the same time.

8 Describe how the apparatus shown in figure 1.11.9 can be used to measure the velocity of a trolley at two different points as it travels down a runway.
a) State what measurements must be taken and explain how the velocities are calculated.

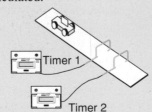

fig 1.11.9
Measuring velocity

b) Describe how the acceleration of the trolley along the track could be measured. List any additional apparatus you would require and explain how you would calculate the acceleration.

9 A body, initially at rest, moves in a single direction with a constant acceleration of 5 m s^{-2} for 10 seconds and then slows down at 2 m s^{-2}, until it comes to rest.
a) What is the maximum speed of this object?
b) At what time after reaching its maximum speed does it come to rest?
c) Draw a velocity-time graph for this motion.
d) What is the total distance it travels?

10 The apparatus shown below is used to record the motion of a trolley up a wooden runway. The trolley is given a push and released at the exact moment the motion sensor starts recording.

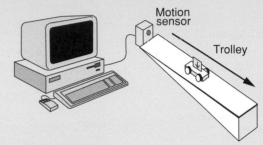

fig 1.11.10 Recording motion

The displacement-time graph recorded is as shown.

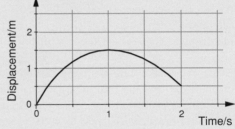

fig 1.11.11

Which of the following graphs shows the most probable velocity-time graph for the same motion? Explain your answer.

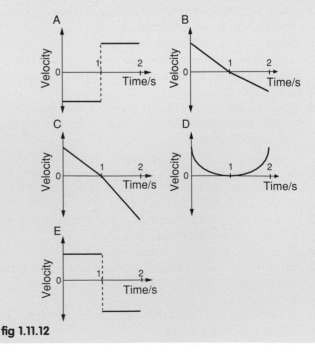

fig 1.11.12

11 The graph of figure 1.11.13 is obtained for a moving object which passes a marker at time $t = 0$.

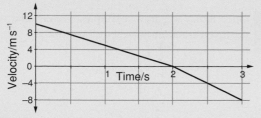

fig 1.11.13

a) How far is the object from the marker after 2 seconds?
b) What is the displacement of the object after 3 seconds?

12 A trolley is travelling in a straight line without changing direction.
Which set of the graphs below shows displacement-time; velocity-time and acceleration-time graphs for this motion?

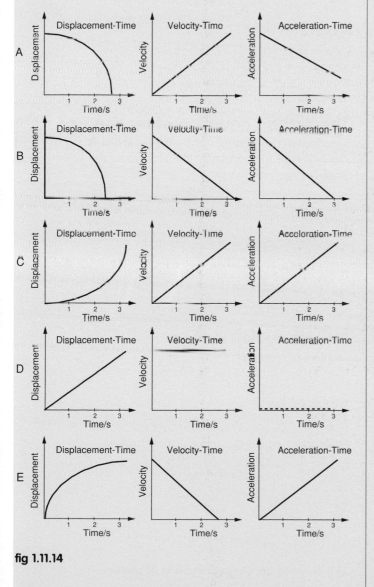

fig 1.11.14

13 A constant force is used to move an object. The velocity-time graph for the motion of the object is shown in figure 1.11.15.

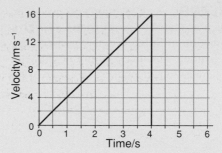

fig 1.11.15

Which row in the following table shows the correct values for the instantaneous speed after 3 seconds and the average speed during the first 3 seconds of the motion?

	Instantaneous speed	Average speed
A	$16.0 \ \mathrm{m\,s^{-1}}$	$16.0 \ \mathrm{m\,s^{-1}}$
B	$16.0 \ \mathrm{m\,s^{-1}}$	$10.7 \ \mathrm{m\,s^{-1}}$
C	$12.0 \ \mathrm{m\,s^{-1}}$	$8.0 \ \mathrm{m\,s^{-1}}$
D	$8.0 \ \mathrm{m\,s^{-1}}$	$24.0 \ \mathrm{m\,s^{-1}}$
E	$12.0 \ \mathrm{m\,s^{-1}}$	$6.0 \ \mathrm{m\,s^{-1}}$

Table 1.11.2

14 The following table shows the velocity of an object for the first 10 seconds of its motion.

Time (s)	0	2	4	6	8	10
Velocity (m s⁻¹)	0	20	40	40	24	8

Table 1.11.3

Which of the following graphs shows how the acceleration varies with time for this motion? Explain your answer.

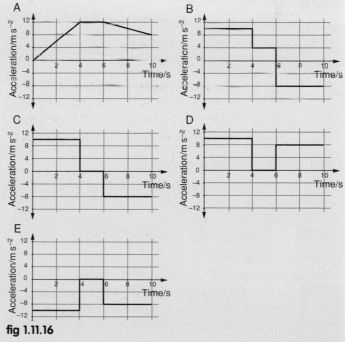

fig 1.11.16

The equations of motion

Introduction

A graph such as that shown in figure 1.12.1 gives information about the motion at different times between the start and the finish of a journey. The object is initially moving at a velocity, u, and accelerates to a velocity, v, in a time of t seconds.

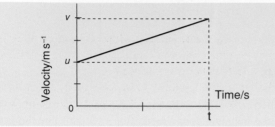

Figure 1.12.1

This graph represents a constant acceleration where the value of the acceleration can be found from;

$$a = \frac{v - u}{t}$$

Rearranging this equation gives;

$$v = u + at$$

This equation allows you to predict the velocity at any given time for a moving object accelerating uniformly from an initial velocity u for a time t. This equation is sometimes called the **first equation of motion**.

The first equation of motion

Q A falling mass accelerates at 9.8 m s^{-2} from rest for 2 seconds before hitting the ground. Calculate the velocity of the mass just before it hits the ground.

A It is best to start by drawing a diagram and marking on the positive vector direction.

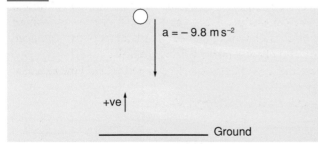

Figure 1.12.2 A falling mass accelerates

If the mass falls from rest, this tells you that the initial velocity, u, was 0 m s^{-1}. So;

$$v = u + at$$
$$v = 0 + (-9.8) \times 2$$
$$v = -19.6 \text{ m s}^{-1}$$

The mass hits the ground with a velocity of 19.6 m s^{-1} downwards

Q An object is projected upwards with an initial upward velocity of 49 m s^{-1}. For what time will it continue to rise if the deceleration during the upwards motion is 9.8 m s^{-2}?

A Once again we draw a simple diagram and mark the positive direction, for vector quantities.

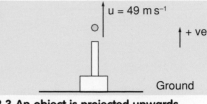

Figure 1.12.3 An object is projected upwards

This means that the initial velocity is 49 m s^{-1} and the deceleration is -9.8 m s^{-2}. When the object stops rising its final velocity v will equal 0 m s^{-1}. Therefore at the highest point;

$$v = 0 \text{ m s}^{-1}$$
$$v = u + at$$
$$0 = 49 + (-9.8) \times t$$
$$t = \frac{-49}{-9.8}$$
$$t = 5 \text{ s}$$

The object rises for 5 seconds before beginning to fall.

The second equation of motion

The displacement during the journey graphed in figure 1.12.1 can be found by calculating the area under the velocity-time graph.

$$Area\ under\ graph = Area\ of\ \boxed{} + Area\ of\ \triangle$$

$$Displacement = ut + \frac{1}{2}(v - u)\,t$$

If this displacement is given the symbol, s, then;

$$s = ut + \frac{1}{2}(v - u)\,t \qquad\qquad equation\ 1$$

But since acceleration is defined as;

$$a = \frac{v - u}{t}$$

We can see that

$$v - u = at \qquad\qquad equation\ 2$$

Substituting equation 2 into equation 1 gives;

$$s = ut + \frac{1}{2}at \times t$$

So

$$s = ut + \frac{1}{2}at^2$$

This is called the **second equation of motion**.

Q A ball is projected upwards with an initial upward velocity of 49 m s^{-1}. The acceleration downwards is 9.8 m s^{-2}. Calculate the time taken for the ball to return to its starting point.

 A We again draw a diagram and mark on the positive direction for vector quantities.

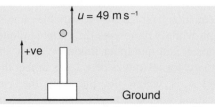

Figure 1.12.4 A ball falls back towards its starting position

When the ball returns to its starting point its displacement is zero therefore:

$$s = ut + \frac{1}{2}at^2 = 0 = 49 \times t + \frac{1}{2}(-9.8) \times t^2$$

$$4.9\,t^2 = 49\,t$$

$$t = 10\ seconds$$

Ten seconds after the ball is thrown upwards it is back at its starting position.

Combining the first and second equations of motion

Q A car, initially at rest accelerates uniformly at 5 m s^{-2}. Find the final velocity after it has travelled 102.4 metres.

A Initially this question seems to provide too little information. We need to find v, but we can't use the first equation of motion as we haven't been given the time that the acceleration lasted, t. However, we can use the second equation of motion to find t.

$$s = ut + \frac{1}{2}at^2$$

$$102.4 = 0 + \frac{1}{2} \times 5\,t^2$$

Rearranging to find t gives;

$$t = \sqrt{\frac{2 \times 102.4}{5}}$$

$$t = 6.4\ seconds$$

Now we can find the final velocity by substituting the value for t into the first equation of motion.

$$v = u + at$$
$$v = 0 + 5 \times 6.4$$
$$v = 32\ m\,s^{-1}$$

In doing this calculation we have combined the first and second equations of motion. The key to solving this problem was to determine the time. The first equation of motion;

$$v = u + at$$

rearranges to give;

$$t = \frac{v - u}{a}$$

Substituting this into the second equation of motion we get;

$$s = ut + \frac{1}{2}at^2$$

$$s = u\left(\frac{v-u}{a}\right) + \frac{1}{2}a\left(\frac{v-u}{a}\right)^2$$

To simplify this equation you need to square and multiply out the brackets. Rearranging then gives;

$$v^2 = u^2 + 2as$$

This is known as the **third equation of motion**. It is most useful in cases where information about the time for which the acceleration lasts is not given in the question.

We can also arrive at the third equation of motion by a simpler mathematical manipulation of the first two equations of motion.

$$v = u + at$$
$$v^2 = (u + at)^2$$
$$v^2 = u^2 + 2uat + a^2t^2$$
$$v^2 = u^2 + 2a\left(ut + \frac{1}{2}at^2\right)$$

but since; $\quad s = ut + \frac{1}{2}at^2$

we can say; $\quad v^2 = u^2 + 2as$

Summary

The three equations we have studied, link the five variables;

$u = Initial\ velocity$ $\qquad t = Time\ acceleration\ lasted$
$v = Final\ velocity$ $\qquad\quad s = Displacement$
$a = Acceleration$

The three equations of motion are;

First equation $v = u + at$

Second equation $s = ut + \frac{1}{2}at^2$

Third equation $v^2 = u^2 + 2as$

Each of these equations contains four of the five different variables. Given three pieces of information you can choose one of these equations to find the fourth quantity, but you need to be careful about which equation you select.

The previous examples will help you understand how to tackle questions.

For the third worked example we can see;

u	v	a	t	s
+49	not used	−9.8	?	0

In this case we want an equation linking u, a, t, and s so it is most appropriate to use the second equation of motion.

$$s = ut + \frac{1}{2}at^2$$

For the final worked example we can see;

u	v	a	t	s
0	?	5	not used	102.4

To find the final velocity in this example you must choose an equation linking u, v, a and s so it is best to use the third equation of motion.

$$v^2 = u^2 + 2as$$

1.13 Equation of motion experiments

Testing v = u + at

EXPERIMENT Using the apparatus shown in figure 1.13.1 the card is held above light gate A and released. The system measures the speed at light gates A and B. Since the card is falling freely due to the force of gravity its acceleration is constant at 9.8 m s^{-2}.

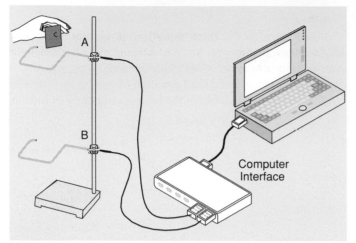

Figure 1.13.1

Method

In this experiment you should hold the card just above light gate A. Prepare the system to measure the speeds at light gates A and B, along with the time to travel from A to B. Release the card and record data in a table similar to table 1.13.1. Repeat the measurements using different starting positions and different distances between the light gates.

Results

	Speed at A	Speed at B	Time to travel from A to B
1			
2			
3			

Table 1.13.1

Analysis

For each of the sets of data, you have measured an initial speed (speed at A) and the time for which the acceleration of 9.8 m s^{-2} lasted, before the final speed (speed at B) was measured. Test $v = u + at$ by using your data to confirm that;

Speed at B = Speed at A + 9.8 × Time to travel from A to B

Graphical analysis

The equation $v = u + at$ describes the relationship between variables v, u and t. In the previous experiment you tried to confirm this equation by showing that a measured value and a value calculated were equal, within the limits of experimental error. An alternative method for verifying relationships or equations involves plotting straight line graphs. Before this can be done we often need to rearrange equations to match the mathematical equation which describes a straight line;

$$y = mx + c$$

In this equation y represents the variable plotted on the y-axis of the graph while x represents the variable plotted on the x-axis. The letter m is a multiplier linking y and x and c is a number added to mx for all values of x. (m is in fact the gradient of the line and c is where the line cuts the y-axis). A simple mathematical equation might link the variables x and y by stating;

$$y = 3x + 7$$

If x can have values between 0 and 7 then the values of y are as shown in the table and plotted on the graph.

x values	y values
0	7
1	10
2	13
3	16
4	19
5	22
6	25
7	28

Table 1.13.2

Figure 1.13.2

The fact that this is a straight line shows that there is a linear relationship between the variables plotted on the y- and x-axes. The gradient of this line is given by the multiplier m in the equation $y = mx + c$. The line cuts the y-axis at the value given by c in this equation.

For the first equation of motion;

$$v = u + at$$

the final velocity v is the result of a constant acceleration, for a variable time, t. We can match this equation to the mathematical equation for a straight line by rearranging;

	$v = u + at$
But as	$y = mx + c$
Rearranging gives	$v = at + u$

This would suggest that collecting values for the velocity after a constant acceleration lasting different times would produce a straight line graph. To achieve this we would have to plot the velocity achieved on the y-axis and the travel time on the x-axis. The gradient of the line will give the acceleration.

You should note that this analysis assumes that the initial velocity is the same for each repetition of the experiment. As the experiment outlined earlier was designed to give *different* initial speeds you could not use the data from that experiment to plot a straight line graph. However, you could use the same apparatus, keeping the height from which the card is dropped constant and collecting data by altering only the position of the *lower* light gate.

Testing $s = ut + \frac{1}{2}at^2$

EXPERIMENT This equation links the displacement with the initial velocity for a moving object which has a constant acceleration for a certain time. An object falling due to the force of gravity falls in a straight line with a constant acceleration. When released from rest, the equation governing its motion simplifies to;

$$h = \frac{1}{2}gt^2 \qquad \text{equation 1}$$

where h is the height fallen in time t and g is the acceleration due to gravity. (Note that h is used in place of s.)

Method

You can use the apparatus shown to measure the time that a ball takes to fall from different heights.

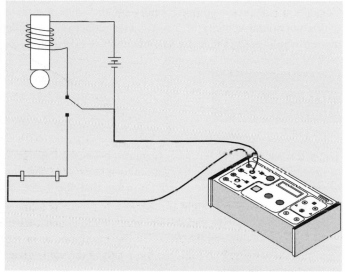

Figure 1.13.3 Apparatus for testing an equation

It is good technique to repeat the timings a few times before altering the height. You should then record the mean fall time for each height in a table similar to table 1.13.3.

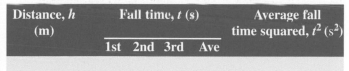

Distance, h (m)	Fall time, t (s)				Average fall time squared, t^2 (s²)
	1st	2nd	3rd	Ave	

Table 1.13.3

Analysis

The mathematical equation for any straight line graph states that;

$$y = mx + c$$

To put the variables for equation 1 into the form required for a straight line graph we must plot the height fallen, h, on the y-axis and the fall time squared (t^2) on the x-axis. (In this case c is given the value 0.) The gradient of the 'best fit' straight line is $\frac{1}{2}g$. You can confirm that the value for the acceleration due to the force of gravity is 9.8 m s^{-2}. (Drawing the 'best fit' straight line on the graph makes allowance for experimental error).

Testing $v^2 = u^2 + 2as$

EXPERIMENT In this experiment you will find the speed of a trolley after it has travelled different distances down an inclined runway. The speed is found by measuring how long it takes for a card of known length, attached to the trolley, to pass through a light beam. If the trolley is released from rest this equation of motion simplifies to;

$$v^2 = 2as$$

Method

To test the validity of this equation you should align the centre of the interrupt card with the start line marked on the runway and measure the speed of the trolley when it has travelled a known distance.

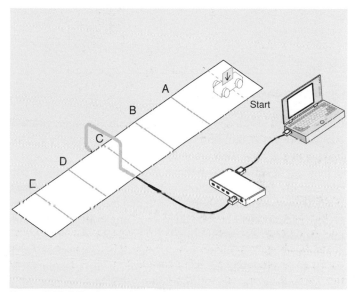

Figure 1.13.4 Testing an equation

It is good technique to repeat the measurement a few times before moving the light gate to a new position and measuring the new speed. You should record your results in a table similar to 1.13.4

Distance, s (m) from start to	Speed (m s^{-1})				Average speed, v, (m s^{-1})	Speed squared v^2 (m s^{-1})²
	1st	2nd	3rd	4th		
A =						
B =						
C =						
D =						
E =						

Table 1.13.4

Analysis

To put the variables from this equation into the form required for a straight line graph we must plot speed squared (v^2) on the y-axis and displacement on the x-axis. The gradient of the best fit straight line graph, which passes through the origin, would then have a value of $2a$. You could check this by measuring the acceleration of the trolley down the runway using two light gates or a single light gate and double interrupt card.

1.14 Projectiles

Introduction

The apparatus shown in figure 1.14.1 can be used to drop ball P vertically while simultaneously projecting ball Q horizontally.

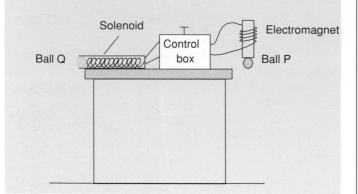

Figure 1.14.1 Studying projectiles

If this motion is videoed and replayed in slow motion, you will see the projected ball and the ball dropped vertically hit the floor at *exactly the same instant*.

It may seem difficult to believe that an archer, firing a bolt horizontally from a crossbow and at the same time dropping the crossbow, would see both the bolt and the crossbow landing on the ground at the same instant! In reality both would not hit the ground at the same time, as it would be difficult to synchronise the release of the bolt with dropping the bow. Also the flights on the bolt are designed to give it 'lift', keeping it in the air longer. In our simple model for the motion of a projectile we have to make some assumptions:

1 **Only the force of gravity acts on the vertical motion.**
2 **No frictional forces impede the horizontal motion.**

If a scale is placed behind the apparatus of figure 1.14.1 we will see that the horizontal motion of the projected ball is a constant velocity. This means that in equal time intervals the ball, projected horizontally, travels the same horizontal distance.

Investigating projectile motion

The photograph of figure 1.14.2 shows the path of a projectile with an initial horizontal velocity.

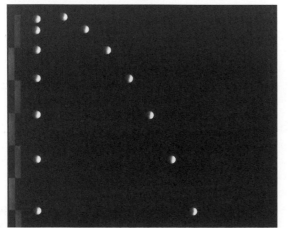

Figure 1.14.2 A multi-image photograph of projectile motion

In this photograph the position of the ball is shown at several different times during its motion.

From this picture you can see that the horizontal distance between images is constant whilst the vertical distance between images increases. The horizontal velocity is therefore constant whilst the dropped ball is accelerating vertically because of the force of gravity.

We can summarise these observations by treating the movement of a projectile as independent horizontal and vertical motions:

1 **The horizontal motion is a constant speed.**
2 **The vertical motion is a constant acceleration.**

Simulating projectile motion

Consider a ball projected horizontally off the edge of a table with a horizontal velocity of $5\ \mathrm{m\,s^{-1}}$. Our model of projectile motion would predict that the ball travels 5 m horizontally in each of the subsequent 3 seconds.

Time (s)	0	0.5	1.0	1.5	2.0	2.5	3.0
Horizontal distance travelled (m)	0	2.5	5	7.5	10	12.5	15

Table 1.14.1

The ball is accelerating vertically from rest because of the force of gravity. We can calculate the vertical displacement using the second equation of motion and knowing that the force of gravity causes an acceleration of $-9.8\ \mathrm{m\,s^{-2}}$.

Calculating the vertical displacements at the times shown in table 1.14.1 gives the values shown in table 1.14.2.

Time (s)	0	0.5	1.0	1.5	2.0	2.5	3.0
Vertical displacement (m)	0	–1.22	–4.9	–11	–19.6	–30.6	–44

Table 1.14.2

Plotting the horizontal and vertical positions at half second intervals shows the curved path of the projectile, figure 1.14.3.

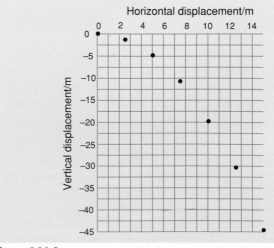

Figure 1.14.3

The similarity between the photograph of figure 1.14.2 and the plotted values of figure 1.14.3 confirms that the movement of a projectile can indeed be treated as independent horizontal and vertical motions.

Projectile motion examples

Q A communications line is fired with a horizontal velocity of 90 m s^{-1} from the deck of an oil rig towards a nearby support vessel. The rig is 48 metres above sea level and the support vessel sits 3.9 metres above the water.

(a) Calculate the time it takes the line to travel from the oil rig to the ship.
(b) What is the distance between the oil rig and the ship?

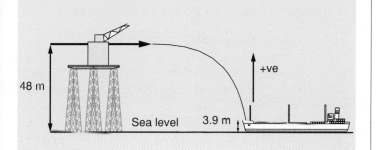

Figure 1.14.4

A (a) As the line travels from the oil rig to the ship it has to travel a vertical distance of;

$$48 - 3.9 = 44.1 \ m$$

The vertical motion of a projectile is an acceleration of 9.8 m s^{-2} caused by the force of gravity. As the line is fired horizontally the initial vertical velocity is zero. Therefore

$$\downarrow s = ut + \frac{1}{2}at^2$$

$$\downarrow -44.1 = 0 + \frac{1}{2} \times (-9.8) \times t^2$$

$$t = \sqrt{\frac{2 \times -44.1}{-9.8}}$$

$$t = 3s$$

(b) We have calculated that it takes 3 seconds for the line to fall the required distance between the rig and the ship. During this time the horizontal speed is constant at 90 m s^{-1}. There is no acceleration in the horizontal direction, therefore;

$$\rightarrow s = ut + \frac{1}{2}at^2$$

$$\rightarrow s = 90 \times 3 + \frac{1}{2} \times 0 \times 3^2$$

$$\rightarrow s = 270 \ m$$

Therefore the ship and the oil rig are 270 metres apart.

Q An aircraft flying horizontally at 60 m s^{-1}, 122.5 metres above the ground releases a package of supplies.

(a) What is the horizontal distance of the release point from the target zone if the package is to score a direct hit?
(b) What is the horizontal velocity of the package just before impact?
(c) What is the vertical velocity of the package just before impact?

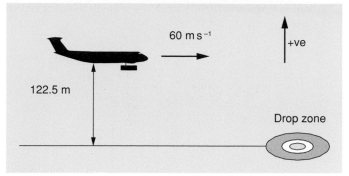

Figure 1.14.5

A (a) As the aircraft is flying horizontally the initial velocity in the vertical direction is 0 m s^{-1}. We must now find the time for the package to fall the vertical distance of 122.5 metres.

$$s = ut + \frac{1}{2}at^2$$

$$-122.5 = 0 + \frac{1}{2} \times (-9.8) \times t^2$$

$$t = \sqrt{\frac{2 \times (-122.5)}{-9.8}}$$

$$t = 5 \ s$$

We have calculated that it takes 5 seconds for the package to reach ground level. As the package is attached to the aeroplane before its release its speed in the horizontal direction is initially 60 m s^{-1}. While it is falling the horizontal speed of the package remains constant at 60 m s^{-1}, therefore;

$$\rightarrow s = ut + \frac{1}{2}at^2$$

$$\rightarrow s = 60 \times 5 + \frac{1}{2} \times 0 \times 5^2$$

$$\rightarrow s = 300 \ m$$

The package must be released when the plane is a horizontal distance of 300 metres from the target.

(b) Projectiles travel at a constant horizontal speed so the horizontal speed on impact is 60 m s^{-1}, the same as throughout the drop.

(c) The package is accelerating vertically throughout the entire motion so we can calculate its vertical velocity using;

$$v = u + at$$
$$v = 0 + (-9.8) \times 5$$
$$v = -49 \ m \ s^{-1}$$

Just before impact the package has a velocity of 49 m s^{-1} downwards as well as a horizontal velocity of 60 m s^{-1}.

Worked example 1

A ball is thrown vertically upwards with an initial speed of $14\ \mathrm{m\,s^{-1}}$.

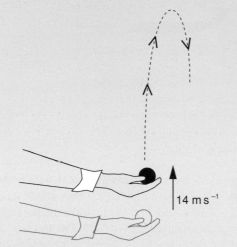

fig 1.15.1 Throwing a ball upwards

a) Describe what happens to the speed of the ball as it rises to its maximum height.

b) How long does the ball take to reach its highest point?

c) How far is the highest point above the release point?

Answers and comments

a) The speed of the ball decreases from the initial value of $14\ \mathrm{m\,s^{-1}}$ as the ball rises to its maximum height.

b) At the highest point the speed of the ball is zero. To calculate the time to rise to the highest point we must use vectors.

u	v	a	t	s
+14	0	−9.8	?	not used

$$v = u + at = 14 + (-9.8) \times t$$

$$t = \frac{14}{9.8} = 1.43\ seconds$$

c) We can again find the maximum height using an equation of motion.

u	v	a	t	s
+14	0	−9.8	1.43	?

Vector +ve ↑

$$s = ut + \frac{1}{2}at^2 = 14 \times 1.43 + \frac{1}{2}(-9.8) \times 1.43^2$$

$$s = 10\ m$$

After 1.43 seconds the ball is 10 m above the point where it was released.

Worked example 2

A pupil investigating projectile motion sets up the apparatus shown in figure 1.15.2

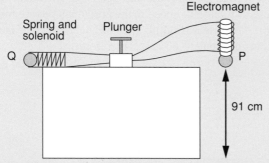

fig 1.15.2 Apparatus for investigating projectile action

Pressing the plunger on the control box releases ball P from the electromagnet and fires ball Q horizontally at a speed of $3.5\ \mathrm{m\,s^{-1}}$.

a) Calculate the time that ball P takes to fall to the floor.

b) How long after the plunger is pressed will ball Q hit the floor?

c) What horizontal distance will ball Q travel before it reaches the floor?

Answers and comments

a) Ball P falls vertically from rest, because of the force of gravity. We can use an equation of motion to find the time taken to fall 91 cm (0.91 m) downwards.

u	v	a	t	s
0	not used	−9.8	?	−0.91

Vector +ve ↑

$$s = ut + \frac{1}{2}at^2$$

$$-0.91 = 0 + \frac{1}{2} \times (-9.8) \times t^2$$

$$t = \sqrt{\frac{2 \times (-0.91)}{-9.8}} = 0.43\ seconds$$

b) The vertical motion of the projected ball is the same as that of the falling ball so ball Q also takes 0.43 s to fall to the floor.

c) The horizontal speed of the projected ball remains constant so there is no horizontal acceleration. We can find the horizontal displacement as follows:

u	v	a	t	s
−3.5	not used	0	0.43	?

Vector +ve →

$$s = ut + \frac{1}{2}at^2 = -3.5 \times 0.43 + \frac{1}{2} \times 0 \times 0.43^2$$

$$s = -1.5\ m$$

Questions

1 A stone is thrown vertically upwards with an initial speed of 28 m s^{-1}.
   a) Calculate the maximum height reached.
   b) Calculate the time taken to reach the highest point.
   c) Calculate the time taken for the stone to return to its starting position.

2 A car starts from rest and accelerates at 2 m s^{-2} for 6 seconds.
   a) Show, by calculation, that the car is travelling at 12 m s^{-1} after 6 seconds.
   b) The car then travels at this speed for a further 30 seconds. How far does the car travel during these 30 seconds?
   c) How far is the car from its staring position after 36 seconds?

3 A car travels at a steady velocity of +20 m s^{-1} for 6 seconds before the brakes are applied to bring it to rest in a further 8 seconds.
   a) Calculate the acceleration of the car after the brakes are applied.
   b) How far does the car travel while it is slowing down?
   c) Draw a graph to show the velocity of the car during the 14 seconds of its motion.
   d) How far is the car from its starting point after 10 seconds?

4 A stone is thrown vertically upwards and its height above the ground at various times is recorded in table 1.15.1.

Time (s)	0	1	2	3	5	6	7	8
Height (m)	0	35	60	75	75	60	35	0

Table 1.15.1

a) Plot a graph of height (y-axis) versus time for this motion.
b) From your graph, determine the maximum height reached.
c) What time does the stone take to fall from the highest point back to the ground?
d) Calculate the acceleration of the stone during its fall.
e) What is the speed of the stone just before it hits the ground?

5 Describe an experiment to measure the acceleration caused by the force of gravity acting on a 1 kg mass dropped from rest. Your answer should include;
   a) A diagram of the apparatus you would use.
   b) A description of the measurements you would take.
   c) A description of the way you would calculate the acceleration due to gravity.

6 A stunt motorcyclist jumping across a river, as shown in figure 1.15.3 lands in the middle of the landing zone which is 19.6 m vertically below the take-off point.
   a) What is the time between take-off and landing?
   b) What is the minimum speed at take-off that will ensure that the motorcyclist lands in the centre of the landing zone?
   c) On one occasion, the motorcyclist has a horizontal speed of 25 m s^{-1} at take-off. How far from the centre of the landing zone does the motorcycle land?

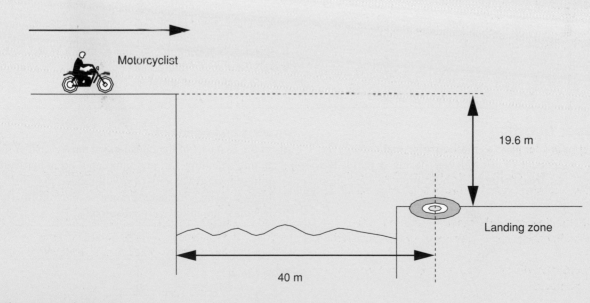

fig 1.15.3 The motorcycle stunt

1.16 Exam style questions

Past paper question

In a 'handicap' sprint race, sprinters P and Q start the race at the same time but from different starting lines on the track. The handicapping is such that both sprinters reach the line XY, as shown below, at the same time

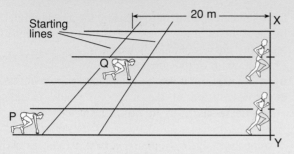

Starting lines

20 m

fig 1.16.1

Sprinter P has a constant acceleration of 1.6 m s^{-2} from the start line to the line XY. Sprinter Q has a constant acceleration of 1.2 m s^{-2} from the start line to XY

a) Calculate the time taken by the sprinters to reach line XY.
b) Find the speed of each sprinter at this line.
c) What is the distance between the starting lines for sprinters P and Q? **SQA 1995**

Answers and comments

a) Both sprinters start at the same time and arrive at the line XY at the same time beacuse they accelerate differently during the race.
For sprinter P we know the following information

	u	v	a	t	s
	0	not used	1.6	?	20

Table 1.16.1

We can now use the second equation of motion to find the time;

$$s = ut + \frac{1}{2}at^2$$

$$20 = 0 + \frac{1}{2} \times 1.6t^2$$

$$t = \sqrt{\frac{2 \times 20}{1.6}}$$

$$t = 5 \ seconds$$

b) To find the speed of each sprinter at line XY we can use the equation $v = u + at$.

For sprinter P $v = 0 + 1.6 \times 5$ $v = 8 \ ms^{-1}$.
for Sprinter Q $v = 0 + 1.2 \times 5$ $v = 6 \ ms^{-1}$.

c) We know that sprinter Q reaches a speed of 6 m s^{-1} from rest having accelerated at 1.2 m s^{-2}. We can calculate how far he travels using;

$$v^2 = u^2 + 2as$$
$$6^2 = 0 + 2 \times 1.2 \times s$$

$$s = \frac{36}{2.4} = 15 \ m$$

If sprinter Q travels 15 m to the line XY and sprinter P travels 20 m to XY there must be 20 – 15 = 5 m between the starting positions of sprinters P and Q.

Past paper question

A long jumper devises a method for estimating the horizontal component of his velocity during a jump. His method involves first finding out how high he can jump vertically.

0.86 m

fig 1.16.2

Answers and comments

a) In the first part of the question we must consider the long jumper standing still and jumping 0.86 m vertically. His initial upward velocity of u will reduce to 0 m s^{-1} at the highest point. We can summarise the information and calculate the initial vertical velocity.

	u	v	a	t	s
	?	0	–9.8	not used	0.86

Table 1.16.2

Vector +ve

$$v^2 = u^2 + 2as$$
$$0 = u^2 + 2 \times -9.8 \times 0.86$$
$$u^2 = 16.856$$
$$u = 4.1 \ m \ s^{-1}$$

He finds that the maximum height he can jump is 0.86 m.
a) Show that his initial vertical velocity is 4.1 m s^{-1}.
He now assumes that when he is long jumping, the initial vertical component of his velocity at take-off is 4.1 m s^{-1}. The length of his long jump is 7.8 m.
b) Calculate the value that he should obtain for the horizontal component of his velocity v_H. **SQA 1994**

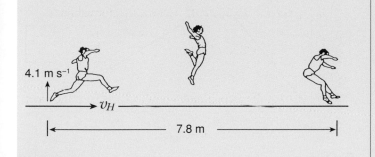

fig 1.16.3

b) We must now find the time of flight, which will be twice the time taken to jump to the highest point. We can find the time to reach the highest point from;

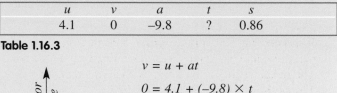

	u	v	a	t	s
	4.1	0	–9.8	?	0.86

Table 1.16.3

$$v = u + at$$
$$0 = 4.1 + (-9.8) \times t$$
$$t = \frac{4.1}{9.8} = 0.42s$$

The total time of flight = 2 × 0.42 = 0.84 seconds

The horizontal component of the jumper's velocity, v_H, is constant.
The jumper travels 7.8 m in 0.84 seconds so;

$$v_H = \frac{7.8}{0.84} = 9.3 \, m \, s^{-1}$$

Questions

1 A boy throws a rock horizontally from the top of a lighthouse with a horizontal speed of 8 m s^{-1}. The rock hits the water 3.5 seconds after being released.

fig 1.16.4 A rock projected from a lighthouse

a) Describe the horizontal component of the rock's velocity during its flight.
b) How far, horizontally, from the base of the lighthouse does the rock enter the water?
c) How far vertically, is the point where the rock was released above the point where it entered the water?

2 A sprinter starting from rest accelerates uniformly over the first 30 m of a 60 m dash. At a point 30 m from the start his speed is 12 m s^{-1}.
a) What is his acceleration during the first 30 m?
b) What time does it take the sprinter to cover the first 30 m?
c) How long does it take the sprinter to complete the final 30 m of the race if he maintains the speed of 12 m s^{-1} right to the finish line?
d) The sprinter takes 0.15 seconds to react to the starting pistol. What is his total time for the 60 m dash?
e) Draw a speed-time graph for this motion starting from the instant that the starting pistol is fired. (Mark numerical values on both axes of your graph.)

3 An object starts from rest and moves due east, accelerating at 3 m s^{-2} for 4 seconds.
a) What is the velocity of this object after 4 seconds?
b) Show that after 4 seconds the object is 24 m due east of its starting position.

c) After 4 seconds the object stops accelerating and travels for a further 36 m before decelerating uniformly to rest. The total time for which the object moves is 10 seconds.
(i) Draw a graph to show the velocity of the object during the 10 seconds of its motion.
(ii) Use your graph to find the displacement of the object's stopping point relative to the start.

4 The speed of a lorry 'freewheeling' along a level road reduces constantly from 4 m s^{-1} to 1.5 m s^{-1} while the lorry travels a distance of 27.5 m.
a) Calculate the deceleration of the lorry.
b) How much further will the lorry travel before coming to rest, if the deceleration remains the same?
c) What time does the lorry take to come to rest from its initial speed of 4 m s^{-1}?

5 Car A is stopped at traffic lights on a dual carriageway. When the light changes to green car A starts to accelerate at 4.3 m s^{-2}.
Just as the lights change car B travelling at a steady 12 m s^{-1}, passes car A.
a) Copy and complete the following table to show how far car B has travelled 1,2,3 and 4 seconds after passing the traffic lights.

	Time (s)	1	2	3	4
Car B	Distance (m)				

Table 1.16.4

b) Calculate the speed of car A 1, 2, 3 and 4 seconds after it starts moving. Enter the values in the following table.

	Time (s)	1	2	3	4
Car A	Speed (ms^{-1})				

Table 1.16.5

c) On the same axes draw speed-time graphs for the motions of both cars during the first 4 seconds.

d) After what time are both cars travelling at the same speed?

e) How far does car A travel during the first 4 seconds?

f) After 4 seconds which car is in front and by how much?

6 A week-day Inter City express train travels the 300 km between London and York in 3 hours.

a) What is the average speed of the train?

b) The Saturday train completes the same journey in 3 hours but spends 45 minutes stopped at stations. A passenger who catches both the week-day and Saturday service feels that the Saturday train is travelling faster, yet the train management maintain that the average speed for the journey is the same each day of the week. Use your knowledge of physics to explain why both the passenger and the train managers are correct.

7 A dragster in a race covers the 440 m from rest in 5.45 seconds.

fig 1.16.5 A dragster

a) Calculate the speed of the dragster at the end of the course if its acceleration is constant throughout the 5.45 seconds.

b) Describe how the speed of the dragster at the end of the race could be measured. List any equipment that you would use and describe what measuurements you would take. Explain how the speed is determined from the measurements that you make.

8 An aeroplane is travelling with a velocity of 120 m s^{-1} due north.

a) The aeroplane enters an area where wind turbulence gives it an additional velocity of 20 m s^{-1} to the east. What is the resultant velocity of the aeroplane as seen by an observer on the ground?

b) During a different part of the flight, a controller in a control tower tracking the plane estimates that its velocity relative to the ground is 150 m s^{-1} north. Describe the velocity of the wind in this region.

9 Two children are sitting in a train which is travelling due north at a speed of 18 m s^{-1}. The children are sitting one either side of the central aisle.

The child on the west side of the aisle rolls a ball 3 m directly across the aisle to his sister. The ball travels at a constant speed and takes 3 seconds to roll from one child to the other.

a) What is the northwards component of the velocity of the ball as it rolls between the children?

b) What is the eastwards component of the velocity of the ball?

c) Calculate the resultant velocity of the ball relative to the ground.

d) What distance, relative to the ground, does the ball travel as it rolls between the children?

10 Figure 1.16.6 shows a velocity-time graph for the motion of a lift in a department store.

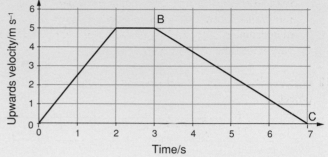

fig 1.16.6

a) Describe the motion of the lift during the first 2 seconds.

b) Calculate the acceleration of the lift during the first 2 seconds.

c) During the section labelled BC the lift is slowing down so that it can stop at the top floor. How far below the level of the top floor does it start to slow down?

d) How far does the lift travel between the start and end of its motion?

11 A motion sensor placed at the top of a runway points at a trolley as it rolls down a slope. Initially the motion sensor is 0.4 m from the trolley.

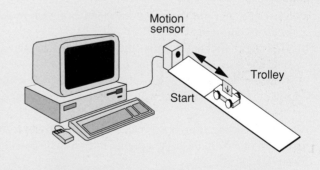

fig 1.16.7 Motion sensor equipment

The trolley is released from rest and its distance from the motion sensor is measured at regular intervals. The results are as tabulated below.

Time (s)	0	1	2	3	4	5
Distance from motion sensor (m)	0.4	0.5	0.8	1.3	2.0	2.9

Table 1.16.6

a) A pupil inspecting this results table cannot decide whether the results show a constant speed or a constant acceleration. Explain why these results cannot represent a constant speed.

b) Plot a graph to show how the displacement of the trolley from its starting point varies with time. How far is the trolley from the motion sensor 3.5 seconds after it is released?

c) The trolley is returned to its starting position and the slope of the runway is increased to give a greater acceleration. The trolley is released and its distance from the motion sensor is again recorded at regular intervals. Draw on your graph to show how the displacement of the trolley from its starting point would vary with time in this case.

12 Figure 1.16.8 shows a velocity-time graph for a moving object.

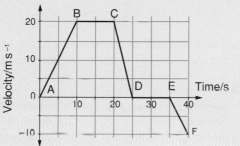

fig 1.16.8

a) Describe the motion of the object over the region DE.
b) How far does the object travel during the region marked CD?
c) What is the object's displacement from the start after 40 seconds?

13 In the apparatus of figure 1.16.9 a motion sensor points at a trolley travelling towards a fixed support.

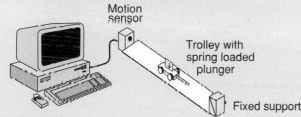

fig 1.16.9

The trolley is fitted with a spring loaded buffer and after hitting the support it returns back up the runway.

The motion sensor regularly measures the distance to the trolley and plots a graph like that shown in figure 1.16.10.

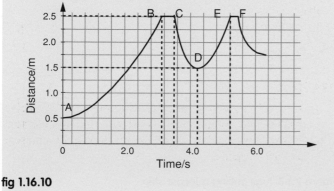

fig 1.16.10

a) Copy figure 1.16.10 and identify the region of the graph where the trolley is first travelling towards the fixed support.

b) Identify two regions on the graph where the trolley is in contact with the fixed support.

c) How far does the trolley return back up the runway after its first contact with the fixed support?

d) Describe the motion of the trolley at the point labelled D on the graph.

e) Calculate the acceleration during the part DE on the graph.

14 A stunt car driver uses a ramp as shown in figure 1.16.11 to jump over a parked bus. The path of the car is as shown.

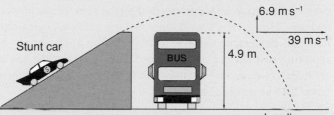

fig 1.16.11 Path of a stunt car

The horizontal and vertical components of the car's velocity at take off are as shown.

a) What is the time taken for the car to reach its highest point after leaving the ramp?

b) What is the time taken for the car to reach the landing area?

c) Calculate the horizontal distance between the landing point and the end of the take-off ramp.

15 A golfer practising her drive stands on the edge of a cliff and hits a golf ball horizontally. The sea is 37 m vertically below the edge of the cliff.

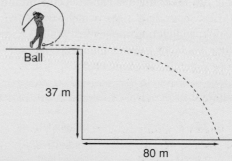

fig 1.16.12 Path of a golf ball

After one particular swing the ball follows the path as shown before landing in the water.

a) How long after it is struck does the ball enter the water?

b) What is the horizontal component of the velocity of the ball during its motion?

c) What is the vertical component of the velocity of the ball just before it enters the water?

d) Calculate the resultant velocity of the ball as it strikes the water.

2.1 Scalars and vectors

Introduction

In the previous chapters you have been introduced to some physical quantities such as mass, time, distance and speed. The size of a physical quantity is expressed as a number and a unit. In stating that the mass of the earth is 6×10^{24} kg or that the earth completes one orbit of the sun in 365 days, we have fully defined the mass and time involved by giving both a number and the appropriate unit. Omitting either the number or the unit leaves the size of the quantity uncertain.

Vectors

In the above examples the mass of the earth and the time for its orbit of the sun were *fully* described by the appropriate number and unit. However, a boat skipper making a distress call would not give the rescue services enough information simply by reporting that he was 300 km from Aberdeen. For the rescue services to locate the boat they would have to know how far to travel *and* in which direction.

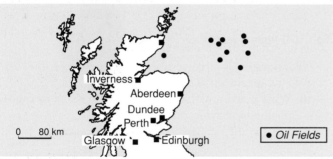

Figure 2.1.1 Locating a boat near Aberdeen

As this example shows, the position of the boat cannot be specified by simply stating a distance and the unit. The direction must also be given for this physical quantity to be fully described. **Quantities where the direction must be stated to convey their full meaning are called vector quantities**.

Quantities like mass and time which are fully described without stating a direction are called **scalars**. Some of the quantities that you will meet in the mechanics section of this course can be classified as shown below.

Scalar quantities	Vector quantities
Time	Velocity
Mass	Displacement
Distance	Acceleration
Speed	Force
Energy	Momentum
Power	Impulse

Table 2.1.1

Combining quantities

Some of the quantities in this table are made up by combining others. For example, acceleration is defined as;

$$a = \frac{v - u}{t}$$

The change in velocity on the top line is clearly a vector quantity whereas the time on the bottom line is scalar. In this case, combining a vector and a scalar gives another vector quantity. In a later chapter you will be introduced to momentum. This is defined by the equation;

$$Momentum = Mass \times Velocity$$

Momentum is the product of a vector quantity and a scalar quantity and is itself a vector quantity.

The kinetic energy of a mass, moving at a certain velocity is defined by the equation;

$$Kinetic\ energy = \frac{1}{2} Mass \times Velocity^2$$

Once again this equation includes both a scalar and a vector quantity but **kinetic energy is itself a scalar quantity.**

Forces

When considering the velocity of a moving object we use + and − signs to indicate movement in directly opposite directions. Similarly the forces acting in the tug-of-war contest shown in figure 2.1.2 act in directly opposite directions so we can say that one force acts in a positive direction while the other acts in the negative direction.

Adding vectors

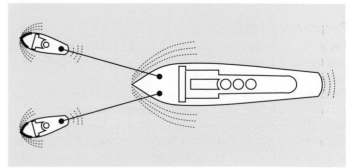

Figure 2.1.2 Tug-of-war

In figure 2.1.3 neither tug boat is pulling in the exact direction in which the ship moves, but clearly the combined effect of both pulling forces moves the ship straight ahead. We must therefore consider how forces in two different directions can add to give a resultant force in a third direction. As you will see n subsequent sections, forces are measured in newtons (N).

Figure 2.1.3 Tugs pulling a larger ship

Forces must be combined using the rules of vector addition. The simplest way to add vectors can be seen from the following example.

 Forces of 50 N and 75 N pull on the same point as shown in figure 2.1.4. Using the method outlined below find the resultant of these two forces.

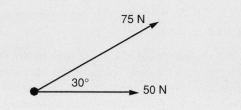

Figure 2.1.4

 1 Draw a sketch of the situation showing the sizes of the vectors and the angles between them, see figure 2.1.5
2 Choose a suitable scale
3 Draw the first vector to the appropriate scale
4 Draw the second vector, starting at the end of the first one and remembering to keep the angle the same as in the question
5 The resultant is found by drawing from the start of the first vector to the end of the last vector
6 Measure the length of the resultant and use the scale to convert back to newtons
7 Remember to measure and state the angle.

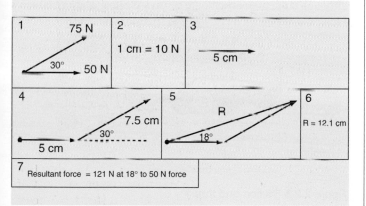

Figure 2.1.5 Drawing a scale diagram

An alternative approach uses the mathematical cosine and sine rules to find the resultant of two vectors.

Resolving vectors

The previous example was solved by adding the two vectors to find the resultant. This resultant is the combined effect of the two forces.

There are situations where it is useful to treat a single force as a resultant and divide it into two components. For example, a force of 9 N due east could be made up from forces of 6 N due east and 3 N due east. Or it could also have been the resultant of a 14 N force east and a 5 N newton force in a westerly direction (see figure 2.1.6)

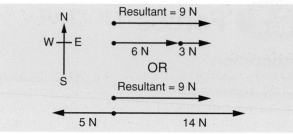

Figure 2.1.6

Resolving is often useful to separate a vector into vertical and horizontal components.

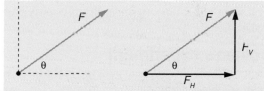

Figure 2.1.7

Figure 2.1.7 shows how forces of 3 N and 4 N can be added to give a resultant of 5 N at an angle of θ to the direction of the 4 N force. Forces of 4 N and 3 N are the horizontal and vertical components respectively of the 5 N force which acts at angle θ. In this simple example we have chosen forces of 3 N, 4 N and 5 N as they make a simple, right angled triangle. In general we can resolve any force, F, into its horizontal, F_H, and vertical, F_V, components as shown in figure 2.1.8.

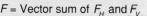

Figure 2.1.8 F = Vector sum of F_H and F_V

$$\cos \theta = \frac{F_H}{F} \qquad \sin \theta = \frac{F_V}{F}$$

$$F_H = F\cos \theta \qquad F_V = F\sin \theta$$

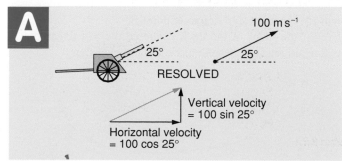 A shell leaves the muzzle of an artillery gun with a velocity of 100 m s^{-1} at an angle of 25° to the horizontal. Calculate values for the initial velocities in the horizontal and vertical directions.

Figure 2.1.9

$$\textit{Horizontal velocity} = 100 \cos 25° = 90.6 \ m\,s^{-1}$$
$$\textit{Initial vertical velocity} = 100 \sin 25° = 42.3 \ m\,s^{-1}$$

By treating these as the initial horizontal and vertical velocities we can use the equations of motion to find the range of the projected shell and the maximum height it reaches.

2.2 Forces

Introduction

A force is a push or a pull acting on an object. If opposite poles of magnets are suspended near each other they will produce an attractive force. Similarly when a bat hits a ball the direction of the ball changes because of the effect of the force. There are many examples of contact and non-contact forces other than those shown in figure 2.2.1.

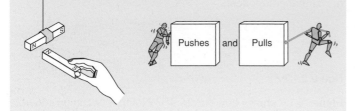

Figure 2.2.1 The effect of forces

When forces act on an object they change either its shape, the direction in which it moves or the speed at which the object moves. It is quite easy to imagine how contact forces can cause changes but more difficult to explain how non-contact forces such as gravity, magnetism or electrostatic forces can affect objects some distance away from the source of the force.

Galileo's experiment

Ancient Greek philosophers believed that a force was needed to make an object move at a constant speed. This work was questioned by Galileo's famous thought experiment. This imagines a ball released at point A on a curved *frictionless* track.

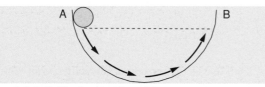

Figure 2.2.2 Galileo's experiment

Since no energy is lost through friction the ball when released will rise up to its start height at point B.
Even if the shape of the track is altered the ball released at A will again reach B, despite having to travel farther.

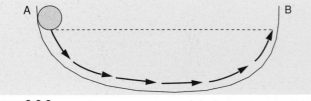

Figure 2.2.3

In the absence of friction, the ball released on one side of the curve will always reach the same height on the other side of the track.

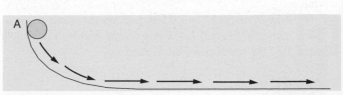

Figure 2.2.4

If the ball is again released at point A as shown in figure 2.2.4 it will try to reach the same height on the other side of the track. However, since the track is level it will continue to roll 'forever' at a constant speed, provided no forces act to change the motion.

Newton's First Law

In 1687, Isaac Newton published his famous work formalising Galileo's ideas. His First Law of Motion can be expressed as;

An object at rest will remain at rest, and a moving object will continue with a constant velocity unless an unbalanced force acts.

This is sometimes referred to as the ***Balanced forces no change in motion law***.

Balanced forces

In some ways, the belief of the ancient Greek philosophers is understandable. In their experience, as in ours, any moving object will eventually stop unless it is being pushed. The push is needed to balance or overcome the frictional forces which, although sometimes small, are always present. But in Galileo's thought experiment there were no frictional forces so the object continued moving even though it was not being pushed.

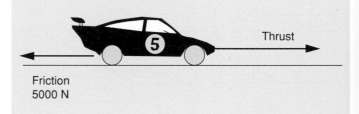

Figure 2.2.5 Friction and thrust

When the car shown in figure 2.2.5 moves at a constant $60\,\mathrm{km\,h^{-1}}$, the frictional forces opposing the motion are 5000 N. To maintain this constant speed the car's engine must produce a forward force, called a **thrust**, of 5000 N. If the car were to travel at $90\,\mathrm{km\,h^{-1}}$, the frictional forces would be larger, so the engine would need to produce a greater thrust to balance the higher frictional forces.
When a car accelerates, the friction and drag forces increase. Eventually, a time will be reached where the car's engine is producing maximum thrust and the car is travelling at its top speed. Car designers wishing for higher top speeds can use larger engines to produce more thrust or they can reduce the friction and drag so that the thrust and the forces opposing the motion only become equal when the car is travelling faster.

Figure 2.2.6 Car in a wind tunnel test

Terminal velocity

Sky divers jumping from an aeroplane accelerate towards the ground and as their speed increases the frictional forces also increase. Eventually a time is reached when the sky diver is travelling at top speed. This is called the **terminal velocity**. Teams of sky divers wishing to do formation stunts try to control their terminal velocity so that all members of the team are falling at the same speed.

When nearing the ground sky divers open their parachutes. This dramatically increases the frictional forces opposing their descent. The increased frictional forces reduce the speed at which they fall. As the speed reduces, the frictional forces also reduce to a point where the frictional forces again balance the downwards force. The sky diver has now achieved a lower terminal velocity. In this way experienced parachutists try to achieve very low landing speeds.

Measuring frictional forces

EXPERIMENT When a trolley is pushed along a level runway and released, it slows down because of the frictional forces.

It is possible to use the force produced by a falling mass to pull a trolley along a level runway.

You can use the motion sensor to monitor the speed of the trolley as it is pulled and you can alter the size of the falling mass to find the force needed to pull the trolley at a constant speed.

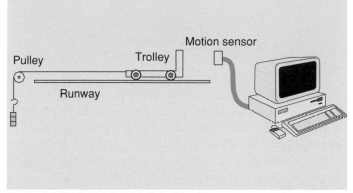

Figure 2.2.8 Measuring frictional forces

When the trolley travels at a constant speed the pulling force and the frictional force must be equal and opposite in direction. The size of the frictional force between the rolling trolley and the runway is therefore equal to the pulling force required for constant speed.

You might wish to extend this experiment to look at the frictional forces involved with different surfaces. You may wish to try polished surfaces, rough wood surfaces or different types of carpet.

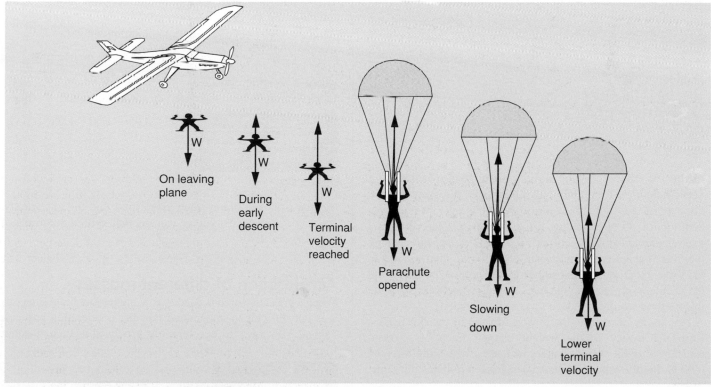

Figure 2.2.7 Velocity during a sky diver's descent

2.3 Force, mass and acceleration

Introduction

When balanced forces act on a moving object, the object moves with a constant velocity. If the forces acting are not balanced, the velocity of the moving object changes.

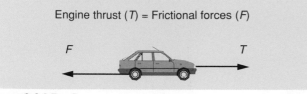

Figure 2.3.1 T = F

Figure 2.3.1 shows that, if the thrust from the car's engine increases, the car speeds up. It accelerates because the thrust is larger than the frictional forces. However if the thrust is removed or the brakes are applied the car slows down. In these cases the **retarding forces** exceed the forward forces so the car decelerates.

Quantitative accelerations or decelerations

By considering the above example we see that unbalanced forces cause accelerations or decelerations. Our everyday experience would tell us that the larger the unbalanced force the larger the acceleration or deceleration. Similarly, it is not too difficult to believe that objects with more mass will require a bigger unbalanced force to produce a certain acceleration.

EXPERIMENT We can investigate the effect of different unbalanced forces on the acceleration of a constant mass, using apparatus like that shown in figure 2.3.2 below.

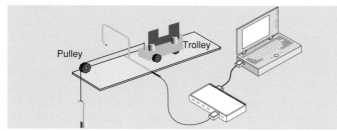

Figure 2.3.2

In this experiment, towing forces are supplied by allowing a mass, hanging over a pulley, to fall towards the ground. Each 50 g mass added to the falling weight hanger provides an additional 0.49 N of towing force. However if the additional masses are added from *outside the system* we are not keeping the overall accelerating mass constant. This would mean that in the same experiment we are altering both the towing force *and* the accelerating mass so the tests would not be fair. Adopting this method would introduce *a systematic error.*

A better way of altering the towing force is to transfer masses from the trolley to the weight hanger. Each 50 g mass transferred to the weight hanger increases the towing force by 0.49 N, but the total mass being accelerated remains the same.

Accelerations and unbalanced forces

Table 2.3.1 shows results obtained when the accelerations caused by a range of pulling forces are measured.

Pulling force/N	0.49	0.98	1.47	1.96	2.45
Acceleration/m s^{-2}	0.54	1.05	1.63	2.2	2.72

Table 2.3.1

These results confirm everyday experience; **as the towing force increases the acceleration increases**. The simplest mathematical relationship linking quantities which increase in this way is called a **direct proportionality**. We can test to see if the acceleration is directly proportional to the towing force in two ways. Firstly we can plot a graph of acceleration versus towing force. If the best fit line is a straight line passing through the origin (0,0), we can conclude that the acceleration, a, is directly proportional (α) to the towing force, F.

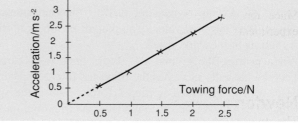

Figure 2.3.3

In mathematical notation we would write;

$$a \propto F \ or \ a = kF$$

where 'k' is some constant numerical value.

The second method of testing for a direct proportionality is to rearrange the above equation to give;

$$\frac{a}{F} = k$$

where 'k' is some constant.

Using the results from the table you can calculate the $\frac{a}{F}$ values.

Pulling force/N	0.49	0.98	1.47	1.96	2.45
Acceleration/m s^{-2}	0.54	1.05	1.63	2.2	2.72
$\frac{a}{F}$ calculated values	1.10	1.07	1.11	1.12	1.11

Table 2.3.2

We can calculate the average value for k and the uncertainty. If the uncertainty is small we can say that, within the limits of experimental error;

$$a \propto F$$

Acceleration of different masses

When a constant force is used to tow bigger and bigger masses it is not difficult to understand that the acceleration reduces. Table 2.3.3 shows the accelerations produced when a constant force of 1.96 N tows a range of masses along a level runway.

Mass/kg	0.90	1.10	1.30	1.50	1.70
Acceleration/m s^{-2}	2.20	1.80	1.52	1.32	1.16

Table 2.3.3

Clearly from these results we can see that as the mass increases the acceleration decreases. The simplest mathematical relationship linking quantities which vary in this manner is called an **indirect proportionality**.

Mathematically we can state an indirect proportionality as;

$$a \propto \frac{1}{m} \quad or \quad a = \frac{k}{m}$$

Again, we can test to see if our data fits this type of relationship by drawing a graph of acceleration, *a* versus 1/mass, 1/*m*. A straight line through the origin would confirm that acceleration is directly proportional to 1/*m*. Alternatively we can rearrange the above equation to show that;

$$ma = k$$

Using the results from the table you can calculate the *ma* products.

Mass/kg	0.90	1.10	1.30	1.50	1.70
Acceleration/m s^{-2}	2.20	1.80	1.52	1.32	1.16
ma calculated values	1.98	1.98	1.98	1.98	1.97

Table 2.3.4

Since the *ma* products are constant, within the limits of experimental error, we can again say;

$$a \propto \frac{1}{m}$$

Newton's Second Law

The two relationships governing acceleration, mass and unbalanced force can be combined into a law. The two relationships are;

$$a \propto \frac{1}{m} \quad and \quad a \propto F$$

Therefore we can say;

$$a \propto \frac{F}{m}$$

Which rearranges to give;

$$F \propto ma$$
$$or \; F = kma$$

In this equation the constant *k* is simply a number. Indeed many years ago it was arranged that this constant should equal 1! This was done when the newton was adopted as the unit of force.

One newton (1N) is defined as the unbalanced force required to make a mass of 1 kg accelerate at 1 m s^{-2}.

Therefore when expressing forces in units of newtons we can write Newton's Second Law of Motion as

$$F = ma$$

Measuring forces

Forces can be measured with a Newton balance which is sometimes referred to as a force meter.

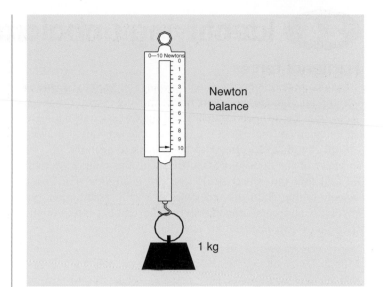

Figure 2.3.4 A Newton balance

When a mass is hung on the hook of the Newton balance the spring extends because the mass is being pulled by an attractive force towards the centre of the earth. When the mass is stationary on the end of the Newton balance, the force exerted upwards by the spring on the mass is equal and opposite to the force of the earth pulling down on the mass. As you will see in section 2.12, the force of the earth pulling on any object is called its **weight**, so the spring in the Newton balance is exerting a force equal to the weight of the object. The spring extends and **Hooke's Law** states that its extension is directly proportional to the applied force. This allows the calibrated scale marked beside the spring in the Newton balance to have equally spaced graduations.

Calibrating a Newton balance

When a mass of 1 kg is falling freely, the pull of the earth causes it to accelerate at 9.8 m s^{-2}. According to Newton's Second Law we can say that a 1 kg mass accelerating at 9.8 m s^{-2} is being pulled by an unbalanced force of 9.8 N. If the same 1 kg mass is then held stationary while attached to the end of the uncalibrated Newton balance, the spring must be exerting an upward force of 9.8 N. The extension which this causes can be marked and the space on the scale between the spring's original length position and the 9.8 N mark, divided into equal graduations.

Watch your weight!

The word 'weight' is widely misused in everyday conversation. Its scientific meaning and its meaning in the minds of the general public are different. Non-scientists use the word weight when they are talking about the mass of an object and will speak of a weight of 'so many kilograms'.

Scientists use the word weight to describe a measurement of the force of attraction between a planet and an object. Weight, being a force, is measured in newtons.

2.4 Identifying unbalanced forces

Frictional forces

To calculate accelerations using $F = ma$, you need to be able to identify all of the forces acting on the moving object. But it can sometimes be difficult to identify all the forces contributing to an unbalanced force. Friction, for example acts on all moving objects, but its exact effect can be difficult to quantify.

In many of the questions associated with this course you will be told that the *effects of friction are negligible*. For heavy objects falling short distances this is a valid assumption, but skydivers and parascenders rely on the braking effect of the frictional forces between the parachute and the air.

Figure 2.4.1 Benefits of friction

There are other situations where large frictional forces are desirable. In football matches goalkeepers wear gloves to give better grip when the ball is wet. Similarly, the manufacturers of car tyres do research to find which types of tyre and tread pattern have the best road holding capabilities. Lorry brakes must generate large frictional forces with the wheels to cause decelerations.

Engineers use different methods to reduce the frictional forces where motion must continue with a minimum input of energy. Many bearings are greased or lubricated to minimise the effect of friction. Dance floors used by ballroom dancers are polished to ease the motion of the competitors. In classroom experiments we can use air tracks or air tables to reduce the effects of friction.

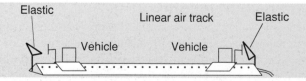

Figure 2.4.2 Linear air track

The vehicles on an air track travel on a cushion of air. This means that hard solid surfaces are no longer touching. The contact is between the metal surfaces and the moving air molecules. However, if the vehicle is given a push and allowed to rebound off elastic bands at both ends, it hardly slows down. Frictional forces still act but the fact that the vehicle travels for

such a long time before stopping shows that the frictional forces are small.

Calculating unbalanced forces

A lorry with a mass of 20,000 kg has six wheels, each of which contributes 1250 N to the overall frictional force. If the engine is producing a thrust of 100,000 N, we can calculate the acceleration.

Figure 2.4.3 Calculating the acceleration of a lorry

To analyse this situation it is helpful to draw a diagram showing the forces acting. This type of diagram is called a **free body diagram**.

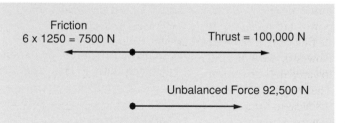

Figure 2.4.4 Free body diagram

Using Newton's Second Law;

$$F = ma$$
$$92,500 = 20,000 \times a$$

Therefore;

$$a = 4.6 \, m \, s^{-2}$$

It is important to identify all the forces acting before calculating the unbalanced force. Once the unbalanced force is found the acceleration is easily calculated from Newton's Second Law.

Tied objects

A car and a caravan with a combined mass of 800 kg travel along a rough road with an acceleration of $3 \, m \, s^{-2}$. If the frictional forces between the wheels and the road total 3,000 N, we can calculate the size of the thrust produced by the car's engine.

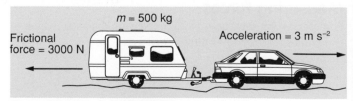

Figure 2.4.5 Car towing caravan

Even though the car and caravan are two separate objects, they are connected and travelling together so their acceleration must be the same.

> *The unbalanced force = Mass × Acceleration*
> *F = 800 × 3*
> *F = 2400 N*

Unbalanced force 2400 N

This is made up from;
    Friction = 3000 N   Thrust

from the diagram we can see that;

> *Thrust – 3000 = 2400 N*

Therefore;

> *Thrust = 5400 N*

In this example the engine produces a force of 5400 N.

Rockets taking off

Any rocket about to take off must produce an upwards force sufficient to lift it from the launch pad. The absolute minimum force required from the rocket's engines must equal the downward forces keeping the rocket on the launch pad. If the thrust produced by the rocket's engine is greater than the downward forces the rocket will accelerate upwards.

At first sight it is not obvious what causes the downward force. However, we must remember that the rocket is being pulled towards the centre of the earth with a force called the weight. The weight of a mass of 1 kg has previously been shown to be 9.8 N so to lift a rocket of mass 10,000 kg from the launch pad requires a minimum force of 98,000 N.

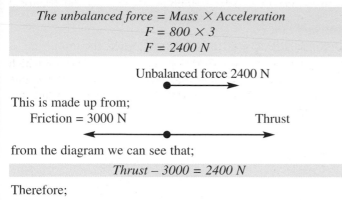
Thrust | Unbalanced force 30,000 N
$m = 10,000$ kg
Weight = 98,000 N

Figure 2.4.6 Forces on a rocket

To make the rocket accelerate upwards at 3 m s^{-2} requires an unbalanced force given by;

> *F = ma*
> *F = 10,000 × 3*
> *F = 30,000 N*

Therefore

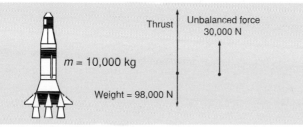
Thrust | Unbalanced force 30,000 N

Weight = 98,000 N

From the free body diagram above we can see that the resultant of the thrust and the weight is 30,000N;

> *Thrust – 98,000 = 30,000 N*
> *Thrust = 30,000 + 98,000*
> *Engine thrust = 128,000 N upwards*

Changing masses and forces

The energy needed to lift the rocket from the launch pad is supplied from the fuel burnt in the engines. As the engines produce a constant thrust by burning fuel, the mass of the rocket reduces. We can explain the subsequent motion by considering Newton's Second Law in the form;

$$a = \frac{F}{m}$$

If the quantity on the top line of this equation remains *constant* while the quantity on the bottom line *reduces* then the acceleration must *increase*.

If enough of a chain dangles over the edge of a bench it will fall to the ground accelerating as it goes. The links falling vertically are providing the towing force, so as more and more of the chain gets pulled over the edge, the towing force increases. This means that the chain of *constant mass* is being pulled by an *increasing force*. Once again we can explain the subsequent motion by considering Newton's Second Law;

$$a = \frac{F}{m}$$

The force on the top line of this equation is increasing while the mass on the bottom line remains constant. Again, the overall effect is an increase in the acceleration.

In Formula 1 racing, the teams have to carefully judge the mass of fuel carried by the car. Carrying smaller masses of fuel means that the acceleration produced by the engine is larger, so the car can complete each lap quicker. However, team managers must consider the length of a race and balance the time lost refuelling with the advantages of minimising the mass of fuel carried.

Figure 2.4.7 Formula 1 car refuelling

2.5 Consolidation questions

Worked example 1

Figure 2.5.1 shows a situation where an object of mass 28 kg is held stationary by the forces acting in ropes X, Y and Z. The sizes of the forces are as shown. (You should ignore the effects of the force of friction.)

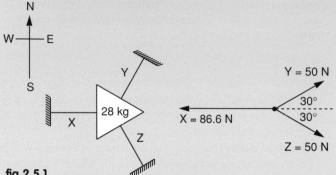

fig 2.5.1

a) What is the resultant of the 50 N forces which act in ropes Y and Z? Explain your answer.

b) The rope pulling due west snaps. Calculate the initial acceleration of the mass.

c) The mass is again secured so that the forces are as shown in figure 2.5.1. What would the acceleration of the object be if the rope Y snapped?

Answers and comments

a) Since the object is stationary, and therefore not accelerating in any direction, the forces acting in any particular direction must balance.
    The 86.6 N force pulling due west must be balanced by a total force of 86.6 N acting due east.

b) If the rope X snaps there is an unbalanced force of 86.6 N due east. We can calculate the acceleration from $F = ma$;

$$a = \frac{F}{m} = \frac{86.6}{28}$$
$$a = 3.09\ m\,s^{-2}$$

c) If rope Y snaps the mass accelerates because of the unbalanced force from the remaining ropes. The resultant force will be 50 N in a direction opposite to the original force in Y.

$$a = \frac{F}{m} = \frac{50}{28}$$
$$a = 1.79\ m\,s^{-2}$$

The acceleration is 1.79 $m\,s^{-2}$ at an angle of 30° south of west, bearing 240.

Worked example 2

A steel ball with a mass of 0.1 kg travelling horizontally at a speed of 8 m s^{-1}, embeds itself in a sandbag which brings it to rest in 0.08 seconds.

a) What is the deceleration of the steel ball as it is brought to rest by the sandbag?

b) What is the size of the average force causing the steel ball to slow down?

c) The size of the force decelerating the steel ball varies with time as shown in figure 2.5.2.

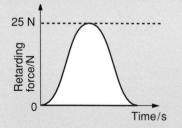

fig 2.5.2

Calculate the maximum deceleration of the steel ball.

Answers and comments

a) The steel ball slows down from a speed of 8 ms^{-1} to rest in a time of 0.08 seconds. We can calculate the acceleration during this time from;

$$a = \frac{v - u}{t} = \frac{0 - 8}{0.08}$$
$$a = -100\ m\,s^{-2}$$

The negative sign means that this is a deceleration of 100 m s^{-2}.

b) We can calculate the average force causing the acceleration from;

$$F = m \times a = 0.1 \times -100$$
$$F = -10\ N$$

The negative sign indicates that this force acts in the opposite direction to the initial motion of the ball.

c) In the earlier parts of this question we have assumed that the deceleration was constant because the force was constant. In this part we are asked to consider the situation where the force is varying.
    The maximum deceleration will occur where the retarding force is greatest. From the graph the largest retarding force is 25 N so the maximum deceleration will be given by;

$$F_{max} = ma$$
$$25 = 0.1 \times a_{max}$$

$$a_{max} = \frac{25}{0.1} = 250\ m\,s^{-2}$$

The maximum retardation of the ball is 250 m s^{-2}.

Questions

1 Which one of the following is a scalar quantity?
   a) Velocity
   b) Acceleration
   c) Force
   d) Momentum
   e) Kinetic energy.

2 At the start of a yacht race one competitor sails 30 km due north in 23 hours before being blown 3 km due south by a gale force wind which lasts for 1 hour.
   a) What is the yacht's displacement at the end of the journey?
   b) What is the average velocity of the yacht for the day?

3 Describe and explain how the resultant of two non-parallel forces acting at a point may be determined.

4 Figure 2.5.3 shows two forces pulling at right angles on an object at point P.

fig 2.5.3

Using a scale drawing find the resultant of the two forces.

5 The resultant of the two forces shown in figure 2.5.4 acts at angle of 30° to the 40 N force.

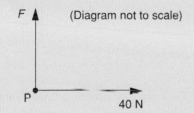

fig 2.5.4

What is the magnitude of the force labelled F?

6 A garden roller is pulled with a force of 200 N acting at an angle of 40° to the level ground.

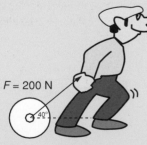

F = 200 N

fig 2.5.5

What is the magnitude of the part of this force which pulls the roller in the forward direction?

7 Figure 2.5.6 shows a graph for the speed of a parachutist during a descent.

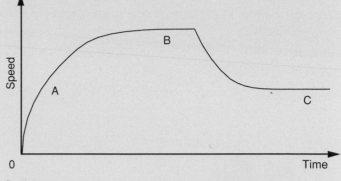

fig 2.5.6

   a) Draw diagrams to show the relative sizes of the vertical forces acting on the parachutist at the points marked A, B and C.
   b) Explain why the steady speed at point C is less than the steady speed at point B.

8 In figure 2.5.7 a force of 8 N causes a mass of 4 kg to accelerate at 0.75 m s⁻² along a rough horizontal surface.

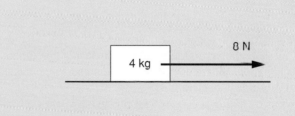

8 N

4 kg

fig 2.5.7

   a) What unbalanced force is needed to cause the acceleration?
   b) What is the magnitude of the forces opposing the 8 N pulling force?
   c) Oil is spread on the ground in an attempt to increase the acceleration. Explain in terms of the horizontal forces acting on the block why the acceleration increases.

9 A mass of 10 kg is dropped over the side of an airship hovering 40 m above the ground.

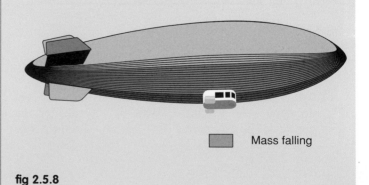

Mass falling

fig 2.5.8

a) What is the initial acceleration of this mass?

b) As the mass falls the frictional forces acting on it increase.

    (i) Explain what happens to the speed of the mass as it falls.

    (ii) Explain what happens to the acceleration of the mass as it falls.

10 A scooter travels along a straight level road at a maximum speed of 25 m s^{-1}.

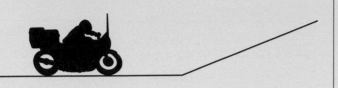

fig 2.5.9

a) The frictional forces acting are 750 N. What is the forward force (thrust) produced by the scooter's engine?

b) Explain why the scooter decelerates as it starts to climb uphill even though the thrust from the engine stays constant.

11 A sports car of mass 550 kg accelerates from rest to a speed of 30 m s^{-1} in 6 seconds.

a) Calculate the acceleration of the car during this time.

b) What is the magnitude of the unbalanced force needed to cause this acceleration?

c) When travelling at 30 m s^{-1} the engine is switched off and the car coasts to a halt. While slowing down the car travels 90 m.

    (i) Calculate the deceleration during the slowing down.

    (ii) What is the magnitude of the frictional forces acting while the car comes to rest?

12 A breakdown truck tows a car of mass 975 kg along a level road at a constant speed. The total frictional force between the wheels of the car and the road is 1275 N.

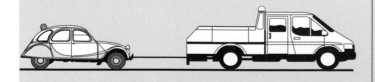

fig 2.5.10

a) What is the size of the force of the truck on the car to maintain this constant speed?

b) The breakdown truck accelerates so that the car accelerates at 0.5 m s^{-2}. What is the size of the additional force on the tow rope to produce this acceleration?

c) What is the total force pulling the car while it is accelerating?

d) Due to a weakness in the rope the maximum towing force that it can exert is 2000 N. What is the maximum acceleration that the car could have before the tow rope breaks?

13 A rope with a breaking strength of 1000 N is used to lift a bucket of mass 80 kg through a vertical height of 80 m onto a ledge.

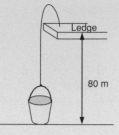

fig 2.5.11

a) What is the weight of the bucket?

b) What is the maximum force that can accelerate the bucket upwards?

c) What is the maximum acceleration which the bucket can have in the upwards direction?

d) What is the minimum time for the bucket to rise from the ground on to the ledge?

14 A rocket with a mass of 1×10^5 kg rests on a launch pad.

fig 2.5.12

a) What force is needed to accelerate the rocket upwards at 1 m s^{-2}?

b) The rocket produces a constant thrust. Explain why the acceleration increases as the motion continues.

15 An aeroplane flying north west at a steady speed at a constant height is acted on by the forces shown in figure 2.5.13.

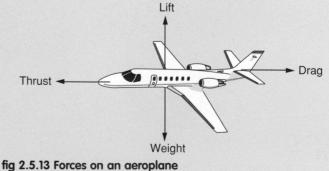

fig 2.5.13 Forces on an aeroplane

a) Explain why the lift force must be exactly equal in size, and act in exactly the opposite direction, to the weight.

b) To rise to a higher altitude the plane accelerates forward. Explain in terms of the forces acting why this causes the plane to rise.

16 An engineer models the forces acting on a front-wheel drive car travelling at a steady speed along a level road as shown in figure 2.5.14.

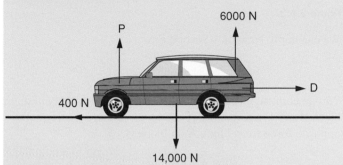
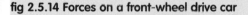

fig 2.5.14 Forces on a front-wheel drive car

P is the force from the road supporting the front of the car while D is the drag force caused by the air resistance.

a) Explain why P must equal 8000 N.

b) Explain why the engineer can assume that the drag forces at this speed total 400 N.

c) The driver now accelerates the car by doubling the forward force. What is the size of the force causing the acceleration?

d) Calculate the acceleration of the car.

e) In her model the engineer assumes that the acceleration due to gravity is 10 m s^{-2}. Show, by calculation, that the mass of the car is 1400 kg.

f) As the speed of the car increases the driver notices that the acceleration reduces. Explain, in terms of the forces involved, why the car will eventually reach a top speed.

17 A pathfinder landing craft descending vertically towards the surface of a planet travels at a constant speed of 2 m s^{-1}.

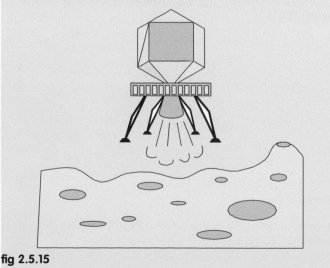

fig 2.5.15

The total mass of the landing craft is 125 kg and the acceleration due to gravity on the planet causes an acceleration of 40 m s^{-2}.

a) Calculate the upward thrust from the engine that will cause the landing craft to descend at a steady speed.

b) The upward force is increased to 5050 N for the final 5 seconds of the descent. What is the deceleration of the landing craft during the final 5 seconds of the descent?

c) Calculate the speed of the landing craft as it touches down.

d) Sketch a graph to show how the velocity of the landing craft changes during the 5 seconds after the thrust of the engine increases to 5050 N.

e) How far above the surface of the planet is the landing craft when the engine force increases to 5050 N?

2.6 Newton's Third Law

Introduction

Newton's First and Second Laws explain the effect of forces acting on a *single* object. The First law describes the **motion if the forces on a single object balance**, while the Second law quantifies the **acceleration of a single object if the forces are not balanced**.

Figure 2.6.1 Car pulling a caravan

If the frictional forces between each of a caravan's two wheels and the road are 1000 N, the car towing the caravan must pull it with a force of 2000 N to maintain a constant speed.

If the forces opposing the motion of the car itself total 4000 N, the car's engine must produce a thrust of 6000 N to move the car and caravan at a constant speed. If the caravan were to break free, a thrust of 6000 N from the engine would make the car accelerate. Clearly, the very act of pulling a caravan has an effect on the car. This should be no surprise as it is in keeping with everyday experience.

This example shows that when object A exerts a pulling force on object B, there is also a pull on object A due to the presence of B. **Newton's Third Law** formalises this situation and states that;

If A exerts a force on B, then B exerts a force on A which is equal in size and in the opposite direction.

This law links the *action* of a force pulling an object with the *reaction* on the body providing the pulling force.

Examples of pairs of forces.

Two masses placed side by side on a frictionless surface are accelerated by a force of 20 N.

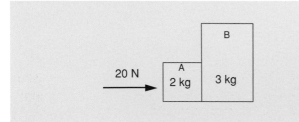

Figure 2.6.2

It is possible to find the force exerted by the 3 kg block on the 2 kg one. Since the blocks are together they have the same acceleration.

Applying Newton's Second Law for the blocks *together*,

$$F = m \times a$$
$$20 = 5 \times a$$
$$a = 4\ m\,s^{-2}$$

Both blocks therefore accelerate to the right at $4\ m\,s^{-2}$. Now considering the forces acting only on the 2 kg mass.

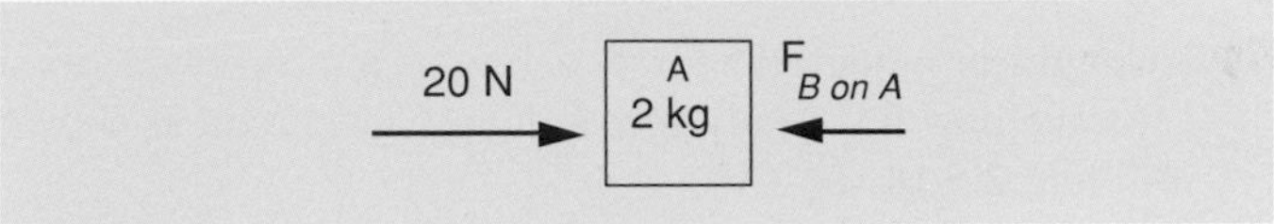

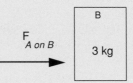

Figure 2.6.3

We can apply Newton's Second Law;

$$20 - F_{B\ on\ A} = 2 \times 4$$
$$F_{B\ on\ A} = 12\ N\ to\ the\ left$$

We can confirm that this is correct by considering that although block B exerts a force of 12 N on block A, there is an equal and opposite force from block A on block B.

$$F_{B\ on\ A} = F_{A\ on\ B}$$
$$F_{B\ on\ A} = 12\ N$$
$$F_{A\ on\ B} = 12\ N\ to\ the\ right$$

Figure 2.6.4

We may now calculate the effect this 12 N force has on block B.

$$F_{A\ on\ B} = 3 \times a$$
$$12 = 3 \times a$$
$$a = 4\ m\,s^{-2}$$

Therefore the 3 kg block accelerates at $4\ m\,s^{-2}$ to the right. This value should be no surprise as the blocks moving together must have the same acceleration.

We may ask whether the force between the two blocks would have been different if the bigger block was placed in front of the smaller one.

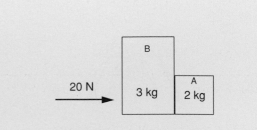

Figure 2.6.5

Once again using Newton's Second Law for the whole system shows that the blocks will accelerate at $4\ m\,s^{-2}$.
Considering the 3 kg block alone;

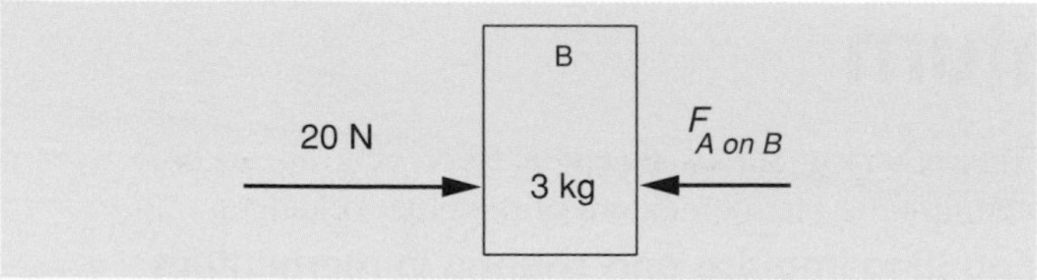

B

20 N → 3 kg ← $F_{A \text{ on } B}$

Figure 2.6.6

$$20 - F_{A \text{ on } B} = 3 \times 4$$
$$F_{A \text{ on } B} = 8\ N$$

In this case the force at the point of contact between the blocks is only 8 N.

Additional example

When using Newton's Third Law we must be careful to identify the objects on which the forces act.

Consider again the example illustrated in figure 2.6.1. If the frictional forces at each of the caravan's wheels are 1000 N the car must be providing the caravan with a towing force of 2000 N. The caravan must also be pulling back on the car with a force of 2000 N. Since the car and caravan are joined by a rigid tow bar the tension in the tow bar is 2000 N.

If the car which has a mass of 800 kg, and the caravan, a mass of 500 kg, now accelerate forwards at 3 m s^{-2}, we can calculate the new tension in the tow bar.

Since the car and caravan are joined they have the same acceleration. So we firstly consider them *together*.

From Newton's Second Law we can say;

$$F = ma$$

$$\frac{Engine}{thrust\ (E_T)} - \frac{Opposing}{forces} = Total\ mass \times Acceleration$$

The forces opposing the motion arise because of friction between the six tyres (two from the caravan + four on car) and the ground. The friction from each tyre is 1000 N. Therefore;

$$E_T - 6000 = 1300 \times 3$$
$$E_T = 3900 + 6000$$
$$E_T = 9900\ N$$

So to provide the acceleration the car's engine must produce a thrust of 9900 N.

We can also use Newton's Second Law for the forces acting on *the caravan alone*.

2000 N

← Frictional forces ● Force from tow bar →

Figure 2.6.7 Free body diagram for the caravan

$$Force\ from\ tow\ bar - 2000 = 500 \times 3$$
$$Therefore\ the\ force\ of\ tension\ in\ the\ tow\ bar = 3500\ N$$

So the caravan is receiving a pulling force of 3500 N from the car.

Consider the situation if the car and caravan separated. The total frictional forces from the car's four tyres is now 4000 N.

$$Engine\ thrust - opposing\ forces = mass \times acceleration.$$
$$Engine\ thrust - 4000 = 800 \times 3$$
$$Engine\ thrust = 6400\ N.$$

This shows that for the car alone to accelerate at 3 m s^{-2} the engine would have to provide a thrust of 6400 N. However, with the caravan attached the engine must provide a thrust of 9900 N. The extra force required because of the caravan is therefore 3500 N. This is exactly equal to the tension force from the car accelerating the caravan at 3 m s^{-2}.

In this example we can say that *if the car exerts a force of 3500 N on the caravan, then the caravan exerts an equal and opposite force of 3500 N on the car.*

Identifying pairs of forces

In the examples studied so far the pairs of equal and opposite force have been fairly obvious. Consider the front tyre on a front-wheel drive car just before the car starts to move. The engine produces a force which pushes the bottom of the tyre backwards. It is the equal and opposite frictional force between the tyre and the road that is responsible for the car's forward motion!

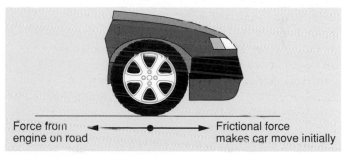

Force from ← ● → Frictional force
engine on road   makes car move initially

Figure 2.6.8 Forces on a car tyre

The moon orbiting the earth experiences a force of attraction which keeps it in orbit. Similarly the moon attracts the earth with an equal and opposite force.

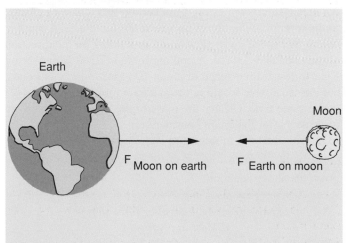

Earth

Moon

$F_{\text{Moon on earth}}$ →   ← $F_{\text{Earth on moon}}$

Figure 2.6.9 Forces on the moon and the earth

2.7 Impulse and momentum

Introduction

Newton's Laws explain the behaviour of objects when forces act on them. They explain the effect of balanced forces or changes in motion caused by unbalanced forces. Newton's Third Law even links the effect of a force on an object that is pulled, with the effect on the object doing the pulling. These laws are still the basis for much of the innovation in modern engineering design. As well as considering the *size* of forces acting on an object an engineer must consider the *time* for which the forces act. A small force acting for a long period can cause the same effect as a larger force that acts only for a short time. In physics the term *impulse* is defined as;

$$Impulse = Force \times Time\ of\ contact$$

When a car is stopped quickly by applying a large braking force for a short time the velocity of the car changes very quickly. The car could also have been brought to a halt using a smaller braking force in contact with the wheels for a longer time. From this example we can see that a force acting for a certain time causes a change in motion.

Newton's Second Law can be stated as;

$$F = ma$$

$$But;\ a = \frac{v - u}{t}$$

$$So;\ F = m\ \frac{v - u}{t}$$

$$Which\ gives;\ F = \frac{mv - mu}{t}$$

$$or;\ Ft = mv - mu$$

From Newton's Second Law we have been able to equate the impulse of a force with the quantity $mv - mu$.

Momentum

An unbalanced force acting for a certain time causes a change in the motion of the object on which the force acts. A moving object will speed up or slow down depending on the direction of the unbalanced force. A braking force of 4500 N may bring a car to rest from a speed of about 100 $km\,h^{-1}$ in 3.5 seconds, but the same force would not stop a train moving at 100 $km\,h^{-1}$ in the same time. Clearly the mass of the object on which the force acts is an important consideration.

In physics the term **momentum** is defined as;

$$Momentum = Mass \times Velocity$$

From the above equation;

$$Ft = mv - mu$$
$$Impulse = Momentum\ after\ force\ acts\ -$$
$$Momentum\ before$$

or;

$$Impulse = Final\ momentum - Initial\ momentum$$

or in a simpler form;

$$Impulse = Change\ in\ momentum$$

This equation shows that if a force acts on an object for a certain time the momentum of the object changes.

Equating impulse and change in momentum

Impulse and momentum are each the product of a vector quantity and a scalar quantity. They are themselves both vector quantities. In section 2.1 both momentum and impulse are shown as vector quantities in table 2.1.1.

If we consider the equation;

$$Impulse = Change\ in\ momentum$$

We must be able to equate the units on either side of the equals sign. Impulse, being the product of a force and time, must have units of newton seconds (N s).

Since momentum is the product of mass and velocity the units of change in momentum must be kilogram metre per second ($kg\,m\,s^{-1}$). 1 $kg\,m\,s^{-1}$ is exactly the same size as 1 N s.

Quantity	Units
Impulse	N s
Momentum	$kg\,m\,s^{-1}$

Table 2.7.1

Strange as it may seem, we can express the impulse of a force as 30 $kg\,m\,s^{-1}$ and similarly the change in the momentum of a moving object could be written as 45 N s.

 A car with a mass of 500 kg is travelling at 30 $m\,s^{-1}$ when a braking force of 4500 N is applied. Calculate the time taken for the car to stop.

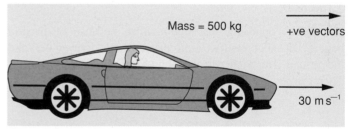

Mass = 500 kg

+ve vectors

30 $m\,s^{-1}$

Figure 2.7.1 Moving car with braking force

A Firstly, when tackling this problem we must remember that we are dealing with vector quantities. The question describes a braking force so this must act in the direction opposite to that in which the car is moving. As a consequence we must consider the velocity as +30 $m\,s^{-1}$ while the braking force, which acts in the negative direction is −4500 N.

Using;

$$Ft = mv - mu$$
$$-4500 \times t = 500 \times 0 - 500 \times 30$$
$$t = \frac{-500 \times 30}{-4500}$$
$$t = 3.3\ seconds.$$

The braking force of 4500 N brings the car to rest in 3.3 seconds.

Impulse and momentum in real situations

The previous example may seem a little contrived. You might feel that the situation has been framed to fit the theory. There are many examples of more realistic situations which can be explained by the concepts of impulse and momentum.

Cricketers or tennis players wishing to play a powerful stroke are told to 'follow through' on their shot. By so doing they apply the maximum force that their bodies can generate for a longer time. This causes a greater impulse and consequently a greater change in momentum, so the ball leaves the bat or racket with a greater velocity.

Figure 2.7.2 Contact between racket and ball

 A cricket ball of mass 0.4 kg, travels with a speed of 25 m s^{-1} towards a batsman. The batsman returns the ball along its original path with a speed of 40 m s^{-1}. Calculate the force on the ball, if ball and bat are in contact for 0.3 seconds.

 We should note that the incoming and outgoing velocities are in opposite directions;

$$\xrightarrow[+ve]{Vector}$$

$$Ft = mv - mu$$
$$F \times 0.3 = 0.4 \times (-40) - 0.4 \times 25$$
$$F = -86.7\ N$$

This negative sign shows that this force is acting in the direction *opposite* to the original motion of the ball. This is therefore the force of the bat on the ball.

This force which we have calculated is an average force which we have assumed will be constant throughout the contact. In reality contact forces can vary as shown in figure 2.7.3.

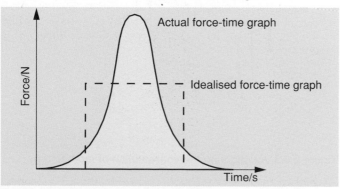

Figure 2.7.3 Variable force-time graph

From this type of graph we can determine the impulse by calculating the area under the curve.

Car safety

Modern car designers are increasingly building advanced safety features into their latest models. Seat belts are the most effective way of restraining the body, but the head will still be jolted during a collision. In the few moments immediately following a frontal impact the driver's head and chest continue their forward motion at the speed with which the car was moving before the impact. If the driver's head hits the steering wheel, dashboard or windscreen of the car, which by this time will have stopped moving, it will be brought to rest by a large impact force acting for a short time. We can explain this from the equation;

$$Ft = mv - mu$$

During a collision the momentum of the driver's body will have been reduced to zero in a very short time. Thus there will be a large force between the driver and the point of impact. This force can cause serious damage to the rib cage, chest or head. If an airbag inflates at the instant the car crashes, the driver's head and chest are brought to rest over a longer period of time and so the force needed to reduce the momentum to zero is less. Another benefit of the air bag is that the impacting force is spread over a greater area so the pressure at the point of contact is smaller and so there is less likelihood of fractured bones.

Does impulse equal change in momentum?

EXPERIMENT In this experiment a vehicle on a linear air track is pulled with a falling mass. Initial and final velocities are measured along with the time that the force takes to change the velocity.

Collect a series of results to test the equation;

$$Ft = mv - mu$$

You should alter the distance between the light gates so that the force takes different times to move the vehicle from speed u to speed v. Care needs to be taken to ensure that the vehicle has passed through the second light gate before the towing force hits the ground.

If you decide to alter the towing force you must remember that the accelerating mass will also change.

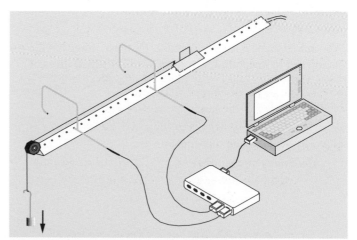

Figure 2.7.4 Investigating impulse and momentum

2.8 Conservation of momentum

Introduction

In a collision between a car and a wall the impulse is numerically the same as the change in momentum. After the collision both the car and the wall are stationary so the initial kinetic energy of the car has been absorbed by the wall and transformed into other forms of energy.

Not all collisions are between one stationary and one moving object, nor do all collisions result in moving objects coming to rest.

A practical model of a collision

Figure 2.8.1 shows a large trolley, fitted with spring loaded bumpers hitting a smaller one which is moving in the opposite direction. After the collision both trolleys move off in the direction in which the heavier one was initially moving.

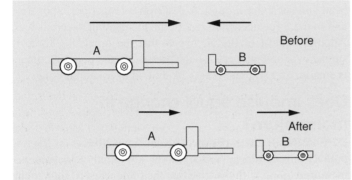

Figure 2.8.1 Before and after collision

While the trolleys are in contact, we can use Newton's Third Law to state that;

The force that trolley A exerts on B ($F_{A\ on\ B}$) is exactly equal in size and acts in the opposite direction to the force that trolley B exerts on trolley A ($F_{B\ on\ A}$).

Therefore;

$$-F_{A\ on\ B} = F_{B\ on\ A}$$

Since the trolleys are in contact with each other for the same length of time we can say;

$$-F_{A\ on\ B} \times t = F_{B\ on\ A} \times t$$

We can also equate the impulse of these forces with the changes in momentum for trolleys A and B which gives;

$$-(m_A v_A - m_A u_A) = m_B v_B - m_B u_B$$

Rearranging this gives;

$$m_A u_A + m_B u_B = m_A v_A + m_B v_B$$

In words we can state this equation as;

Total momentum before the collision	=	Total momentum after the collision

Conservation of momentum

The equation

$$m_A u_A + m_B u_B = m_A v_A + m_B v_B$$

Which can be expressed as;

Total momentum before the collision	=	Total momentum after the collision

is often called the **principle of conservation of momentum** and applies if no forces from outside the system act during the collision.

In deriving the equation we stated that the force that object A exerts on object B is exactly equal in size and acts in the opposite direction to the force that B exerts on A.

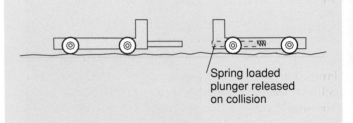
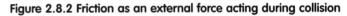

Figure 2.8.2 Friction as an external force acting during collision

Figure 2.8.2 shows a collision between trolleys travelling over a rough surface. The frictional forces act as external forces so the principle of conservation of momentum does not apply.

Testing conservation of momentum

Apparatus, similar to that shown in figure 2.8.3 could be used to verify the principle of conservation of momentum. The air track reduces the effects of friction. In each case you could use vehicles of equal mass or you might like to try situations where one vehicle has a greater mass than the other.

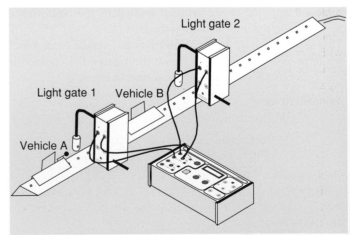

Figure 2.8.3 Investigating the conservation of momentum

You should try the following three situations;

1. The vehicles are fitted with elastic bands or repelling magnets

In one set of experiments where the vehicles were fitted with elastic bumpers the results shown in table 2.8.1 were obtained. In these experiments the target vehicle is initially stationary.

Mass of incoming vehicle (A)	Mass of target vehicle (B)	Velocity of vehicle A before	Velocity of vehicle A after	Velocity of vehicle B after
0.41	0.41	0.59	0	0.59
0.21	0.21	0.61	0	0.61
0.41	0.21	0.55	0.22	0.66
0.21	0.41	0.61	−0.20	0.41

Table 2.8.1

We can test the first set of results using;

$$m_A u_A + m_B u_B = m_A v_A + m_B v_B$$
$$\text{Total momentum before} = 0.24\ kg\,m\,s^{-1}$$
$$\text{Total momentum after} = 0.24\ kg\,m\,s^{-1}$$

From this we can conclude that the momentum after the collision and the momentum before the collision are the same. You should confirm that the principle holds for other sets of results.

2. The vehicles move off together after the collision

The results for a series of experiments where one vehicle is fitted with a pin which embeds itself in a cork target on the other vehicle are shown in table 2.8.2.

Mass of incoming vehicle (A)	Mass of target vehicle (B)	Velocity of vehicle A before	Velocity of vehicle A after	Velocity of vehicle B after
0.41	0.41	0.64	0.32	0.32
0.41	0.41	0.82	0.41	0.41
0.41	0.21	0.55	0.36	0.36
0.21	0.41	0.59	0.20	0.20

Table 2.8.2

Since the vehicles stick together after the impact they move off with a common velocity. This fact may seem obvious but can easily be forgotten.

Again we could test the principle of conservation of momentum using;

$$m_A u_A + m_B u_B = m_A v_A + m_B v_B$$
$$\text{Total momentum before} = 0.26\ kg\,m\,s^{-1}$$
$$\text{Total momentum after} = 0.26\ kg\,m\,s^{-1}$$

Once again we can conclude that momentum is conserved in collisions where the vehicles stick together after impact. You should confirm that the principle holds for the other three sets of results.

3. Repelling forces

The third type of situation where we can use the principle of conservation of momentum is not really what we would think

of as a collision. The vehicles, fitted with repelling magnets, are initially held close together on the air track and allowed to separate.

Mass of vehicle (A)	Velocity through light gate 1	Mass of vehicle B	Velocity through light gate 2
0.41	−0.24	0.41	0.24
0.41	−0.16	0.21	0.32
0.41	−0.19	0.41	0.19
0.21	−0.35	0.41	0.18

Table 2.8.3

When testing these results it is important to remember that momentum is a vector quantity. Since the vehicles are moving in different directions as they separate you must specify which direction is to be regarded as positive. Again we could test the principle of conservation of momentum. Since both vehicles are stationary before they separate we can say;

$$m_A v_A + m_B v_B = 0$$
$$\text{Total momentum before} = 0$$
$$\text{Total momentum after} = 0$$

This situation is sometimes called **an explosion**. We would expect momentum to be conserved because as the vehicles separate, the force exerted by A on B is exactly equal in size but acts in a direction opposite to the force of B on A and both forces act for the same time. You should confirm that the principle holds for the other three sets of results.

 Q A 20 g ball is fired horizontally from a 1.5 kg launcher which is sitting at rest on a frictionless track. The velocity of the ball is 7.5 m s^{-1}.

Calculate the speed of recoil of the launcher.

A Since momentum is conserved we can say;

$$m_L v_L + m_B v_B = 0$$

If we define the direction of the ball as positive for vector quantities we can say;

$$1.5 \times v_L + 0.020 \times 7.5 = 0$$

Therefore;

$$v_L = -0.1\ m\,s^{-1}$$

The launcher recoils at 0.1 m s^{-1}.

To calculate the recoil force if the launching mechanism and ball are in contact for 2 milliseconds, we must consider the impulse on the ball;

$$Ft = mv - mu$$
$$F \times 2 \times 10^{-3} = 0.02 \times 7.5 - 0$$
$$F = 75\ N$$

The launcher exerts a force of 75 N on the ball and therefore by Newton's Third Law the recoil force of the ball on the launcher is also 75 N.

Worked example 1

Masses of 2 kg and 5 kg, placed side by side on a frictionless surface are accelerated by a force of 14 N.

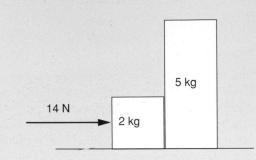

fig 2.9.1

a) Show, by calculation, that the acceleration of the combined masses is 2 m s^{-2}.
b) What is the size of the force required to give a mass of 5 kg an acceleration of 2 m s^{-2}?
c) What is the size of the force provided on the 5 kg mass by the 2 kg mass?

Answers and comments

a) The combined total of the masses is 7 kg. Since no friction acts in the system, the unbalanced force accelerating the masses is 14 N. To find the acceleration caused by an unbalanced force of 14 N we can use Newton's Second Law.

$$F = ma$$
$$14 = 7 \times a$$
$$a = 2\ ms^{-2}$$

b) The force required to make a mass of 5 kg accelerate at 2 m s^{-2} can again be found from Newton's Second Law;

$$F = ma$$
$$F = 5 \times 2$$
$$F = 10\ N$$

c) It is the push from the 2 kg mass that accelerates the 5 kg mass. Therefore, since the 5 kg mass is accelerating at 2 m s^{-2}, the 2 kg mass must be pushing on the 5 kg mass with a force of 10 N.

Worked example 2

A ball with a mass of 20 g is fired horizontally into a catcher mounted on top of a vehicle. The vehicle is resting on an air track.

fig 2.9.2

The vehicle and the catcher have a combined mass of 0.38 kg and move along the air track at a steady speed of 1.2 m s^{-1} after the ball has entered the catcher.

a) State the law of conservation of momentum.
b) Figure 2.9.2 shows the type of apparatus which could be used to investigate this interaction in the laboratory. Explain why the air track is used.
c) What is the total momentum of the ball, catcher and vehicle when they are moving along the runway?
d) Calculate the velocity of the ball before it entered the catcher.

Answers and comments

a) The law of conservation of linear momentum states that, in the absence of external forces, the total momentum after an interaction is equal to the total momentum before the interaction.
b) The law of conservation of momentum only applies in the absence of external forces. In the system being considered friction is an external force, so the air track is used to make the frictional forces as small as possible.
c) Momentum is defined as;

$$Momentum = Mass \times Velocity$$

After the ball is caught the vehicle and catcher (combined mass 0.38 kg) and the ball (mass 0.02 kg) travel at the common speed of 1.2 m s^{-1}.

$$Total\ momentum = Total\ mass \times Velocity$$
$$Total\ momentum = (0.38 + 0.02) \times 1.2$$
$$Total\ momentum = 0.48\ kg\,m\,s^{-1}$$

d) From the principle of conservation of momentum we can say that;

$$Total\ momentum\ before\ catch = Total\ momentum\ after$$
$$Total\ momentum\ before\ catch = 0.48\ kg\,m\,s^{-1}$$
$$Mass\ of\ ball \times Velocity\ of\ ball = 0.48$$
$$0.02 \times Velocity\ of\ ball = 0.48$$
$$Velocity\ of\ ball = \frac{0.48}{0.02}$$
$$Velocity\ of\ ball = 24\ m\,s^{-1}.$$

The ball has a velocity of 24 m s^{-1} and is moving in the same direction as after it is caught.

Questions

1 A trolley with a mass of 2 kg travelling at 3 m s⁻¹ collides with and sticks to another trolley which has a mass of 1.5 kg. Before the collision the trolley of mass 1.5 kg is stationary.
   a) Calculate the momentum of each trolley before the collision.
   b) Calculate the total momentum of the trolleys before the collision.
   c) Calculate the common speed with which the joined trolleys move after the collision.

2 A trolley of mass 2.5 kg, initially travelling to the right at 4 m s⁻¹ collides with a trolley of mass 2 kg travelling in the opposite direction at 4.5 m s⁻¹.

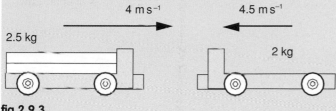

fig 2.9.3

   a) The trolleys stick together after the collision. What is their common velocity after the collision?
   b) Calculate the kinetic energy of the trolley moving to the right before the collision.

3 A ball with a mass of 250 g, held stationary underneath a motion sensor, is released. The velocity of the ball at various points during the motion is measured and the data plotted on a graph as shown.

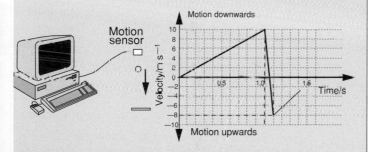

fig 2.9.4

   a) Calculate the momentum of the ball just as it is about to hit the floor.
   b) Calculate the momentum of the ball just as it leaves contact with the floor after the first bounce.
   c) Calculate the change in momentum during the first bounce.
   d) What is the impulse during the first bounce?
   e) Calculate the average resultant force that acts during the first bounce.

4 Two trolleys, initially at rest and touching, are held stationary on a horizontal surface.

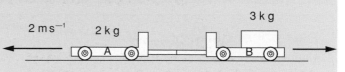

fig 2.9.5

A spring mechanism is released and the trolleys explode apart with trolley A moving to the left at 2 m s⁻¹.
   a) Calculate the velocity of trolley B after the trolleys separate.
   b) A student studying this situation draws figure 2.9.6 to show how he believes that the force exerted by trolley A on trolley B changes as the trolleys separate. What quantity is represented by the hatched area in the graph?

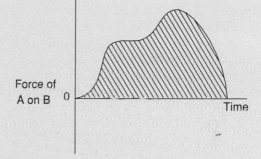

fig 2.9.6

   c) Sketch a copy of figure 2.9.6 and on the same axes show how the force exerted by trolley B on trolley A changes as the trolleys separate.

5 During a car collision the seat belt exerts different forces on the driver at different times. An engineer uses a simple model to produce data for the force-time graph shown in figure 2.9.7.

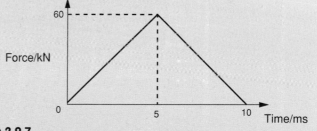

fig 2.9.7

   a) Calculate the impulse of the force from the seat belt on the driver during the collision.
   b) The driver has a mass of 75 kg. Calculate the speed at which the driver is travelling in the car just before the collision brings the car to a halt.
   c) What is the maximum force exerted by the seat belt on the driver.
   d) Many modern cars are fitted with air bags as a safety feature. Explain how an airbag can reduce the likelihood that the driver will sustain injuries to his head and chest during a collision.

Past paper question

During a test on car safety, two cars as shown below are crashed together on a test track

18.0 m s⁻¹   10.8 m s⁻¹

Car A   Car B

fig 2.10.1

a) Car A, which has a mass of 1200 kg and is moving at 18.0 m s⁻¹, approaches car B, which has a mass of 1000 kg and is moving at 10.8 m s⁻¹, in the opposite direction. The cars collide head on, lock together and move off in the direction of car A. Calculate the speed of the cars immediately after the collision.

b) During a second safety test, a dummy in a car is used to demonstrate the effects of a collision. During the collision the head of the dummy strikes the dashboard at 20 m s⁻¹ as shown below and comes to rest in 0.02 seconds. The mass of the head is 5 kg.

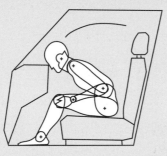

fig 2.10.2

(i) Calculate the average force exerted by the dashboard on the head of the dummy during the collision.

(ii) The test on the dummy is repeated with an airbag which inflates during the collision. During the collision, the head of the dummy again travels forward at 20 m s⁻¹ and is brought to rest by the airbag.

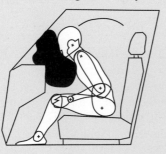

fig 2.10.3

Explain why there is less risk of damage to the head of the dummy when the airbag is used.

SQA 1996

Answers and comments

a) To calculate the speed of the cars when they are locked together we can use the principle of conservation of momentum. Expressing this as an equation gives;

$$m_A u_A + m_B u_B = m_A v_A + m_B v_B$$

Since the cars join together they must move at the same velocity, v, after the collision. Therefore;

$$m_A u_A + m_B u_B = (m_A + m_B)v$$

Since momentum is a vector quantity, before we substitute values into the equation, we must choose one direction as being positive for vector quantities.

$$1200 \times 18 + 1000 \times (-10.8) = (1200 + 1000) \times v$$
$$21{,}600 - 10{,}800 = 2200 \times v$$

Vector +ve →

$$v = \frac{10{,}800}{2200}$$
$$v = 4.91 \ m \, s^{-1}$$

After the collision both cars move to the right at 4.91 m s⁻¹.

b) (i) To calculate the force we can use the impulse and momentum equation using the same vector convention as in part **a**):

$$Force \ on \ head \times time = change \ in \ momentum \ of \ head$$
$$Ft = mv - mu$$
$$F \times 0.02 = 5 \times 0 - 5 \times (-20)$$
$$F = \frac{+100}{0.02}$$

$$F = +5000 \ N$$

The minus sign shows that this force acts from left to right. This then is the force of the dashboard on the head.

(ii) The airbag brings the head to rest by applying a force for a certain time. Since the initial and final speeds are the same with or without the airbag, the change in momentum will be the same. With the airbag the force acts for a longer time so its magnitude is smaller than 5,000 N. The smaller force is less likely to cause an injury. In addition the airbag spreads the force over a larger area so there is less pressure at the point of contact. Again this reduces the risk of damage.

Questions

1 A force acting on a ball of mass 1 kg for 10 milliseconds accelerates it from rest to a speed of 20 m s^{-1}.
   a) Calculate the change in momentum of the ball.
   b) Calculate the average force causing this change in momentum.

2 A pupil trying to analyse the forces acting while he is skateboarding draws the force-time graph shown in figure 2.10.4.

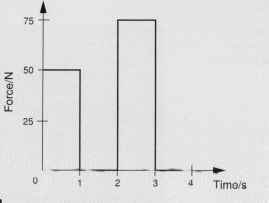

fig 2.10.4

The pupil is initially standing at rest with one foot on the skateboard and the other in contact with the ground. In the first second he pushes forward with a steady force of 50 N.
   a) The pupil and skateboard have a combined mass of 75 kg. Calculate the speed of the skateboarder 1 second after the start of his motion.
   b) From time t=1 to t=2 seconds he has both feet on the board and is moving at a constant speed. During the next second he again pushes on the ground but with a steady force of 75 N. Calculate his speed 3 seconds after the start of the motion.
   c) Draw a speed-time graph for the speed of the skateboarder during the 3 seconds of the motion.

3 In an air track experiment in the laboratory, trolleys travel towards each other as shown in figure 2.10.5.

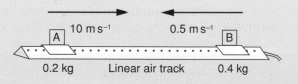

fig 2.10.5

   a) The masses and initial speeds of the trolleys are as shown on the diagram. With the aid of a diagram, describe how these speeds could be measured.
   b) After colliding the trolleys stick together. Calculate their common velocity after the collision.
   c) Figure 2.10.6 shows how the force exerted by trolley A on trolley B changes during the collision. Copy the figure and add a line to show how the force exerted by trolley B on trolley A alters during this collision.

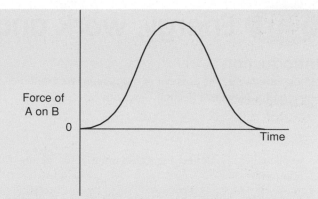

fig 2.10.6

4 In a cricket match, a ball of mass 0.45 kg is bowled towards a batsman at a speed of 20 m s^{-1}. The batsman returns the ball back along its original path at 25 m s^{-1}.
   a) Calculate the initial momentum of the ball.
   b) Calculate the final momentum of the ball.
   c) What is the change in momentum of the ball?
   d) The ball is in contact with the bat for 0.15 seconds. What is the average force acting during the batsman's stroke?
   e) The cricket coach tells the batsman to 'follow through more' on his shots. Explain using the concepts of impulse and momentum, how following through on his shots will result in the ball reaching the boundary quicker.

5 A ball of mass 0.25 kg is dropped from rest onto a steel plate 2.4 m below.
   a) What is the velocity of the ball just before hitting the plate?
   b) What is the momentum of the ball just before hitting the plate?
   c) After its contact with the steel plate the ball bounces to a height of 1.2 m. Calculate the velocity with which the ball left the plate in order to bounce to a height of 1.2 m.
   d) What is the change in momentum of the ball while it is in contact with the steel plate?

6 In a laboratory experiment, trolleys of mass 1 kg and 0.8 kg travelling towards each other on an air track collide.

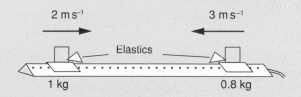

fig 2.10.7

The speeds of the trolleys before the collision are as shown in the diagram.
   a) Calculate the total momentum of the moving trolleys before the collision.
   b) State the principle of conservation of momentum.
   c) After the collision the 1 kg vehicle returns along its original path at 0.5 m s^{-1}. Calculate the velocity of the 0.8 kg vehicle after the collision.

2.11 Energy, work and power

Introduction

When introducing the concept of momentum we combined the equation for Newton's Second Law with;

$$a = \frac{v - u}{t}$$

to mathematically define the effect of a force acting for a certain time;

$$Ft = mv - mu$$

Work and kinetic energy

We can use the third equation of motion;

$$v^2 = u^2 + 2as$$

to find an alternative expression for acceleration; one which does not involve time. Rearranging this equation gives;

$$a = \frac{v^2 - u^2}{2s}$$

If we now substitute this into Newton's Second Law;

$$F = ma$$

$$a = \frac{v^2 - u^2}{2s}$$

$$F = m \times \frac{v^2 - u^2}{2s}$$

This equation simplifies to give;

$$F \times s = \frac{1}{2} mv^2 - \frac{1}{2} mu^2 \qquad \text{equation 1}$$

The quantities on the right hand side of this equation may be familiar since kinetic energy is defined from;

$$Kinetic\ energy = \frac{1}{2} \times Mass \times Velocity^2$$

From equation 1, we can see that a force acting over a certain distance not only changes the velocity of a moving object from u to v, but also changes the kinetic energy.

$$Force \times Distance = Change\ in\ kinetic\ energy$$

Work

The scientific meaning of the word 'work' is much more specific than the way it is used in day to day speech. If the velocity of an object is changed as a result of a force acting, the force is said to be *doing work* on the object. We can quantify the amount of work done by saying that;

$$Work\ done = Force \times \frac{Distance\ moved\ in\ the}{direction\ of\ the\ force}$$

$$W = F \times s$$

The work done by a force is measured in units of newton metres (N m). By comparing this equation with equation 1 we can state;

$$Work\ done = Change\ in\ kinetic\ energy$$

We have already seen how work can be measured in units of newton metres, but since we can equate work done and kinetic energy it must also be possible to express work in units of joules (J). **Work done is a measure of the energy transferred**.

Work and energy are closely linked concepts. It is important however to distinguish between them. Work is done by a force *only* when an object is moving. If the object is stationary there is no change in its kinetic energy therefore no work is being done.

**Figure 2.11.1
Weight-lifter
doing no work!**

While the weight-lifter has the bar in motion he is doing work but while he is holding it aloft he does no work! This seems strange because the weight-lifter begins to feel tired and requires energy to keep the bar raised. We can explain this by noticing that while he is doing no work moving the weights *he is using energy to keep his muscles under sufficient tension force to support the bar.*

Kinetic energy and collisions

In section 2.8 you saw that for a system where no external forces act, momentum is conserved. You might now want to know if kinetic energy is also conserved in these situations.

Table 2.8.1 shows the data collected when vehicles on an air track, fitted with elastic bumpers or repelling magnets collide. Calculate the kinetic energy before and after the collision. Fill in a table like that shown in 2.11.1.

	Total kinetic energy before	Total kinetic energy after
Expt.1		
Expt.2		
Expt.3		
Expt.4		

Table 2.11.1

These calculations should indicate that in this situation kinetic energy is conserved.

When the vehicles stick together after the collision there is less kinetic energy after the collision than before. The data from table 2.8.2 can be analysed to confirm this point.

	Total kinetic energy before	Total kinetic energy after
Expt.5		
Expt.6		
Expt.7		
Expt.8		

Table 2.11.2

In the third type of situation outlined in section 2.8 the vehicles, fitted with repelling magnets are held closely together and then allowed to separate.

Before separation the vehicles are stationary so the total kinetic energy is zero. After separation both are moving so each has some kinetic energy. Since kinetic energy is *not* a vector quantity, the total kinetic energy is not zero. Some energy has been added to the system to make the vehicles move. Therefore, when the vehicles separate *kinetic energy is not conserved*.

Elastic and inelastic collisions

Our consideration of the air track experiments has shown that there are situations where both momentum and kinetic energy are conserved. Such collisions are called **elastic**. In other situations while momentum is conserved kinetic energy is not. These collisions are called **inelastic**.

	Momentum	Kinetic energy
Elastic	Conserved	Conserved
Inelastic	Conserved	Not conserved
Explosion	Conserved	Not conserved

Table 2.11.3

We need to be careful in our use of the concept of elastic collisions. In gases, the molecules must collide elastically with each other and with the walls of the container. If the kinetic energy after these collisions was less than before the collision, this would mean that the gas molecules would slow down, leading to a shrinking in the volume of the gas. Clearly this does not happen, so the collisions between gas molecules must be elastic.

When a hammer strikes a lump of lead the collision is clearly inelastic. The kinetic energy after the collision may be less than before but the *total energy* of the system will remain unchanged. The kinetic energy has been transferred into other forms of energy. As well as the sound energy produced with each blow there will be some heat. After a few strikes the lead will begin to feel warm.

While the terms elastic and inelastic are used for specific situations where kinetic energy is or is not conserved, we must **always remember that the *total energy* within any system remains unchanged.**

Power

In lifting a weight of 5000 N from the ground to a height of 2.5 m, the work done by a weight-lifter can be calculated using;

$$W = F \times s$$
$$W = 5000 \times 2.5$$
$$W = 12,500 \ J$$

Exactly the same amount of work is done when a delivery van driver lifts 50 individual 100 N weights separately through the same height of 2.5 m into his van. In both cases the total energy transferred to the weights is the same, but clearly the weight-lifter has a more difficult task. This is because he is required to provide all the energy over a much shorter period of time. The rate at which the energy is supplied is greater for the weight-lifter.

In physics, **power** is defined as;

$$Power = \frac{Energy \ transferred}{Time} = \frac{Work \ done}{Time}$$

Power can be thought of as the **energy transformed per second**. Alternatively we can say that power is **the rate of doing work** or **the rate of transferring energy**.

Power is measured in units of joules per second ($J \, s^{-1}$) or watts (W). A 100 W light bulb transfers 100 J of electrical energy into other forms of energy each second.

For many years power was measured in 'horse power'. This is a non-standard unit but it is still often used to specify the power of engines. 1 horse power is equivalent to 746 W.

Systems only convert part of the input energy into useful work. The percentage efficiency of the system is defined as;

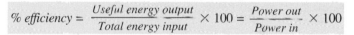

$$\% \ efficiency = \frac{Useful \ energy \ output}{Total \ energy \ input} \times 100 = \frac{Power \ out}{Power \ in} \times 100$$

Work done and kinetic energy

EXPERIMENT In this experiment the vehicle on the linear air track is pulled through a measured distance, *s*, by a falling mass. The initial and final velocities can be measured for a range of distances.

The work done by the force in moving the mass a distance, *s*, can be calculated, as can the change in the kinetic energy. You should be able to confirm that;

Work done = Change in kinetic energy

Note that in calculating the kinetic energy you must use the mass of the vehicle *plus* the mass of the falling weight since the force of gravity is accelerating both of them.

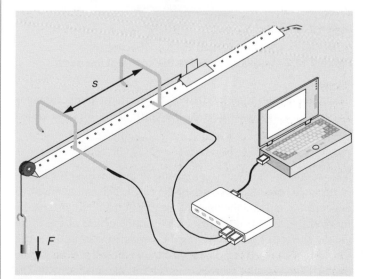

Figure 2.11.2 Investigating work and kinetic energy

2.12 Weight and mass

Introduction

Charged balloons or suspended magnets exert a force on each other even if they are not in contact. These forces are explained as being caused by the electrostatic or magnetic properties of the objects involved. Another *non-contact* force keeps the planets in motion around the sun. This attractive force cannot be explained by the magnetic or electrostatic properties of the planets and the sun. In 1677, Newton proposed the Law of Gravitation stating that all bodies exert attractive forces on each other simply because of their *mass*.

Figure 2.12.1 Planets orbiting around the sun

Just as the force between oppositely charged spheres becomes larger as the charge increases, so the gravitational attraction between two masses is bigger for objects of greater mass. The force of attraction between two masses of 70 kg 1 m apart is only about 3.3×10^{-7} N. This is a very small force indeed compared to the gravitational attractive force between a mass of 70 kg and the earth. This is large because the mass of the earth is 6×10^{24} kg.

Figure 2.12.2 Masses attracted to the centre of the earth

The gravitational force exerted by the earth on another mass is directed towards the centre of the earth. **This force is called the weight**.

Weight

The weight of an object is a measure of the force with which it is attracted towards the centre of the earth. This attractive force depends not only on the mass of the object itself but also on the mass of the earth and the distance from the centre of the earth. On the surface of planets having the same diameter as the earth, the weight of a mass of 1 kg will depend on the mass of the planet. The weight will be largest on the planet with the greatest mass. Although the mass of an object of mass 1 kg never changes, its weight depends on more than simply the

quantity of material from which it is made. We must be careful not to confuse the words mass and weight, which in day to day conversation are used interchangeably.

Figure 2.12.3

Acceleration due to gravity

If a mass of m kg is suspended by a thread from a calibrated Newton balance the force of attraction towards the centre of the earth is recorded on the scale. This force is called the weight of the object. Since the object is stationary the downwards force (the weight) is exactly balanced by an upwards force provided by the tension in the stretched spring inside the Newton balance.

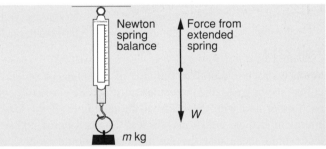

Figure 2.12.4 Forces on a Newton balance

If the thread is cut, the upwards force is removed so the weight, acting as an unbalanced force, causes the mass to accelerate downwards. The size of the downward acceleration caused by gravity is given the symbol, g.
From Newton's Second Law;

$$Unbalanced\ force = Mass \times Acceleration$$
$$W = m \times g$$

Which rearranges to give;

$$g = \frac{W}{m}$$

Falling masses

If a mass of $2m$ kg is suspended from the Newton balance shown in figure 2.12.4 the scale will record a weight *twice* the previous value. The reading will be $2W$ newtons. This shows that the force attracting this mass towards the centre of the earth is now twice as big. Does this bigger force cause a bigger acceleration?
We can calculate the acceleration using Newton's Second Law.

The larger mass is allowed to fall freely;

$$Unbalanced\ force = Mass \times Acceleration$$
$$2W = 2m \times Acceleration$$

Which gives;

$$Acceleration = \frac{2W}{2m}$$

Therefore;

$$Acceleration = \frac{W}{m}$$

The acceleration of the bigger mass is the same as that for the smaller mass. The acceleration due to gravity is the **ratio of the weight to the mass** and does not depend on the mass itself. This means that a heavy and a light object, released from the same point at the same instant will land on the ground at the *same* time. The work done in lifting a mass, *m*, attached to a spring balance as shown in figure 2.12.4 through a vertical height, *h*, is given by;

$$Work\ done = F \times h$$

But since $F = W$ and $W = mg$ we can say;

$$Work\ done = mgh$$

By doing this amount of work the mass has gained potential energy. Therefore;

$$Potential\ energy\ gained = mgh$$

Gravitational field strength

You should notice from the above equation that the acceleration caused by the force of gravity, while being measured in units of $m\,s^{-2}$, can also be measured in units of newtons per kilogram ($N\,kg^{-1}$).

$$Acceleration = \frac{W}{m}$$

For any planet, as the above equation shows, the ratio of the weight of an object to its mass is constant. This ratio is called the **gravitational field strength**. The gravitational field strength near the surface of the planet earth is $9.8\,N\,kg^{-1}$. This means that the earth exerts a force of 9.8 N for each 1 kg of an object's mass. If this mass is released, the weight acts as an unbalanced force and causes an acceleration of $9.8\,m\,s^{-2}$.

The gravitational field strength on the surface of the moon is about $1.6\,N\,kg^{-1}$. A lunar landing craft of mass 500 kg will have a weight of 800 N on the moon.

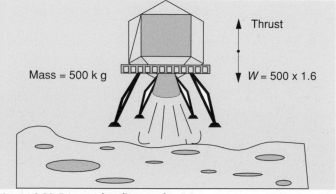

Figure 2.12.5 Lunar landing craft

When the engines of the landing craft exert an upward force of 1280 N while taking off from the moon, the acceleration can be found using;

$$Unbalanced\ force = Mass \times Acceleration$$
$$F = m \times a$$
$$1280 - 800 = 500 \times a$$

Therefore;
$$a = 0.96\ m\,s^{-2}$$

The craft has an acceleration of $0.96\,m\,s^{-2}$.

Real and apparent weights

In a lift travelling upwards or downwards at *a constant speed*, the force exerted upwards by the floor on a passenger's feet is equal and opposite to the passenger's weight. If this person were standing on a set of bathroom scales measuring weight, the reading would be the same as if the lift were stationary.

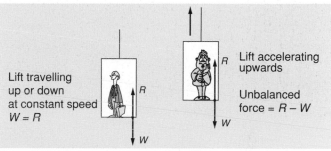

Figure 2.12.6 Forces on a person in a lift

Accelerating upward

If the lift is *accelerating upwards* there must be an unbalanced force in the upwards direction. This means that the floor of the lift is exerting an upwards direction force *greater* than the weight of the passenger. The size of this upwards force is indicated by the reading on the scales so the scales show a reading higher than the expected weight.

Decelerating down

If the lift is decelerating to a stop at the *bottom* of its *downward* journey the resultant force is also in the upwards direction and the scales would read *higher* than expected.

Deceleration upwards

If the lift is *decelerating to a stop* at the *top* of its *upward* motion the resultant force is downwards. This means that the weight is *greater* than the force exerted by the lift on the passenger. So the scales read *less* than the passenger's weight.

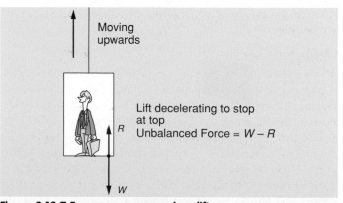

Figure 2.12.7 Forces on a person in a lift

2.13 Exam style questions

Questions

1 Figure 2.13.1 shows a graph obtained when a motion sensor is used to monitor the velocity of a ball as it bounces.

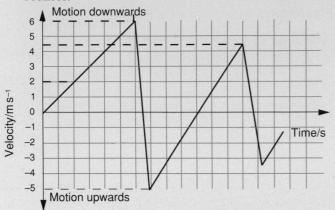

fig 2.13.1

a) The ball has a mass of 0.5 kg. Show that the ball has 9 J of kinetic energy just before it touches the ground at the start of the first bounce.

b) A pupil studying this situation states that the kinetic energy gained by the ball as it falls is a result of the potential energy it loses.

From what height must the ball be released to have 9 J of kinetic energy as it hits the floor?

c) Using other information from the graph calculate the height to which the ball bounces after the first bounce.

2 A lift of mass 100 kg carries three passengers of average mass 80 kg upwards through a vertical height of 5 m in 49 seconds.

a) What is the combined mass of the lift and its occupants?

b) What is the weight of the lift and its occupants?

c) How much work must be done by the motor in lifting the lift and passengers through 5 m?

d) Calculate the power delivered from the lift motor to the lift.

3 A motion sensor is placed at the top of a runway as shown in figure 2.13.2. When released, the trolley accelerates for 3 seconds before reaching the bottom of the slope.

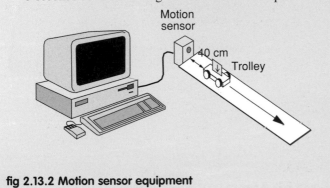

fig 2.13.2 Motion sensor equipment

a) Sketch a graph to show how the displacement of the trolley from the motion sensor varies during the 3 seconds of its motion. (You need not show scales on the axes.)

b) With the aid of a diagram, describe how you would measure the acceleration of the trolley as it travels down the slope.

c) The trolley is found to accelerate along the runway at 0.78 m s^{-2}. Draw a graph to show how the velocity of the trolley changes as it travels down the runway. (Include numerical scales on the axes of the graph.)

In a similar experiment a spring loaded buffer is placed at the bottom of the slope. The trolley is released from the same position as before and has the same acceleration as before.

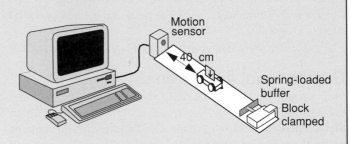

fig 2.13.3 Motion sensor equipment

d) Show, by calculation, that the trolley has a velocity of 2.34 m s^{-1} down the slope, just before it hits the buffer at the bottom.

e) The trolley leaves the buffer with a speed of 1.5 m s^{-1} back up the slope towards the starting position. The mass of the trolley is 0.4 kg. Calculate change in the momentum of the trolley.

4 A car of mass 600 kg travelling at 20 m s^{-1} is brought to rest by a force applied from the brakes.

a) Calculate the kinetic energy of the car just before the brakes are applied.

b) How much energy must be transformed into other forms of energy by the brakes?

c) Draw a diagram to show the direction of motion of the car and the direction in which the braking force acts.

d) The car travels 54 m while decelerating. Calculate the size of the average braking force.

5 A boy of mass 50 kg stands on a force meter placed in the floor of a lift.

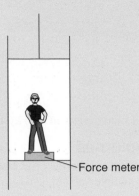

fig 2.13.4 Weight in a lift

a) Show, by calculation, that the force meter will give the weight of the boy standing in the stationary lift as 490 N.

b) What is the weight indicated by the force meter when the lift is travelling upwards at a steady speed? Explain your answer.

c) The lift accelerates upwards at 1 m s⁻². Calculate the size of the unbalanced force needed to cause this acceleration.

   (i) Draw a diagram to show the forces acting on the boy as the lift accelerates upwards at 1 m s⁻².

   (ii) What is the weight indicated by the force meter when the lift is accelerating upwards at 1 m s⁻²?

6 A squash racket with a mass of 0.5 kg hits a ball as shown in figure 2.13.5.

**fig 2.13.5
Squash racket
hitting a ball**

At the point of contact, the racket is moving horizontally at a speed of 30 m s⁻¹ while the ball of mass 75 g is travelling at 20 m s⁻¹ in the opposite direction.

After the contact, the ball returns along its incoming path at 60 m s⁻¹. What is the velocity of the racket after the collision? (You can assume that the external force propelling the racket stops at the moment of impact.)

7 In a sandblasting application, a jet of air is used to carry 25 kg of fine sand particles per second onto the target area of worn stone. The average speed with which the particles approach the wall is 22 m s⁻¹.

a) The sand particles fall vertically to the ground immediately after hitting the wall. Calculate the average force exerted by the sand on the wall during the blasting process.

b) In a different design of sand blaster the sand does not fall vertically but bounces back from the wall. This blaster also delivers particles with a speed of 22 m s⁻¹ at a rate of 25 kg s⁻¹. In which of the designs is the larger force applied to the wall? Explain your answer.

8 A pupil is observing a trolley which has been pushed towards a barrier. The bumper on the front of the trolley is spring loaded so the pupil makes the following statements;

fig 2.13.6

I   The total momentum before the collision equals the total momentum afterwards.

II   During the collision the force from the trolley on the barrier is equal in size but in the opposite direction to the force from the barrier on the trolley.

III Kinetic energy is conserved during this collision.

Which of these statements is/are correct?

A I only
B III only
C I and II only
D I and III only
E I, II and III

9 A 0.2 kg steel ball travelling at 4 m s⁻¹ strikes another initially stationary ball of mass 0.6 kg.

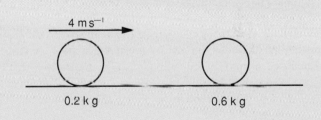

fig2.13.7

If the collision is completely elastic and the 0.2 kg ball returns along its original path, which of A-E below could show possible velocities for the balls?

	Velocity of 0.2 kg ball	Velocity of 0.6 kg ball
A	−2 m s⁻¹	2 m s⁻¹
B	−6 m s⁻¹	2 m s⁻¹
C	−4 m s⁻¹	4 m s⁻¹
D	4 m s⁻¹	−4 m s⁻¹
E	−2 m s⁻¹	6 m s⁻¹

Table 2.13.1

3.1 Density

Introduction

To change 1 kg of ice at 0 °C into water requires 334,000 J of energy. To heat 1 kg of water from 0 °C to 100 °C requires 419,000 J, while to turn the 1 kg of boiling water into steam requires 2,260,000 J. During these changes neither the chemical composition nor the mass of the water changes even though the vibrational energy of the atoms and the bonding between them does. Heating 1 kg of water at its boiling point produces a large volume of steam because all of the bonds holding the molecules together in the liquid have been broken. **The same mass of any material, when in different states, can occupy different volumes.** It is also quite common for the same volume of different materials to have different masses.

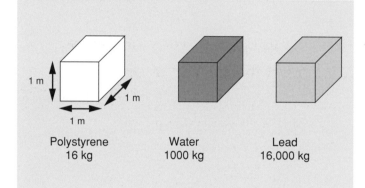

Figure 3.1.1 1 m³ blocks of different materials

To compare the blocks in figure 3.1.1 we should notice that each has a volume of 1 m³. The expanded polystyrene block has a mass of 16 kg, while the lead block has a mass of 16,000 kg. One cubic metre of water has a mass of 1000 kg.

Defining density

The density of a material can be given the symbol, ρ, and is defined as;

$$Density = \frac{Mass}{Volume}$$

$$\rho = \frac{m}{V}$$

The density of a material might be thought of as the mass of a 1 m³ block of the material.
The units of density can be found from the equation;

$$\rho = \frac{m}{V}$$

$$Density = \frac{kg}{m^3}$$

$$Density = kg\,m^{-3}$$

 Find the mass of water needed to fill a 0.5 m³ tank. The density of water is 1000 kg m⁻³.

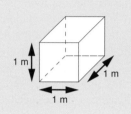

$$\rho = \frac{m}{V}$$

$$Mass = Density \times Volume$$
$$Mass = 1000 \times 0.5$$
$$Mass = 500\ kg$$

The 0.5 m³ tank will hold a mass of 500 kg of water.

Meaningful volumes

The answer in the previous example might have seemed a little high. You might be tempted to think that a 1 m³ tank has a volume of 1000 cm³. This is *incorrect*. Figure 3.1.2 shows a tank which has a volume of 1 m³.

Figure 3.1.2

The volume of the tank can be found using;

$$Volume = Length \times Breadth \times Height$$
$$Volume = 1\ m \times 1\ m \times 1\ m$$
$$Volume = 1\ m^3.$$

To find the volume of this tank in cm³ we must convert *all* of the dimensions into cm;

$$Volume = Length \times Breadth \times Height$$
$$Volume = 100\ cm \times 100\ cm \times 100\ cm$$
$$Volume = 1,000,000\ cm^3.$$

A volume of 1 m³ is therefore equivalent to 1,000,000 cm³.

Q Which of the following has the greater volume (a) 2.5 kg of water or (b) 41 kg of lead?

A To find the volume of each of these samples we must use the equation;

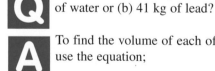

$$\rho = \frac{m}{V}$$

or

$$V = \frac{m}{\rho}$$

For water	*For lead*
$V = \dfrac{2.5\ kg}{1000\ kg\,m^{-3}}$	$V = \dfrac{41\ kg}{16000\ kg\,m^{-3}}$
$V = 2.5 \times 10^{-3}\ m^3$	$V = 2.6 \times 10^{-3}\ m^3$

The 41 kg lead block, despite being a lot heavier, has only a slightly greater volume than 2.5 kg of water.

Explaining density

We can explain why a substance has its *lowest density* in its *gaseous state* by considering the relative distance between the molecules in the different states. In gases the molecules are further apart than in solids or liquids so a certain mass of gas will occupy a greater volume than the same mass of the substance in its liquid or solid states. 1 g of water will occupy a volume of 1 cm^3, but when boiled will produce approximately 1500 cm^3 of steam. Steam is therefore considerably less dense than water.

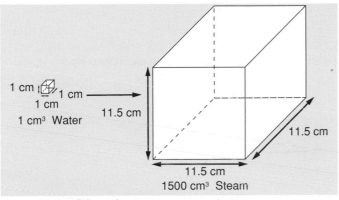

Figure 3.1.3 Solids and gases

To explain why different solid materials have different densities we need to consider the composition of the materials. Solid materials are made up from atoms and molecules bonded closely together. These atoms or molecules are held in position by balanced attractive and repulsive forces. The actual mass of the atoms and the way that the atoms and molecules are packed into the available space determine the density.

Measuring density

To find the density of a substance we need to know the mass of a particular volume. Finding the mass of an object is not difficult and we can use a range of different ways to find the volume.

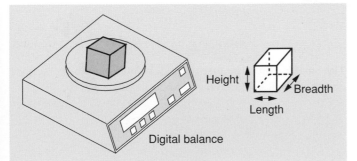

Figure 3.1.4 Volume of a regular object

For a regular object such as that shown in figure 3.1.4 the volume can be found from;

$$Volume = Length \times Breadth \times Height$$

Once the mass and volume are known the density can be found using the equation;

$$Density = \frac{Mass}{Volume}$$

When a **solid** does not have regular sides it is not possible to find the volume by multiplying the length, breadth and height. The volume of an irregular object can be found by immersing the object in water as shown in figure 3.1.5. When the irregular object is totally immersed the water level rises. The size of the increase indicates the volume of the immersed object.

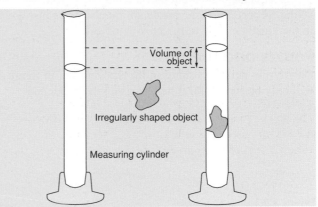

Figure 3.1.5 Measuring the volume of an irregular object

Measuring the density of air

EXPERIMENT Since the density of air is small, a very accurate balance is needed to measure the change in mass when the air is extracted from a flask by a vacuum pump.

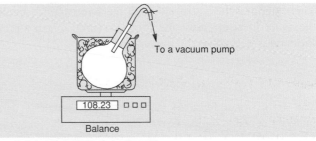

Figure 3.1.6 Air density by extraction

With a good vacuum pump it is reasonable to assume that all of the air in the flask is removed, therefore the volume of the removed air is equal to the flask's volume. The mass of the air removed can be found by weighing the flask before and after the pump is switched on. The density of the air can then be determined.

An alternative way of finding the mass of a known volume of air is to pump more air into a container.

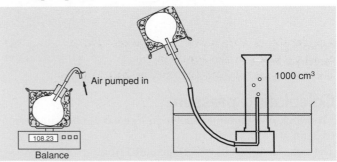

Figure 3.1.7 Air density by pumping

Once the increase in mass is found the air is released and its volume measured by the displacement of water. Knowing the mass of a certain volume of air allows us to calculate its density.

Introduction

We have already discovered that unbalanced forces acting on an object change the motion of the moving object. Such forces acting for a certain time create an impulse and cause a change in the kinetic energy of the moving object.

Forces and area

Clearly the person pushing in the drawing pin in figure 3.2.1 will only be successful if he pushes on the wider end.

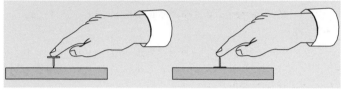

Figure 3.2.1

Pushing with a force of F newtons on the wide end will make the drawing pin force its way into the wood. However, pushing with exactly the same force on the other end of the drawing pin would not force the pin into the wood. It might even be quite painful!

This example shows that although the force applied might be the same in both cases, the force per unit area is different. Pressure is defined as the force per unit area;

$$Pressure = \frac{Force}{Area}$$

The ability of a force to make an object penetrate another material depends on the size of the force and the area. Figure 3.2.2 shows the type of tyres used by forestry vehicles which have to travel over soft ground. Although adding wider wheels increases the weight of the vehicle slightly the extra area in contact with the ground means that the pressure on the soft ground reduces.

**Figure 3.2.2
Forestry vehicles
with wide tyres**

Calculating pressure

The block of aluminium shown in figure 3.2.3 has dimensions of 2 cm × 3 cm × 10 cm.

The volume of this block is calculated from;

$$Volume = Length \times Breadth \times Height$$
$$Volume = 0.1 \times 0.02 \times 0.03$$
$$Volume = 6 \times 10^{-5} \ m^3$$

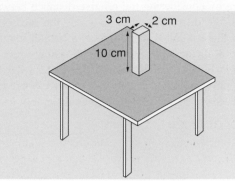

Figure 3.2.3 Pressure exerted by an aluminium block

The density of aluminium is 2700 $kg \, m^{-3}$. The mass of this block is calculated from;

$$\rho = \frac{m}{V}$$

$$m = pV$$
$$Mass = 2700 \times 6 \times 10^{-5}$$
$$Mass = 0.162 \ kg$$

This block will exert its largest pressure on the table when it is resting on its smallest side. The pressure exerted on the table can be calculated from;

$$Pressure = \frac{Force}{Area}$$

The force exerted on the table by the block is equal to the weight of the block which can be calculated from;

$$W = mg$$
$$Weight = 0.162 \times 9.8$$
$$Weight = 1.59 \ N$$

So;

$$Pressure = \frac{1.59}{0.02 \times 0.03}$$

$$Pressure = 2650 \ N m^{-2}$$

The smallest pressure that the block can exert on the surface occurs when it is placed on its largest side;

$$Pressure = \frac{1.59}{0.1 \times 0.03}$$

$$Pressure = 530 \ N m^{-2}$$

Pressure units

Pressures calculated using a force measured in newtons and an area measured in square metres will be in units of newtons per square metre ($N m^{-2}$). There are also a number of other commonly used units for pressures; atmospheres (atm.), millimetres of mercury (mmHg) and millibars are the most common. While these are common in every day usage, the pascal (Pa) is the internationally recognised unit of pressure where 1 Pa is exactly the same as 1 $N m^{-2}$.

3.3 Pressure in liquids

Introduction

To examine oil rig supports, exploration companies use divers to inspect the regions near the surface. For deep water exploration mini-submarines or other remote vehicles are preferred. At points deep in the ocean the *pressure of water* is too great for divers to work safely.

Pressure and depth

The pressure experienced by divers is greater at points further beneath the surface of the water. Figure 3.3.1 shows how water emerges from holes made at different heights in a tin can.

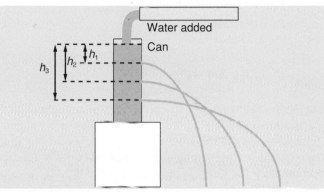

Figure 3.3.1 Water squirting out of three holes in a can

The water flowing into the can maintains a constant 'head' above each of the holes. The water from the bottom hole squirts out furthest because the pressure acting at this depth is equal to the pressure of the atmosphere *plus* the pressure due to the height, h_3, of water. The pressure forcing the water out of the top hole is equal to the atmospheric pressure *plus* the pressure due to height h_1. Since h_3 is greater than h_1, the pressure at the bottom is greater than the pressure at the top.

Pressure acts in all directions

A diver using breathing apparatus has to expand his lungs to inhale the breathing mixture from his tanks. At greater depths this will become increasingly difficult because the movement of the lungs will be opposed by larger forces due to the higher pressure of the water.

Pressure depth equation

Using the apparatus shown in figure 3.3.2 we can determine exactly how the pressure varies with depth.

Placing the open end of the tube at a certain depth in the water will cause an *upward pressure* on the air in the tube equal to atmospheric pressure *plus* the pressure due to that particular depth of water. The pressure sensor can be set to give a reading of zero at the surface of the water in the cylinder. Placing the open end of the tube at different depths and recording the

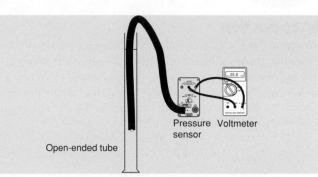

Figure 3.3.2 Apparatus to show how pressure and depth are related

pressure due to that depth of water will give a series of readings which confirm that the pressure caused by the water, *P*, is directly proportional to depth, *h*.

$$P \propto h$$

Liquid pressure formula

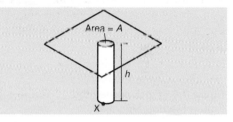

Figure 3.3.3

In figure 3.3.3 the column of water has a base of area *A* and a height *h*. The pressure at the bottom of the column due to the water can be found using;

$$Pressure = \frac{Force}{Area}$$

The force on the bottom of the column is due to the weight of the water above.

$$Weight = Mass \times g$$

We can find the mass of the liquid knowing its density from;

$$Density = \frac{Mass}{Volume}$$

$$m = V \times \rho$$
$$m = (A \times h) \times \rho$$

So;

$$Weight = (A \times h) \times \rho \times g$$

$$Pressure = \frac{(A \times h) \times \rho \times g}{A}$$

$$Pressure = h \times \rho \times g$$

This analysis confirms the experimentally determined direct proportionality between pressure and depth.

3.4 Buoyancy and upthrust

Introduction

In figure 3.4.1 a cube of side 5 cm having a weight of 11 N is suspended from a Newton balance. The block is being pulled towards the centre of the earth with a force of 11 N.

When the block is lowered into a beaker of water the reading shown on the spring balance reduces to 9.8 N. The resultant pull downwards now is only 9.8 N.

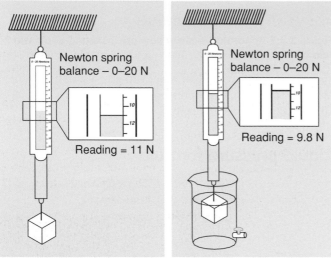

Figure 3.4.1 Block suspended from a Newton balance...

Figure 3.4.2... lowered into water

The apparent reduction in the weight of the cube is caused by the water. To explain this difference we say that the water is providing an upward force of 1.2 N. This force is called a **buoyancy force** or **upthrust**. When the tap on the bottom of the beaker is opened to let the water out the reading on the Newton balance returns slowly to 11 N.

Explaining upthrust

In the last section you saw that;

1 The pressure at a point in a liquid increases with the depth of the point in the liquid.
2 Pressure at a point acts equally in all directions.

In figure 3.4.3 the cube immersed in water will have a pressure acting downwards on the top and a pressure acting upwards on the bottom.

Since the bottom of the cube is deeper in the water than its top, the **pressure acting upwards on the bottom, P_{btm} exceeds the pressure acting downwards on the top, P_{top};**

Pressure difference, ΔP
$$= P_{btm} - P_{top}$$
$$= h_{btm}\rho g - h_{top}\rho g$$
$$= (h_{btm} - h_{top})\,\rho g$$

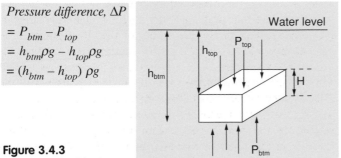

Figure 3.4.3 Forces on the block

The distance $(h_{btm} - h_{top})$ is simply the height, H, of the immersed block, so;

$$\Delta P = H\rho g$$

Pressure is defined as;

$$Pressure = \frac{Force}{Area}$$

The areas of the top and bottom of the cube are equal. Therefore the force on the bottom acting upwards exceeds the force acting downwards on the top. The resultant upwards force is called the buoyancy force or upthrust.

$$Pressure\ difference = \frac{Upthrust}{Area}$$

$$Upthrust = Pressure\ difference \times Area$$
$$Upthrust = \Delta PA$$
$$Upthrust = H\rho gA$$

This shows that the upthrust depends only on the dimensions of the block, the density of the liquid in which it is immersed (remember that ρ is the density of the liquid, *not* the block) and the constant, *g*. **Interestingly, the upthrust does *not* depend on how deep the block is immersed in the liquid!**

Floating and sinking

If the cube shown in figure 3.4.2 is released from the Newton balance it will accelerate towards the bottom of the container because there is *still a resultant force in the downwards direction*. The acceleration will not be as great as if the block were freely falling in air because the unbalanced force downwards is the difference between the weight and the upthrust.

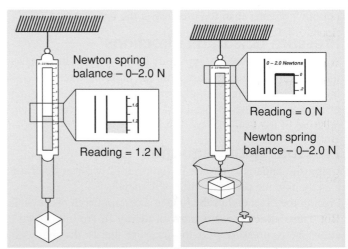

Figure 3.4.4 Repeating the experiment with lower density material

In figure 3.4.4 the cube attached to the end of the spring balance has the same size as that in figure 3.4.2 but is made of a material with a lower density. Its weight indicated on the spring is 1.2 N. When it is immersed in water the reading on the spring balance reduces to 0 N. In this case the upthrust provided by the water is equal to the weight of the cube. There is no unbalanced force and so the cube, if released from the end of the string, will not accelerate towards the bottom of the beaker.

Figure 3.4.5 shows a cube of the same size as those used previously, but made of a material with an even lower density so its weight is less than 1 N. When this block is totally immersed in water the upthrust of 1.2 N is greater than the weight of the cube. In this case the block will move up towards the surface of the liquid.

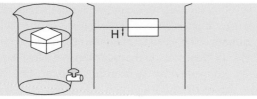

Figure 3.4.5 Floating block

In figure 3.4.5, the block is floating but only part of it is below the surface. The upthrust, given by;

$$Upthrust = H\rho g A$$

will be less because the H value in the above equation will not be the total height of the block but will only **be the distance of the bottom of the block below the surface.** In this way the upthrust can be reduced to a point where it equals the weight of the block.

Archimedes' principle and flotation

When an object is totally or partially immersed in a liquid the level of the liquid in the beaker rises by an amount equal to the volume of the object immersed.

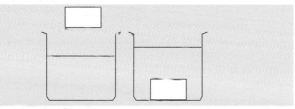

Figure 3.4.6 Sinking block

When an object is wholly or partially immersed in a liquid the upthrust provided by the liquid is given by;

$$Upthrust = H\rho g A \qquad equation\ 1$$

where H is the height and A the area of the base of the object. But;

$$Height \times Area\ of\ Base = Volume\ of\ object$$

So the upthrust from equation 1 can be expressed as;

$$Upthrust = V\rho g \qquad equation\ 2$$

But from the definition of density we can state;

$$Density\ of\ liquid = \frac{Mass\ of\ liquid}{Volume\ of\ liquid}$$

$$Mass\ of\ liquid = Density \times Volume$$

Therefore from equation 2;

$$Upthrust = Mass\ of\ liquid\ displaced \times g$$

or;

$$Upthrust = Weight\ of\ liquid\ displaced$$

When any object is immersed, it experiences an upthrust equal to the weight of the liquid it displaces. This is Archimedes' principle.

$$Upthrust = Weight\ of\ liquid\ displaced$$

For a floating object;

$$Upthrust = Weight\ of\ the\ object.$$

So;

$$Weight\ of\ liquid\ displaced = Weight\ of\ floating\ object$$

Therefore we can say that a floating object displaces its own weight.

Ships and submarines

When it was first proposed that ships should be made from metals rather than wood, many people thought that the ships would sink. It was common experience that lumps of wood would float but metal would sink.

Large steel ships float because they are shaped in such a way that they displace a weight of water equal to their own weight when only *part* of the ship is below the water line.

When the ship is loaded it will be heavier so will have to sit lower in the water to displace its own weight. Ships are marked with a series of *Plimsoll lines* to show where the level of the water line should be with and without cargo.

Figure 3.4.7 Plimsoll lines can be seen below the ship's name

When a submarine is cruising at a constant depth, the upthrust must equal the weight of the submarine.

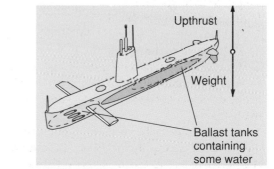

Figure 3.4.8 Submarine

When diving to greater depths the submarine must take water into its ballast tanks. In this way the weight of the submarine is increased, so that the weight is greater than the upthrust. The resultant force is in the downwards direction so the submarine accelerates downwards. When the desired depth has been reached water is expelled from the ballast tanks to the point where the weight of the submarine and the upthrust are again equal.

To rise, the submarine must pump water out of the ballast tanks so that the upthrust is greater than the weight. The resultant force is in the upwards direction and so the submarine will accelerate towards the surface.

Worked example 1

A student wants to find the density of a glass stopper. She puts the stopper on a balance and records its mass as 40.5 g.

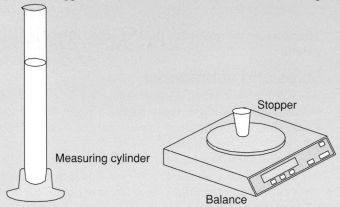

fig 3.5.1 Finding the density of a stopper

a) Describe how she could use the other apparatus shown in figure 3.5.1 to determine the volume of the glass stopper.
b) She determines that the volume of the stopper is 16 cm³.
   (i) What is the volume of the stopper in m³?
   (ii) What is the mass of the stopper in kg?
c) From the information given, determine the density of the glass block in units of kg m⁻³.
d) Calculate the volume of another piece of the same glass which has a mass of 205 g.

Answers and comments

a) To find the volume of the stopper she could use the measuring cylinder. She should record the volume of water in the cylinder and then drop the stopper into the water. The level of the water in the cylinder will increase and indicate a new higher volume. The volume of the stopper is the difference between this and the original volume recorded.

b) i) In this part of the question we must be careful to convert correctly from cm³ into m³. In 1 m³ there are 1×10^6 cm³. So
$$1 \ cm^3 = 1 \times 10^{-6} \ m^3$$
$$16 \ cm^3 = 16 \times 10^{-6} \ m^3$$

   ii) The mass of the stopper is quoted in the question as 40.5 g. This can be expressed as 40.5×10^{-3} kg.

c) The density of the glass can be calculated from
$$\rho = \frac{m}{V} = \frac{40.5 \times 10^{-3}}{16 \times 10^{-6}}$$
$$\rho = 2530 \ kg \, m^{-3}$$

d) We can calculate the volume by rearranging the density equation to give;
$$V = \frac{m}{\rho} = \frac{0.205}{2530}$$
$$V = 8.1 \times 10^{-5} \ m^3$$
$$V = 81 \ cm^3$$

Worked example 2

Figure 3.5.2 shows an attempt to make a device to measure atmospheric pressure.

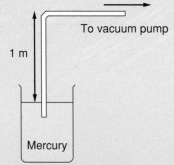

fig 3.5.2 Measuring atmospheric pressure

A glass tube 1 m long and open at both ends has one end placed into a beaker full of mercury. A vacuum pump is attached to the other end of the tube.
a) Explain why switching on the pump causes the mercury to rise up the tube.
b) The mercury rises up the tube until it is 76 cm above the level of the mercury in the beaker. Explain why the mercury level stops rising even when the pump is still switched on.

Answers and comments

a) When the pump is off atmospheric pressure pushes down equally on all parts of the mercury surface. When the pump is switched on the air above the part of the surface within the tube is removed and so the pressure in the glass tube is lowered. The atmospheric pressure acting over the rest of the surface forces the mercury to rise up the tube.

b) When the mercury column is 76 cm tall there is a pressure acting at the base of the column due to a column of mercury 76 cm tall. When the mercury stops moving up the tube the pressure at point X due to the column, is equal to the atmospheric pressure.

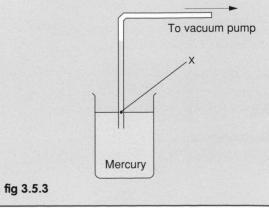

fig 3.5.3

Questions

1 A block of metal of density 2500 kg m⁻³ is 2 m tall and stands on a square base of side 0.4 m as shown in figure 3.5.4.

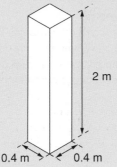

fig 3.5.4 Block of metal

  a) What is the area of the base of the block?
  b) Calculate the volume of the block.
  c) Calculate the mass of the block.
  d) What is the weight of the block?
  e) What pressure does the block exert on the surface on which it is placed?

2 A block of metal has sides of 5 cm, 3 cm and 10 cm. The block is made of a material of density of 19×10^3 kg m⁻³.
  a) Calculate the mass of the block.
  b) Calculate the lowest pressure that it can exert when resting on a flat surface.
  c) Calculate the largest pressure that this block can exert when resting on the flat surface.

3 A ship's anchor has a mass of 495 kg and is made from steel which has a density of 8000 kg m⁻³.

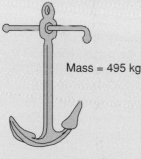

Mass = 495 kg

fig 3.5.5

  a) Calculate the volume of metal in the anchor.
  b) Calculate the weight of the anchor.
  c) The anchor is thrown into water which has a density of 1000 kg m⁻³. Calculate the weight of water displaced by the anchor.
  d) Draw a diagram to show the forces acting on the anchor when it is in the water
  e) Draw a diagram to show the forces acting on the anchor when it is out of the water.

4 **a)** Explain what is meant by pressure.
  b) The density of sea water is 1150 kg m⁻³. Calculate the pressure at a point 30 m below the surface of sea water due to the water alone.

  c) Calculate the pressure at the same depth when the atmospheric pressure of 101 kPa at the surface is included.
  d) Calculate the depth at which the total pressure is 200 kPa.

5 Figure 3.5.6 shows a block of length 15 cm which has a square base of side 4 cm. The block is made from a material of density 19×10^3 kg m⁻³. This block is suspended from a Newton balance.

fig 3.5.6

  a) Calculate the volume of the block in m³.
  b) Calculate the mass of the block.
  c) Show that the weight of the block is 44.7 N.
  d) When the block is fully submerged in a beaker of water the reading on the Newton balance reduces to 42.3 N. Explain why the reading indicated on the Newton balance is less than 44.7 N.

6 A brewer uses a device called a hydrometer to indicate the composition of a liquid. The hydrometer is placed in a liquid and floats at different depths in liquids of different densities. When placed in a liquid with the required density the line marked 'OK' floats level with the surface of the liquid.

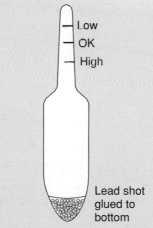

Low
OK
High

Lead shot glued to bottom

fig 3.5.7 A hydrometer

  a) Explain why the line marked 'high' floats level with the surface of the liquid when the hydrometer is placed in a liquid of higher density.
  b) During one particular process in the brewing industry the density reduces slowly. Explain how a hydrometer could be used to monitor the progress of this process.

Past paper question

Figure 3.6.1 shows the effect of increasing the force on a compression spring. This type of spring is used in the design of a safety device for a gas cylinder as shown in figure 3.6.2.

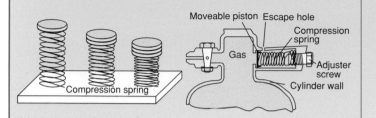

fig 3.6.1 Compression spring   **fig 3.6.2 Safety device**

a) The pressure of the gas in the cylinder is 5×10^5 Pa at a temperature of $20\,°C$. The area of the piston is 2.5×10^{-4} m^2.
  (i) What is the size of the force exerted by the gas on the piston?
  (ii) Explain how the device operates, if the gas pressure in the cylinder exceeds a safety limit.

b) The adjuster is screwed inwards. What would be the effect on the value of the pressure safety limit? Justify your answer. **SQA 1991**

Answers and comments

a) (i) The pressure exerted by the gas in the cylinder acts on the piston as shown in figure 3.6.2. We know both the size of the gas pressure and the area on which it acts. We can calculate the force by rearranging;

$$Pressure = \frac{Force}{Area}$$

$$Force = Pressure \times Area$$
$$Force = 5 \times 10^5 \times 2.5 \times 10^{-4}$$
$$Force = 125\ N$$

  (ii) Figure 3.6.1 shows that when the force on the spring increases it compresses. If the pressure of the gas in the cylinder increases it will exert a larger force on the movable piston. This will compress the spring. If the pressure exceeds the safety limit, the force on the piston will compress the spring sufficiently to expose the escape hole. Gas will therefore escape. The pressure in the cylinder will fall below the safety limit and the spring will extend so that the piston covers the hole and stops the escape of gas.

b) The pressure safety limit would be increased. Greater forces and consequently greater pressures would be needed to compress the spring sufficiently to expose the escape hole.

Past paper question

A sonar detector with a mass of 60 kg is used for monitoring the presence of dolphins. It is attached by a vertical cable to the sea bed so that the detector is held below the surface of the sea.

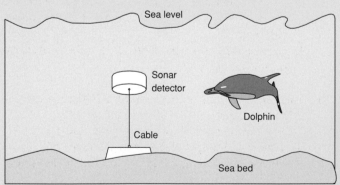

fig 3.6.3 Sonar detector

a) Explain the cause of the buoyancy force on the detector.
b) Draw a diagram showing the buoyancy force and other forces acting on the detector.
c) If the buoyancy force has a value of 31,500 N what is the value of the tension in the cable attached to the sea bed? **SQA 1991**

Answers and comments

a) The pressure at any depth in a liquid is directly proportional to the depth. Therefore the pressure acting upwards on the bottom of the detector exceeds the pressure acting downwards on the top. This means that the force pushing upwards on the bottom is greater than the force pushing downwards on the top. The resultant of these forces acts upwards. This is called the buoyancy force.

b) The tension in the cable acts on the detector. Even though the sonar detector is in the water it is still being pulled towards the centre of the earth. This force is called the weight. The mass of the detector is 60 kg so we can calculate its weight using;

$$W = m\,g$$
$$W = 60 \times 9.8$$
$$W = 588\ N$$

c) Since the forces balance we can say that

$$Buoyancy\ force = Tension + Weight$$
$$Tension = Buoyancy - Weight$$
$$Tension = 31{,}500 - 588$$
$$Tension = 30{,}912\ N.$$

Questions

1 A tin with 5000 cm³ of paint has a mass of 6 kg. The mass of the empty tin plus the lid is 0.75 kg.
   a) Calculate the volume of paint in the tin in m³.
   b) Calculate the density of the paint in the tin.
   c) The paint tin is made from a material whose density is 7600 kg m⁻³. Calculate the volume of metal used to make the paint tin.

2
   a) State how the pressure in a liquid depends on the depth of the liquid and the density of the liquid.

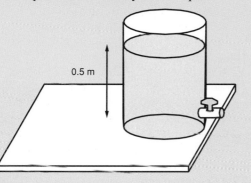

0.5 m

fig 3.6.4

   b) A tank with a base area of 4 m² is drained by a pipe at the base. The horizontal pipe has a cross-sectional area of 0.01 m². The tank is filled with water of density 1000 kg m⁻³ and the surface of the water is 0.5 m above the outlet pipe. Calculate the pressure of the water at a point 0.5 m below the surface.
   c) The tap on the outflow pipe is opened. Calculate the force with which the water is initially pushed out of the pipe.
   d) As water from the tank drains, the jet of water from the open pipe lands on the ground closer to the tank. Explain why this happens.

3
   a) Describe an experiment to measure the density of air.
   b) Calculate the mass of air in a room with dimensions 5 m × 4 m × 3 m. The density of air at atmospheric pressure is 1.2 kg m⁻³.
   c) What is the maximum mass of air that can be removed from a 1000 cm³ flask which is initially open to the atmosphere?
   d) The air pressure at the base of a mountain is found to be 1.01 × 10⁵ Pa. At the top of the mountain the air pressure is measured as 0.8 × 10⁵ Pa. Calculate the height of the mountain assuming that the density of the air at any height is 1.2 kg m⁻³.

4 In an experiment to calibrate a pressure gauge, a 2 kg mass is placed on top of the plunger of a syringe. The area of the syringe is 5 cm².

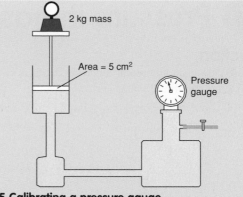

2 kg mass

Area = 5 cm²

Pressure gauge

fig 3.6.5 Calibrating a pressure gauge

   a) Calculate the area of the plunger of the syringe in m².
   b) What is the weight of the 2 kg mass?
   c) Calculate the pressure of the weight on the syringe.
   d) With no weight on the syringe the pressure gauge shows a reading of 100 kPa. What reading should be marked on the scale when 2 kg is placed on top of the syringe?
   e) What reading should be marked on the scale when 2.8 kg is placed on top of the syringe?

5 The lower surface of the escape valve on a pressure cooker has an area of 1 mm². The mass of the escape valve is 50 g.

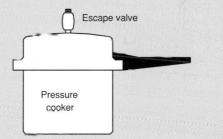

Escape valve

Pressure cooker

fig 3.6.6

   a) What is the weight of the escape valve?
   b) Calculate the area of the lower end of the escape valve in m².
   c) By how much is the pressure in the cooker above atmospheric pressure when the escape valve is activated.
   d) Explain why it is considered dangerous to place other objects on top of the escape valve.

6 A balloon with an envelope volume of 700 m³ is filled with hydrogen of density 9 × 10⁻² kg m⁻³.
   a) What is the mass of hydrogen in the balloon?
   b) What is the weight of the hydrogen in the balloon?
   c) The density of air is 1.2 kg m⁻³. Calculate the weight of 700 m³ of air.
   d) A balloon floats if it displaces a weight of gas greater than its own weight. Calculate the approximate mass of cargo which a 700 m³ balloon filled with hydrogen could lift.

3.7 Pressure in gases

Introduction

The pressure acting on an area at a certain depth is caused by the height of the column of water above.

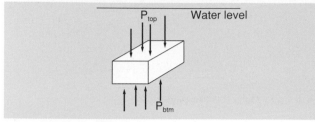

3.7.1 Pressure under water

In general we can say that the pressure, *P*, caused by a fluid at depth *h*, when the fluid is at rest, is given by;

$$P = h\rho g$$

where ρ is the density of the fluid.

Atmospheric pressure

The atmospheric pressure at the surface of the earth is around 1.01×10^5 Pa. This is caused by the weight of a column of air pushing down on an area of the earth's surface.

Accurate measurements record the density of air in the laboratory as 1.3 kgm^{-3}. If we assume that the air is of this density at all altitudes we can use the equation

$$P = H\rho g$$

to work out the height of the column, *H*, needed to create a pressure of 1.01×10^5 Pa.

$$H = \frac{P}{\rho \times g}$$

$$H = \frac{1.01 \times 10^5}{1.3 \times 9.8}$$

$$H = 7930 \ m$$

This analysis shows that the height of the atmosphere should be about 7.9 km. This basic model needs to be extended because Mount Everest is around 9 km high and some climbers have climbed Everest without oxygen.

We can explain this by realising that the density of air is in fact *not* constant at all altitudes. The atmosphere gradually gets less dense with increasing height above sea level so the column of air needed to create a pressure of 1.01×10^5 Pa at sea level is considerably taller than 7.9 km. However, climbers do have to make allowances for the 'thinner' air at higher altitudes.

Figure 3.7.2 Alison Hargreaves was the first woman to climb Everest without oxygen

The effects of atmospheric pressure

We do not normally feel the effect of atmospheric pressure because the pressure inside our bodies is the same as the pressure outside. The effect of atmospheric pressure can be shown by creating spaces from which the air has been removed. Such a space is called a **vacuum.**

The pressure of the air inside the metal container shown in figure 3.7.3 is initially the same as the pressure outside.

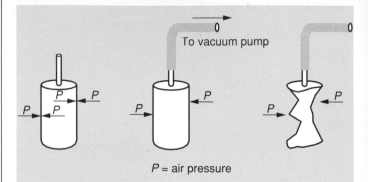

Figure 3.7.3 Collapsing can

When the vacuum pump is used to *reduce the pressure inside the container* the *greater pressure on the outside* acting inwards, forces the can to collapse.

When we use a straw to take drinks from a container our lungs expand as the air from the straw is drawn in. Atmospheric pressure pushing down on the surface of the liquid in the container is now greater than the pressure of the air in the straw. This imbalance forces the liquid from the container into the mouth.

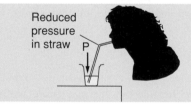

Figure 3.7.4 Sucking from a straw

Heating air

When air in an open space is heated the space between the molecules increases and so the density reduces. This warm air rises and is replaced by colder air, causing convection currents. If the air being warmed is contained within a hot air balloon of constant volume the warm air will provide an upwards force which tries to lift the balloon. The heated air is providing a lifting force by displacing colder, denser air. In the chapter on buoyancy and upthrust in liquids we showed that;

Upthrust = Weight of liquid displaced

If the lifting force created as warm air displaces cold, is greater than the weight of the balloon and the basket, the balloon will rise. Archimedes' principle can be applied to gases as well as liquids.

**Figure 3.7.5
Floating balloon**

Altering the pressure and volume of gases

When a fixed mass of gas is enclosed within the type of syringe shown in figure 3.7.6 the pressure inside the syringe is the same as the pressure on the outside.

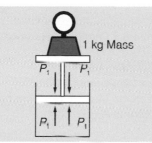

Figure 3.7.6 Piston

The piston of the syringe can move up and down freely without any of the gas inside the syringe escaping. Adding weights on top of the piston increases the external pressure on the gas and compresses it into a smaller volume. As the gas compresses the **internal pressure within the piston increases to a point where it now equals the increased external pressure.**

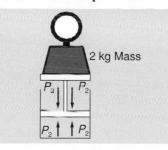

**Figure 3.7.7
Increased pressure**

From this analysis we can conclude that **increasing the pressure on a fixed mass of gas reduces the volume.**

When investigating the relationship between the pressure and volume of an enclosed mass of gas we need to be careful to keep the temperature of the gas constant. In an experiment using apparatus such as that shown in figure 3.7.8 a *fair test* will involve measuring the volume for a range of pressures. To alter the temperature or the mass of the gas *as well as* changing the pressure would not be a fair test of the relationship between the volume and pressure of the enclosed gas.

Pressure-volume variation for an enclosed gas

EXPERIMENT In the apparatus shown in figure 3.7.8, a fixed mass of air is enclosed in the space above a column of high density oil. When the oil is pressurised by pumping air into the

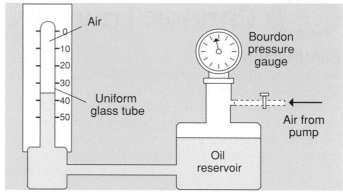

Figure 3.7.8 Boyle's Law apparatus

reservoir the column of oil is forced up the uniform glass tube so that the volume occupied by the fixed mass of air decreases. Because the glass tube containing the gas has a uniform cross-sectional area, measuring the length of the air column can give a good indication of the volume of the gas.

Releasing the air slowly from the reservoir above the oil reduces the pressure and the volume occupied by the fixed mass of air increases. The volume occupied by the enclosed air at different pressures can be found.

Boyle's Law

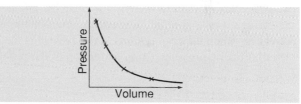

Figure 3.7.9

The simplest form of mathematical relationship linking quantities which vary as shown in figure 3.7.9 is an inverse proportionality. To test if pressure and volume values collected in an experiment as shown in figure 3.7.8 vary in this way, you should plot a graph of pressure versus 1/volume. A graph of this type is shown in figure 3.7.10.

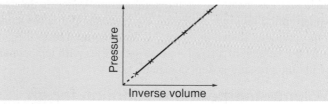

Figure 3.7.10

A straight line as shown in figure 3.7.10 confirms the inverse proportionality between the pressure of a fixed mass of gas and its volume. This relationship is known as Boyle's Law.

$$P \propto \frac{1}{V}$$

An alternative way of confirming the inverse proportionality between the pressure and volume of the enclosed gas is to calculate the product of the pressure and volume values found in the experiment. If the quantities are inversely proportional;

$$Pressure \times Volume = Constant$$

3.8 Charles' Law

Introduction

We know from the last section that the volume of a fixed mass of gas, maintained at a constant temperature, is inversely proportional to the applied pressure. Figure 3.8.1 shows a fixed mass of gas enclosed in a syringe where the piston can move freely. In this situation the pressure is constant. Only atmospheric pressure acts on the gas in the syringe.

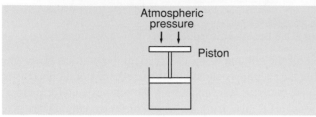

Figure 3.8.1 Syringe

Heating the enclosed gas

When the gas in the syringe shown in figure 3.8.1 is heated its molecules will travel faster and so the pressure inside the syringe will increase. If the pressure of the gas inside the syringe increases, the piston will move out so that the gas occupies a larger volume thereby lowering the pressure inside the syringe. For a specific temperature, the volume will stop increasing when the pressure inside the syringe is the same as the atmospheric pressure acting on the top of the piston.

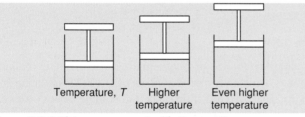

Figure 3.8.2 Piston movement with temperature

The diagram of figure 3.8.2 shows how the volume of an enclosed gas changes as the temperature changes. In each situation the pressure acting on the enclosed gas is the same. The apparatus shown in figure 3.8.3 can be used to investigate how the volume of a fixed mass of gas varies with temperature while the pressure is kept constant.

Volume-temperature variation for an enclosed gas

EXPERIMENT A fixed mass of air is enclosed in a capillary tube by a plug of water which can move freely as the trapped air changes volume. The capillary tube has a uniform cross-sectional area so the length of the air column can be regarded as a measurement of the volume of the enclosed air.

When the capillary tube is immersed in a beaker of water which is slowly heated the volume of the enclosed air can be found at a number of temperatures.

In this experiment it is essential that the water is heated to a desired temperature and that heating is then stopped. Even when heating stops the length of the air column continues to increase as the trapped air warms to the same temperature as the surroundings. Once the trapped air has reached the temperature of the surrounding water the length of the air column and the temperature can be recorded.

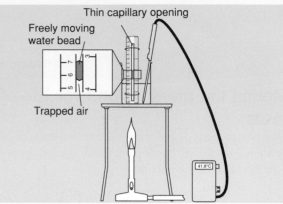

Figure 3.8.3 Charles' Law apparatus

Charles' Law

As the temperature of the enclosed gas in figure 3.8.3 increases the volume of the gas increases. A volume-temperature graph is shown in figure 3.8.4.

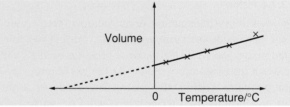

Figure 3.8.4

The simplest type of mathematical relationship linking variables which increase together is a direct proportionality. For a direct proportionality the volume-temperature graph would have to be a straight line passing through the origin. Although the graph *is* a straight line the volume of a fixed mass is not zero at a temperature of 0 °C.

For a direct proportionality the volume would need to double as the temperature doubles. Clearly, if we consider the temperature increasing from 1 °C to 2 °C there is not a doubling of the length of the air column.

The volume of a fixed mass of gas varies in a *linear* manner with the temperature in °C but we cannot say that there is a direct proportionality.

Absolute temperature

For a direct proportionality the volume of the gas would have to be zero when the temperature was zero. This does not happen if temperatures are measured on the Celsius scale but when we use the **absolute or Kelvin scale of temperatures the volume of a fixed mass of gas at constant pressure is directly proportional to the absolute temperature**. This relationship is known as **Charles' Law**. More details of the concept of absolute temperature are given in the section 3.10.

3.9 Pressure–temperature variation

Introduction

EXPERIMENT Using the apparatus shown in figure 3.9.1 we can record the pressures exerted by an enclosed gas as it is heated to different temperatures.

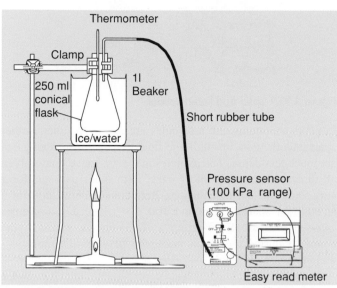

Figure 3.9.1 Investigating change in pressure of a heated gas

A fixed mass of gas, contained in the conical flask is heated indirectly by warming the surrounding water. When taking readings it is a good idea to heat the water slowly and to stop heating when the temperature has risen by roughly 10 °C. The temperature and pressure will continue to rise for a few moments while the gas temperature settles to the same temperature as the water, If you wait too long before taking readings both the temperature and pressure will start to fall, so it is good practice to record the pressure at the highest temperature reached before starting to reheat the water to an even higher temperature. With this method it is unlikely that the table of results will show the temperature rising in intervals of exactly 10 °C but this is not important as you will be able to plot the accurate readings that you have taken.

Analysing experimental results

The results shown in table 3.9.1 were obtained from the experiment outlined above.

Temperature/°C	20	29	42	51	63	75	90
Pressure/kPa	100	103	107	110	115	119	124

Plotting a pressure–temperature graph gives figure 3.9.2.

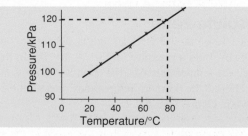

Figure 3.9.2

Although, the pressure and temperature vary in a linear manner, there is not a direct proportionality between pressure and the Celsius temperature because the straight line does not pass through the (0,0) origin.

Mathematically this type of graph can be described by the equation;

$$y = mx + c$$

If pressure, P is the variable on the y-axis and the Celsius temperature, T, is plotted on the x-axis, then;

$$P = mT + c$$

where m is the gradient of the line and c the intercept on the y-axis. We can find numerical values of m and c from the graph.

$$Intercept = 93 \times 10^3 \ Pa$$

$$Gradient = \frac{(120 - 93) \times 10^3}{79 - 0}$$

$$Gradient = 342$$

For this sample of gas we could restate the above equation as;

$$P = 342\ T + 93 \times 10^3$$

We can confirm that this equation describes the experiment by checking that the experimental results fit the model. If the model is correct we can use it to predict the pressure exerted if the gas is warmed to a temperature of 150 °C.

$$P = 342 \times 150 + 93 \times 10^3$$
$$P = 144.3 \ kPa$$

More interestingly we can use this model to predict the temperature needed for the gas pressure to be zero!

$$0 = 342\ T + 93 \times 10^3$$

$$Temperature = -\frac{93 \times 10^3}{342}$$

$$Temperature = -272\ °C$$

This values agrees with the accepted value of –273 °C (0 °K).

Absolute temperature

For a direct proportionality the pressure of the gas would have to be zero when the temperature is zero. This does not happen when the temperatures are measured on the Celsius scale of temperature. When they are measured on the absolute or Kelvin scale, **the pressure exerted by a fixed mass of gas, at constant volume, is directly proportional to the absolute temperature.**

The early experiments on this topic were done by Guillaume Amontons who in 1699 proposed that *equal changes in the temperature of a fixed volume of air resulted in equal variations in pressure.* This law is referred to in most texts as the Pressure or Third Gas Law.

3.10 Absolute temperature and kinetic theory

Introduction

In the previous chapters we have considered how gases behave as their surrounding conditions are changed. We will now look at how these macroscopic effects can be explained in terms of changes on a microscopic scale.

We already know that matter is made up from tiny particles called atoms. These atoms can join together to form molecules.

Brownian motion

In 1827 the Scottish botanist Robert Brown observed the random and erratic movement of pollen grains suspended in water. He suggested that the motion of the large visible particles was a result of collisions with smaller moving, invisible molecules and this movement is now called **Brownian motion**. This is considered as strong evidence for the theory that water and other liquids are composed of molecules which are in a continuous state of random motion.

We can also observe this random motion in gases using the apparatus shown in figure 3.10.1.

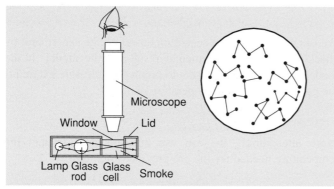

Figure 3.10.1 Brownian motion apparatus

A small quantity of smoke is introduced into a glass cell which is covered with a transparent slide. When the microscope is focused properly it is possible to see specks moving randomly through the cell. The specks are smoke particles which are reflecting the light into the lens of the microscope. The specks, which are big enough to be seen, are moving randomly because they are being bombarded by the surrounding air molecules. For a smoke particle to move, the air molecules must be bombarding it at that instant more on one side than on the other. This results in erratic motion. **From this we can conclude that the small, invisible air molecules are themselves in a state of random motion.**

This idea gives rise to the kinetic theory of gases which we can use to explain the relationships between the pressure, volume and temperature of an enclosed gas.

Kinetic theory

The kinetic model describes solids as being held together by balanced attractive and repulsive forces between neighbouring atoms or molecules. The bonded atoms or molecules cannot move freely but can vibrate. If the solid is heated the vibration

of the atoms or molecules increases. In liquids the atoms or molecules not only vibrate but can move around. This allows liquids to flow.

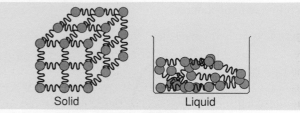

Figure 3.10.2 Solid and liquid model

In gases the atoms and molecules are about 10 times further apart than they are in solids and liquids. This means that gases are much less dense. The particles in the gas move randomly in all directions, coming into contact with other particles for a very short time during collisions. Between collisions they travel at constant speeds of around 500 m s^{-1}. The gas molecules cause a pressure by exerting a force on the walls of the container during the elastic collisions between the molecules and the walls.

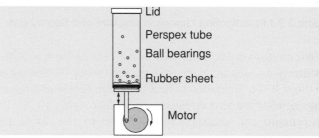

Figure 3.10.3 Mechanical model of kinetic theory of gases

The apparatus shown in figure 3.10.3 can be used as a mechanical analogy to illustrate the behaviour of particles in a gas. As the arm attached to the motor hits the rubber membrane the ball bearings are forced to move in a random manner. The pressure is caused by the particles hitting the walls and lid of the container. Making the piston vibrate faster causes the ball bearings to move faster and the particles will hit the walls more often and with greater force. With the particles moving faster the pressure increases. **Heating an enclosed gas forces the atoms or molecules to move more quickly and consequently the pressure exerted by the gas on its container increases.** We have been able to use the kinetic model to explain how the pressure of a fixed mass of gas, kept at constant volume, varies with temperature.

Absolute zero

At sea level in the UK, air has a density of approximately 1.3 kg m^{-3} and a pressure of $1.01 \times 10^5 \text{ Pa}$. In the experiment where the temperature of a fixed mass of air at constant volume was altered, the pressure-temperature variation recorded was as shown in figure 3.10.4.

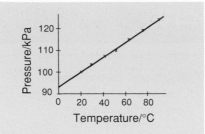

Figure 3.10.4

By redrawing this graph and extrapolating (figure 3.10.5) we can find the temperature at which the pressure exerted by the gas is zero. **Since gas pressure is caused by the movement of the atoms or molecules we can say that if the pressure is zero the molecules have stopped moving.**

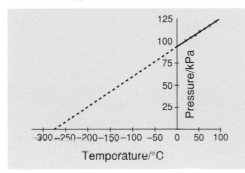

Figure 3.10.5

Accurate measurements show that the pressure is zero when the temperature is approximately −273 °C. This temperature is called **absolute zero** and in theory, is the temperature at which the **gas molecules are assumed to have no kinetic energy** so no more energy can be taken from them.

The absolute scale of temperature

An **absolute or thermodynamic scale** of temperature considers the kinetic energy of the gas atoms or molecules as a measurement of the temperature. The absolute scale of temperature therefore starts at −273 °C. This temperature is called 0 Kelvin or 0 K. The Kelvin scale of temperature is named in honour of Lord Kelvin who devised it.

Each temperature interval of one degree on the celsius scale is *exactly the same* interval as 1 K. The relationship between temperatures on both scales is given by;

$$T_K = T_{°C} + 273$$

On the Kelvin scale the freezing point of water is 273 K and the boiling point of pure water 373 K.

The pressure-temperature graph shown in figure 3.10.5 can be redrawn using absolute temperatures (figure 3.10.6).

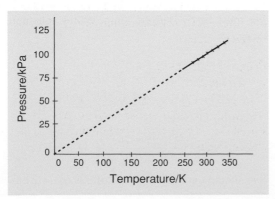

Figure 3.10.6

Since the graph of pressure versus absolute temperature is a straight line passing through the origin we can state that the pressure exerted by a fixed mass of gas at constant volume, is directly proportional to the absolute temperature. Mathematically this can be stated as;

$$P \propto T_K \ or \ \frac{P}{T_K} = constant$$

The other gas laws

When using the absolute scale of temperature we can express Charles' Law as **the volume, V, of a fixed mass of gas at constant pressure is directly proportional to the absolute temperature, T_K;**

$$V \propto T_K \ or \ \frac{V}{T_K} = constant$$

Boyle's Law links the pressure and volume of a fixed mass of gas at constant temperature and is expressed as;

$$P \propto \frac{1}{V} \ or \ PV = constant$$

These three gas laws can be summarised into a general gas equation. This can be used for situations where the mass of the gas is fixed but the pressure, volume and temperature change.

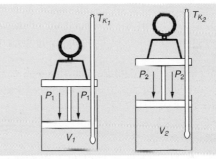

Figure 3.10.7 Diagram to show Gas Law

An enclosed mass of gas is contained in a syringe where the piston is free to move. At a Kelvin temperature T_{K1} the gas has a volume of V_1 and is at a pressure of P_1. If the conditions are changed so that the absolute temperature becomes T_{K2}, the volume, V_2, and the pressure, P_2, the gas laws require that,

$$\frac{P_1 V_1}{T_{K1}} = \frac{P_2 V_2}{T_{K2}}$$

A little further...

Ideal gases

The kinetic model for the behaviour of gases becomes less accurate as the temperature of the enclosed gas approaches 0 K. Theoretically at 0 K the gas should have no volume but this only happens for **ideal gases**. In reality as a gas cools towards 0 K the volume reduces because the gas atoms or molecules pack closer together. But the smallest space that the gas can occupy is equal to the volume of the molecules themselves and not zero. Other gases change state at very high pressures and therefore also do not behave as ideal gases. **Most gases at moderate temperatures and pressures show behaviour similar to that assumed of an ideal gas.**

Worked example 1

A car tyre contains a fixed mass of air. The pressure of the air in the tyre is 305 kN m^{-2} when the air in the tyre is at a temperature of 16 °C.

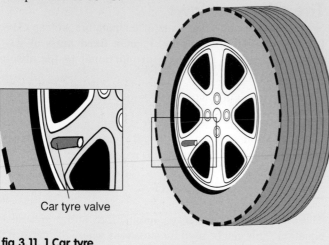

Car tyre valve

fig 3.11. 1 Car tyre

a) What is the temperature of the air on the Kelvin scale?
b) After a rallying event the temperature of the air in the tyre has increased to 28 °C. Calculate the pressure of the air in the tyre.
c) The owner of the car decides that this is a dangerously high pressure and decides to open the valve. Explain how this reduces the pressure of air in the tyre.

Answers and comments

a) To convert temperatures from one scale to another we can use the equation;

$$T_K = T_{°C} + 273 = 16 + 273$$
$$T_K = 289\ K$$

b) To answer this question we must assume that the volume of the tyre does not change. This allows us to use the relationship;

$$P \propto T_K$$

or;

$$\frac{P}{T_K} = constant$$

$$\frac{P_1}{T_{K_1}} = \frac{P_2}{T_{K_2}}$$

substituting gives

$$\frac{305}{289} = \frac{P_2}{273 + 28}$$

$$P_2 = \frac{305}{289} \times 301$$

$$P_2 = 318\ kPa$$

c) The pressure within the tyre is directly proportional to the mass of the gas in the tyre. If half the gas is released the pressure halves. Releasing some of the gas reduces the number of molecules colliding with the inner surface of the tyre. So the total outward force and consequently the pressure reduces.

Worked example 2

In a chemistry experiment a student collects 35 cm^3 of carbon dioxide gas.

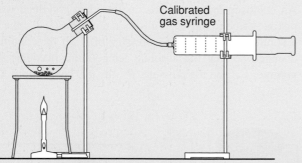

Calibrated gas syringe

fig 3.11.2 Gas collection

The temperature in the room is 20 °C and on the day that the experiment is performed the atmospheric pressure is 105 kPa. The student reads that gaseous volumes are usually quoted as being measured at *standard temperature and pressure*. Standard temperature is 0 °C and standard pressure is 100 kPa.
a) Explain why international journals quote the volumes of gases collected from experiments at standard temperature and pressure.
b) Calculate the volume of gas that she collects at standard temperature and pressure.

Answers and comments

a) The volume occupied by a particular mass of gas depends on the temperature and pressure at which it is collected. In different parts of the world a chemical reaction involving the same quantities of the same reactants will produce the same mass of gas but the volume occupied by this mass may be different depending on the location at which it is collected.
b) To convert the volume of a particular mass of gas collected under one set of conditions into a volume that the same mass of gas would occupy at another set of conditions we must use the equation;

$$\frac{P_1 V_1}{T_{K_1}} = \frac{P_2 V_2}{T_{K_2}}$$

We must firstly convert temperatures into units of kelvin, using;

$$T_K = T_{°C} + 273$$
$$T_{K_1} = 20 + 273 = 293\ K;\ T_{K_2} = 0 + 273 = 273\ K$$
$$\frac{P_1 V_1}{T_{K_1}} = \frac{P_2 V_2}{T_{K_2}} = \frac{105 \times 35}{293} = \frac{100 \times V_2}{273}$$

$$V_2 = \frac{273 \times 35 \times 105}{293 \times 100} = 34.2\ cm^3$$

She should quote the volume of gas as 34.2 cm^3.

Questions

1 Figure 3.11.3 shows a column of dry air at a temperature of 20 °C.

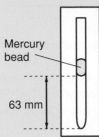

Mercury
bead

63 mm

fig 3.11.3 Column of air

a) Calculate the length of the air column if the air in the column is heated to 90 °C.

b) Sketch a graph of length of column versus temperature to show how the length of the column will change as the air in the column cools from 90 °C to 20 °C.

2

a) Explain what is meant by the absolute zero of temperature.

b) Describe a simple experiment which you could perform in the school laboratory to estimate the value of absolute zero on the Celsius scale of temperature. Draw a diagram of the apparatus that you would use. Describe how you would take the measurements and how you would estimate the value of absolute zero.

c) A sealed container with a volume of 500 cm³ contains a certain mass of gas at a temperature of 27 °C. To what temperature must the gas in the container be heated for the pressure to double?

3

a) State Boyle's Law linking the pressure exerted by a fixed mass of gas and the volume occupied by the gas when the gas is kept at a constant temperature.

b) A thick-walled steel cylinder used for storing compressed air has a safety valve which releases some of the air when the pressure in the cylinder exceeds 1×10^6 Pa. The air in the cylinder is normally stored at 18 °C and a pressure of 0.8×10^6 Pa. At what temperature will the safety valve activate?

4 Before starting a long journey a motorist finds that the pressure of the air in her car tyres is 3×10^5 Pa. She estimates that the temperature of the air in the tyre is 17 °C. At the end of the journey she finds that the tyre pressure has risen to 3.12×10^5 Pa. Assuming that the volume of the tyre has not altered calculate the temperature of the air in the tyre at the end of the journey.

5 Describe an experiment to examine the relationship between the volume and temperature of a fixed mass of dry air. Your answer should include;

a) A labelled diagram of the apparatus you would use.

b) A list of the measurements you would take and a description of how you would analyse the results.

c) A student records the following results to such an experiment.

Volume/cm³	7.0	7.6	8.2	8.6	8.8
Temperature/K	288	313	338	353	363

Table 3.11.1

Plot a volume-temperature graph for these results.

d) From your graph determine the volume of the air at 0 °C.

e) Explain the type of variation shown in your graph in terms of the kinetic theory of gases.

6 The aerosol container shown in figure 3.11.4 has a volume of 150 cm³. This aerosol contains a gas at a pressure of 3.5×10^5 Pa.

fig 3.11.4 Aerosol

a) What will be the volume of the gas contained in the aerosol if it is allowed to expand at an atmospheric pressure of 101 kPa? (You may assume that the temperature of the gas does not alter.)

b) Explain in terms of the kinetic theory why the trapped gas exerts a pressure on the walls of the aerosol.

c) Most aerosols have a warning printed on the can saying that it is dangerous to put the can into a fire. Explain in terms of the kinetic theory of gases why it is dangerous to dispose of aerosol cans in this way.

7 The volume of a sealed weather balloon is 30 m³ at ground level where the atmospheric pressure is 100 kPa and the temperature is 7 °C. What is the volume of the balloon at a height of 1500 m where atmospheric pressure is 80 kPa and the temperature is 37 °C?

Worked example

A pupil uses the apparatus shown below to investigate the properties of a simple gas.

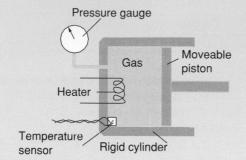

fig 3.12.1 Investigating a gas

The volume of the sample can be changed by moving the piston. The temperature of the sample of gas can be increased by using the heater.

At the start, the pressure of the gas is 400 kPa and its volume is 1000 cm³. During the investigation, the pressure and volume of the gas change as indicated by sections AB and BC on the graph below.

During section AB, the temperature of the gas is constant at 300 K.

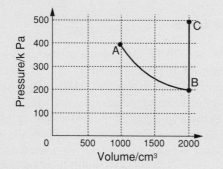

Figure 3.12.2

a) Calculate the volume of the gas when the pressure is 250 kPa during the stage AB.

b) State what happens to the pressure, volume and temperature of the gas over the section of the graph which starts at B and finishes at C.

c) What is the temperature of the gas, in kelvin, corresponding to point C on the graph?

SQA 1993

Answers and comments

a) We are told that during the stage AB the temperature of the fixed mass of gas is constant. We are therefore considering a pressure, volume relationship at constant temperature so we can apply Boyle's Law;

$$Pressure \times Volume = Constant$$
$$P_1 V_1 = P_2 V_2$$

From the diagram we can see that the volume is 1000 cm³ when the pressure is 400 kPa. We can take these values as V_1 and P_1. Substituting values

$$400 \times 1000 = 250 \times V_2$$

$$V_2 = \frac{400 \times 1000}{250}$$

$$V_2 = 1600 \ cm^3$$

b) From the axes on the graph we can see that during section BC the volume stays constant while the pressure increases. To cause a change in pressure without changing the volume the temperature must have increased.

c) To calculate the temperature at C we can use the pressure-temperature law. The value of T_1 is 300 K and at point B the pressure P_1 is 200 kPa.

We want to calculate the temperature when the pressure, P_2 is 500 kPa.

$$P_1 \propto T_k$$

or;

$$\frac{P_1}{T_K} = Constant$$

or;

$$\frac{P_1}{T_{K_1}} = \frac{P_2}{T_{K_2}}$$

$$\frac{200}{300} = \frac{500}{T_2}$$

$$T_2 = \frac{500 \times 300}{200}$$

$$T_2 = 750 \ K$$

Questions

1 A pupil is asked to complete the following statements of two of the gas laws by filling in the correct words in the spaces W, X, Y and Z.

The pressure of a fixed mass of an ideal gas kept at constant ---W--- is ---X--- proportional to the volume.

The pressure of a fixed mass of an ideal gas kept at constant ---Y--- is directly proportional to the ---Z--- temperature.

Which of A–E below shows the correct choices for W, X, Y and Z.

	W	X	Y	Z
A	volume	directly	temp	Celsius
B	temp	inversely	volume	Kelvin
C	volume	inversely	temp	Celsius
D	temp	inversely	volume	Celsius
E	temp	directly	volume	Kelvin

Table 3.12.1

2 The following statements compare the Celsius and Kelvin scales of temperature. Which statement is true?

A They measure different physical quantities.

B A temperature rise of 10 K is less than a rise of 10 °C.

C Doubling the Kelvin temperature will have more effect on the size of a balloon filled with helium, than doubling the Celsius temperature (both at constant pressure).

D Kelvin temperatures are numerically lower than Celsius ones.

E Kelvin temperatures can be negative numbers.

3 The air in the 500 cm³ flask shown in figure 3.12.3 is heated from 27 °C to 390 K before heating is stopped.

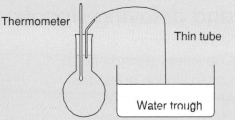

Figure 3.12.3

a) What is the temperature of 27 °C on the Kelvin scale of temperature?

b) The air in the flask has a volume of 500 cm³ at 27 °C. Calculate the volume that the same mass of air would occupy at the same pressure and at a temperature of 390 K.

c) Approximately what volume of air will leave the flask when it is heated to 390 K?

d) Approximately how much water will be drawn back into the flask as the system cools back down to 27 °C? Explain your answer.

4 A cylinder with a volume 0.17 m³ containing 20 kg of compressed air is stored at a temperature of 7 °C. The pressure of the gas in the cylinder is 9.5×10^5 Pa.

a) The compressed air is used in a chemical process which requires air at a temperature of 7 °C to be fed in at a pressure of 150 kPa and at a rate of 1000 cm³ per minute.

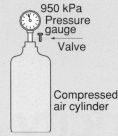

Figure 3.12.4

(i) Calculate the volume that the air in the cylinder would occupy at a pressure of 150 kPa.

(ii) What volume of air can one cylinder supply to the chemical process?

(iii) Calculate the length of time that one cylinder can supply the chemical process.

b) The escape valve on the cylinder activates when the pressure inside the cylinder exceeds 1×10^6 Pa. Calculate the maximum temperature at which the full cylinders could be stored before the escape valve activates.

5 In an investigation of the pressure–volume relationship for a fixed mass of gas at constant temperature a pupil connects a syringe containing 20 cm³ of air to a digital pressure meter as shown in figure 3.12.5.

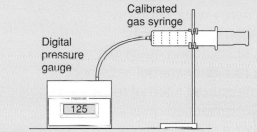

Figure 3.12.5

She applies different forces to the piston of the syringe and records a series of pressure and volume readings.

Pressure/kPa	100	200	150	300
Volume/cm³	15	7.4	10	5

Table 3.12.2

a) Without drawing a graph use a mathematical method to establish the relationship between pressure and volume for the trapped air.

b) Calculate the pressure that would be required to make the volume of the trapped air equal 12 cm³.

c) The barrel of the syringe has an internal diameter of 1 cm.

(i) Calculate the area of the end of the plunger.

(ii) Calculate the force applied by the girl when the trapped air has a volume of 5 cm³.

d) Sketch a graph to show how the force applied by the girl varies with volume. Both axes should show numerical scales.

Electricity & electronics

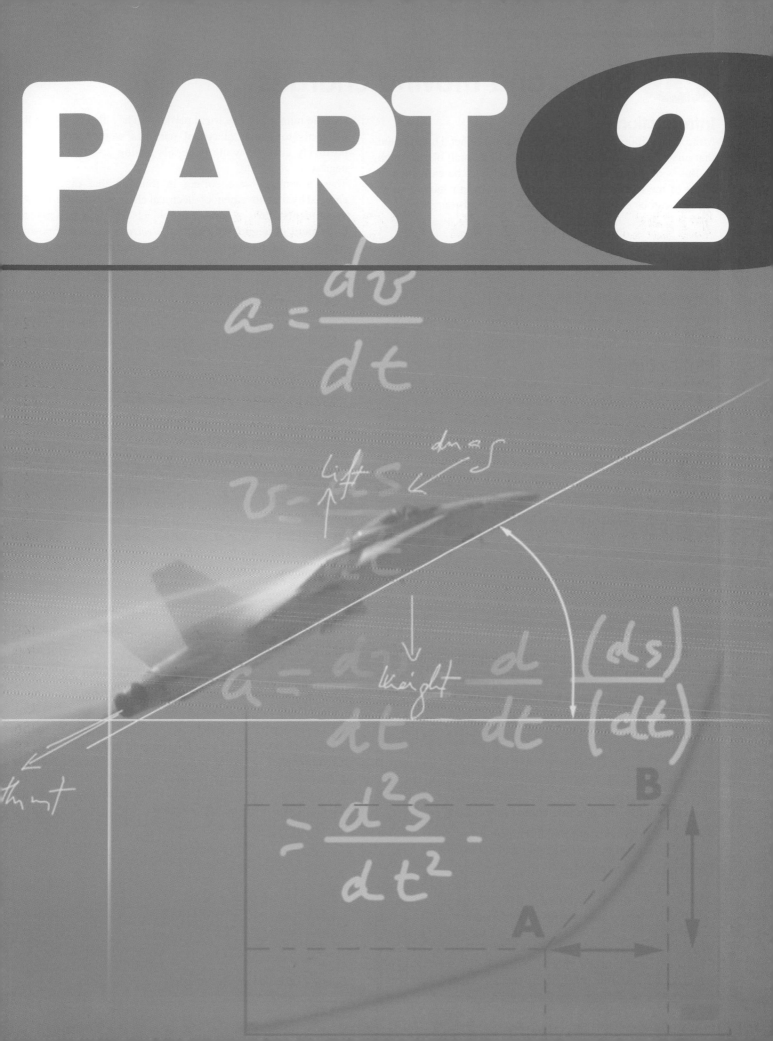

4.1 Static and moving charges

Introduction

More than 2500 years ago the Greek philosopher Thales noted that an amber rod rubbed with fur could lift small objects such as flakes of straw. This observation could not be explained in terms of the forces known at the time but from what we know today it can be explained in terms of electrical charge.

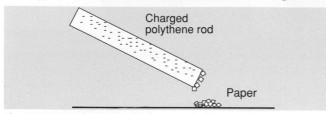

Figure 4.1.1 A charged polythene rod picks up paper

Static charges

When the polythene rod in figure 4.1.1 is rubbed with a woollen cloth it becomes electrically charged and will pick up small pieces of cotton wool or paper. When the charged polythene rod is placed near another charged polythene rod which is suspended freely, there is a *force of repulsion* between the two. Each rod exerts a force of repulsion on the other.

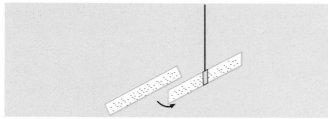

Figure 4.1.2 Similarly charged rods repel each other

When a rod of cellulose acetate is rubbed with a cloth it too will pick up small pieces of fluff or paper. However, if this rod is placed beside the suspended polythene rod, the rods will attract. Each rod now exerts an attractive force on the other.

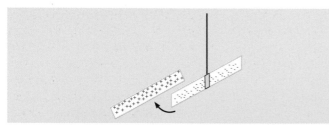

Figure 4.1.3 Rods with opposite charge attract each other

If both the suspended and held charged strips were made of cellulose acetate they would again repel each other.

Explaining static charge

From the experiments shown on figures 4.1.2 and 4.1.3 we can see that there are two different type of charge. Benjamin Franklin used the words 'positive' and 'negative' to distinguish between them. For example, polythene becomes negatively charged when rubbed, while cellulose acetate becomes positively charged when rubbed. Scientists in the eighteenth century believed that electrically neutral materials have a specific number of electrical particles. They also proposed that materials to which electrical particles were added became positively charged while materials having less particles were negatively charged.

These scientists were correct in thinking that objects were charged by the movement of electrical charges. However modern experiments have shown that objects are charged by the **movement of negatively charged *electrons*** and not of positive particles as implied by the thinking of the eighteenth century.

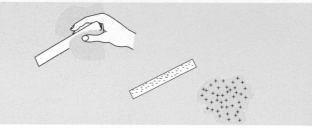

Figure 4.1.4 Transfer of electrons from cloth to rod

In figure 4.1.4, both the polythene rod and cloth are initially uncharged. When the polythene rod is rubbed with the cloth some of the *electrons* on the cloth are transferred to the rod which becomes *negatively charged*. The cloth, due to the loss of electrons, becomes positively charged.

When the cellulose acetate is rubbed with an uncharged cloth some of the electrons from the rod are transferred onto the cloth, making the *rod positively charged* and the cloth negatively charged.

Figure 4.1.5 Transfer of electrons from rod to cloth

Figure 4.1.6 shows a piece of apparatus commonly used to indicate the presence of charge. The gold leaf electroscope has a central metal stem and a metal cap. Beside the stem there is a thin gold leaf. When electrical charges are put onto the cap of the electroscope they spread throughout the metal. The force of repulsion between the stem and the leaf causes the leaf to diverge from the stem.

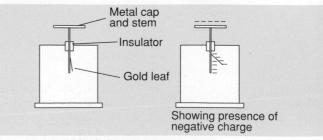

Figure 4.1.6 Gold leaf electroscope

Making charges move

The photograph of figure 4.1.7 shows a girl standing on a platform with her hand on the dome of a Van der Graaff generator. Negative charges from the girl are attracted to the Van der Graaff which is positively charged. The similarly charged strands of hair repel each other and this causes her hair to stand on end.

When she takes her hand off the Van der Graaff generator and stands on the floor her hair returns to its natural position. This is because there is a transfer of electrons from the earth to the girl which neutralises the charge on her body. In this case the earth acts as a source of electrons cancelling the effect of the positive charge on the girl.

Figure 4.1.7 Negative charge moves from student to Van der Graaf generator

Figure 4.1.8 Lightning

Figure 4.1.8 shows a very dramatic example of charges on the move! If the clouds become very very negatively charged the charges can flow to earth producing bolts of lightning. In this case the earth acts as a 'sink' accepting electrons from negatively charged clouds.

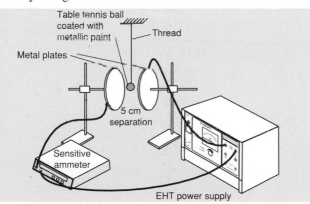

Figure 4.1.9 Experiment with charge

In figure 4.1.9 a d.c power supply is used to give electrons to one metal plate and remove them from the other.

The table tennis ball placed between the plates is coated with metallic, conducting paint and when touched momentarily against the negative plate the ball becomes negatively charged. It will then be repelled from the negative plate and attracted towards the positive plate. On touching the positive plate it will give up its negative charge and become positively charged instead.

The unbalanced force on the ball now moves it towards the negatively charged plate, where it will pick up negative charges again and move towards the positive. The ball will therefore shuttle backwards and forwards between the plates.

Figure 4.1.10 Ping pong!

If a very sensitive charge measuring device is connected between one of the plates and the power supply it registers a flow of charge. The ball is transferring small quantities of charge from one terminal of the power supply to the other across the gap. The faster the ball moves the greater the charge transferred per second, and so the bigger the reading on the ammeter.

The continuous movement of charge.

When the power supply and ammeter used in the above experiment are connected to a lamp as shown in figure 4.1.11, the lamp lights up and the ammeter registers a reading just as it did in the shuttling ball experiment.

From the similarity of these two experiments, we conclude that **a current in a circuit is a flow of charges from one terminal of a power supply to the other.**

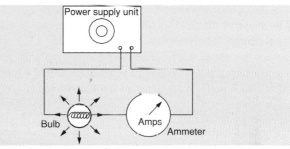

Figure 4.1.11

We know that it is **negatively charged electrons which move in metallic conductors creating a current.** Since much work on electricity was done before electron movement was accepted, many texts still adhere to the idea that *conventional current is the movement of positive charges from the positive pole of the battery to the negative.* Other texts use the arrows on a circuit diagram to indicate the direction in which the *electrons* move.

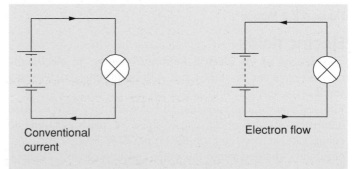

Figure 4.1.12 Conventional current and electron flow

4.2 Electric fields and currents

Introduction

In the shuttling ball experiment, the ball moves between the plates because it experiences a force due to the presence of nearby electrical charges. The charged ball experiences a force from the charged plates even when the ball and the plates are not in contact. *Forces acting at a distance* can be explained by the concept of a **field of force** or more simply a **field.**

Magnetic fields

You may have encountered the idea of a field in work with magnets.

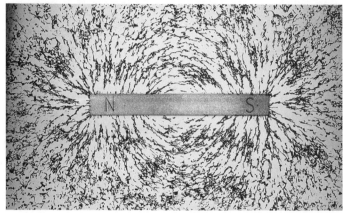

Figure 4.2.1 Iron filings showing the field around a bar magnet

The shape of the field surrounding a magnet can be found using iron filings. The magnets shown in figure 4.2.2 will either attract or repel each other depending on the polarity of neighbouring poles. When two similar poles are brought near each other their fields interact causing the magnets to repel each other.

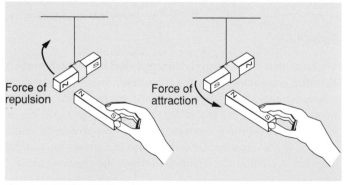

Figure 4.2.2 Magnets

Electric fields

The pattern of the electric field surrounding an electrically charged plate can be shown, not with iron filings, but by using small seeds suspended in a light oil.
The seeds are initially uncharged but when they are sprinkled onto the oil, the plates cause some of their electrons to congregate at one end of the seed. This end is therefore negatively charged while the other end is positively charged. The seeds line up end to end.

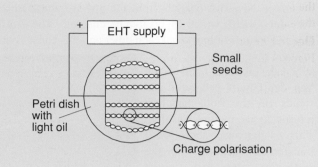

Figure 4.2.3 The pattern of a field made by electrically charged plates

Electric field patterns

When the charged conductors are parallel, the seeds line up as shown in figure 4.2.4. Figure 4.2.5 shows how this pattern is represented diagrammatically. Two parallel conductors form a *uniformly parallel electric field at right angles to the plates.* This is called a **uniform electric field**. At its edges the field curves slightly outwards away from the stronger field located in the region between the plates.

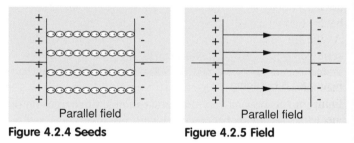

Figure 4.2.4 Seeds   **Figure 4.2.5 Field**

In figure 4.2.6 the metal conductors are a pin and a surrounding circular loop. The conductors have created a **radial field** which is represented by the type of diagram shown in figure 4.2.7.

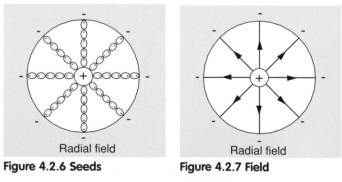

Figure 4.2.6 Seeds   **Figure 4.2.7 Field**

The arrows on the field diagrams show the direction that a small positive charge would move if released in the field. This direction is called the **field direction**. In the areas near the centre of the field in figure 4.2.7 the field lines are closer together than they are near the outer rim. This shows that the field is stronger at the centre than at the rim. By this we mean that the force on a charge placed at the centre is greater than the force on the same charge near the rim.

In figure 4.2.5 the field lines are parallel and therefore are not closer at any one point than at any other. This means that the field strength is the *same at all points* in the uniform field so the force on a charge placed at any point in the uniform field is the same.

Electric field strength

Figure 4.2.9 shows a uniform field where the force experienced by a charge is greater than the force on the same charge placed in the uniform field of figure 4.2.8.

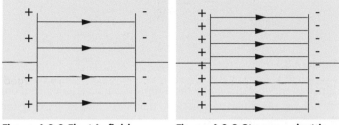

Figure 4.2.8 Electric field   **Figure 4.2.9 Stronger electric field**

The greater field strength in figure 4.2.9 is created by increasing the size of the output voltage of the power supply connected across the plates.

Electric fields in practice

Electric fields are created in electrostatic precipitators which are used to reduce the quantity of flue ash that would otherwise be discharged into the atmosphere by coal fired power stations. As the flue gases with particles of ash rise up the chimney, a series of central electrodes ionise air molecules which attach themselves to the ash. The now positively charged particles are then deposited on the oppositely charged electrodes on the lining of the chimney. The electrodes on the lining are then mechanically shaken to remove the ash.

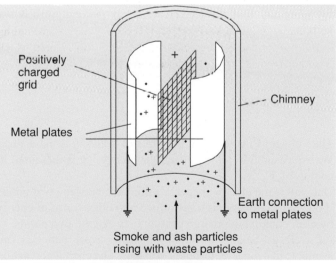

Figure 4.2.10 An electrostatic precipitator

Currents in conductors

In a lightning strike, as in the electrostatic precipitator *very high voltages* are required to make the charges move. However, charges move in a metal conductor even when there is only a *small voltage*. To explain this we must consider how electrons are bound within materials.

In lightning and in the electrostatic precipitator the natural forces holding the positive and negative parts of the air molecules together must be overcome by the forces caused by the external electric field. Since the forces holding the air molecules together are large, high voltages are needed.

Some of the electrons within the electrical conductor shown in figure 4.2.11, although *bound* within the materials as a whole, are free to move when they experience a force from even a small electric field.

In **perfect insulators** all of the electrons are bound to specific nuclei. Adding or removing electrons at one place *does not* cause a flow of electrons.

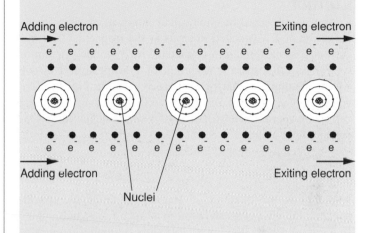

Figure 4.2.11 Electron model

A conduction model

We have already considered the idea of conventional current. Using this model positive charge carriers leave the positive side of the battery or power supply and move towards the negative terminal. With this model the direction in which the positive charge carriers move in a circuit is the same as the direction in which positive charge carriers will move when placed in an electric field.

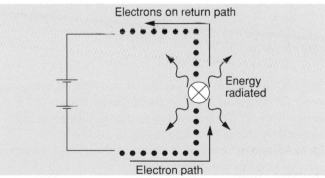

Figure 4.2.12

In figure 4.2.12 the energy radiated by the lamp as heat and light is provided by the electrical energy from the battery.

If we base our conduction model on the movement of electrons, energy is still required from the battery or power supply to make the electrons move through the lamp. With this model the **electrons move from the negative terminal of the battery to the positive terminal transferring energy into heat and light as they pass through the lamp.**

4.3 Current and voltage

Introduction

The lamp in figure 4.3.1 lights because electrical energy from the battery is transferred to the lamp. Our conduction model would lead us to believe that the electrons moving in the conductors in a circuit transfer energy through a circuit to a transducer such as a lamp, LED or buzzer.

In this section we will look in more detail at the rules governing the flow of charges in a circuit and the energy carried by the charges.

Current

Current is measured in units called **amperes, A.** The precise definition of 1 A is stated in terms of the magnetic effect of a current passing through a conductor. As a working definition we can say that there is a current of **1 ampere in a circuit when 1 coulomb of charge passes any point in 1 second.** The coulomb, C is the unit of charge, with 1 C being the charge carried by approximately 6.25×10^{18} electrons. In symbols the current, I, in a circuit is defined as;

$$I = \frac{Q}{t}$$

where Q is the quantity of charge, in coulombs, moving between the ends of a conductor in a time of t seconds.

In a circuit where 24 C of charge flow through a lamp in 30 seconds the current is;

$$I = \frac{Q}{t} = \frac{24}{30}$$
$$I = 0.8A$$

Similarly if there is a current of 5 A in a heater we can calculate the quantity of charge passing through the heater in 2.5 minutes.

$$I = \frac{Q}{t}$$
$$Q = I \times t = 5 \times (2.5 \times 60)$$
$$Q = 750 \, C$$

Our definition states that current is the flow of charge per second. A current of 1 A is a flow of charge at a rate of 1 C per second. This definition allows us to state that a current of 2.5 A is caused by charge flowing at a rate of $2.5 \, C \, s^{-1}$.

The definition of current also allows us to express coulombs in terms of amperes and seconds using;

$$Q = I \times t$$

$$Charge = Current \times Time$$
$$\text{in coulombs} \quad \text{in amperes} \quad \text{in seconds}$$

Just as earlier we were able to equate amperes with coulombs per second, this equation allows us to equate charge with the product of amperes and seconds. A charge of 250 C could be regarded as 250 A s.

Voltage

In figure 4.3.1 charge flows around a circuit. Some of the chemical energy stored in the battery is transformed into heat and light radiated by the lamp.

Figure 4.3.1

Since energy is radiated by the lamp we can conclude, from our model of electrical conduction, that the charges leaving the lamp have less energy than those entering.

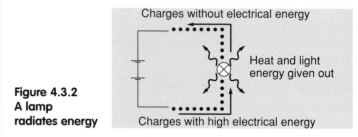

**Figure 4.3.2
A lamp
radiates energy**

The electrons are moving from a place of high energy to a place of lower energy. **This electrical energy difference across the lamp is maintained by the battery.** Each coulomb of charge moving from the high energy side of the battery to the low energy side carries a certain quantity of energy from the battery to the lamp. **The voltage across the lamp, V, is the electrical energy, W, transferred by 1 coulomb of charge. We can therefore say that the electrical energy per coulomb is the voltage.**

$$Voltage = \frac{Electrical \; energy}{Charge}$$

$$V = \frac{W}{Q}$$

The voltage, V, is measured in units of volts, V, if the energy, W, is measured in joules, J, and the charge, Q, is measured in coulombs, C.

One volt is the electrical energy difference between two points if one joule of energy is transformed into other forms of energy by each coulomb of charge passing between the two points.

As well as having different physical sizes, batteries are available in a range of different voltages. The chemical reaction in a 9 V battery can provide 9 J of energy to each 1 C of charge that the battery supplies to an external component.

If an electron with a charge of -1.6×10^{-19} C is placed at rest beside the negatively charged plate of figure 4.3.3 it will accelerate towards the positively charged plate.

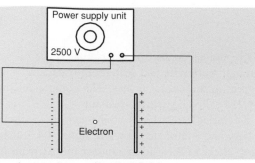

Figure 4.3.3 An electron in an electrical field

The kinetic energy of the electron will increase as it moves in the field. The electron's kinetic energy increases because of the work done by the electric field. We can find the increase in the kinetic energy of the electron using;

$$V = \frac{W}{Q}$$

$$W = QV = 1.6 \times 10^{-19} \times 2500$$
$$W = 4.0 \times 10^{-16} J$$

If the electron is initially at rest we can calculate its speed, v, on arriving at the positive plate because we know that the mass of the electron is 9.1×10^{-31} kg.

Work done = kinetic energy gained

$$4 \times 10^{-16} = \frac{1}{2} m \times v^2$$

$$4 \times 10^{-16} = \frac{1}{2} \times 9.1 \times 10^{-31} \times v^2$$

$$v = \sqrt{\frac{2 \times 4 \times 10^{-16}}{9.1 \times 10^{-31}}}$$

$$v = 3 \times 10^7 \, m\,s^{-1}$$

Measuring voltage

You will have used a voltmeter to measure the voltages in a circuit. You are now in a position to understand a little more about what the voltmeter is actually measuring. In figure 4.3.4 the voltmeter is measuring the difference in energy between the charges entering and leaving the lamp.

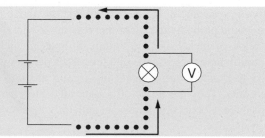

Figure 4.3.4 Measuring voltage across a lamp

The voltmeter display shows the energy difference per coulomb. **Since the voltmeter is measuring a difference in energy it must be connected across the lamp as shown.**

1

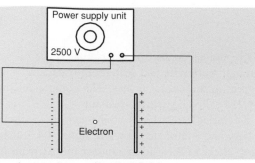

Figure 4.3.3 An electron in an electrical field

The kinetic energy of the electron will increase as it moves in the field. The electron's kinetic energy increases because of the work done by the electric field. We can find the increase in the kinetic energy of the electron using;

$$V = \frac{W}{Q}$$

$$W = QV = 1.6 \times 10^{-19} \times 2500$$
$$W = 4.0 \times 10^{-16} J$$

If the electron is initially at rest we can calculate its speed, v, on arriving at the positive plate because we know that the mass of the electron is 9.1×10^{-31} kg.

Work done = kinetic energy gained

$$4 \times 10^{-16} = \frac{1}{2} m \times v^2$$

$$4 \times 10^{-16} = \frac{1}{2} \times 9.1 \times 10^{-31} \times v^2$$

$$v = \sqrt{\frac{2 \times 4 \times 10^{-16}}{9.1 \times 10^{-31}}}$$

$$v = 3 \times 10^7 \, m\,s^{-1}$$

Measuring voltage

You will have used a voltmeter to measure the voltages in a circuit. You are now in a position to understand a little more about what the voltmeter is actually measuring. In figure 4.3.4 the voltmeter is measuring the difference in energy between the charges entering and leaving the lamp.

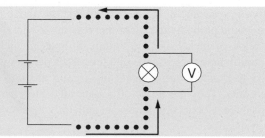

Figure 4.3.4 Measuring voltage across a lamp

The voltmeter display shows the energy difference per coulomb. **Since the voltmeter is measuring a difference in energy it must be connected across the lamp as shown.**

4.4 Resistance

Introduction

Charge moves more readily through some materials than through others. Those materials where the electrons are free to allow the conduction process are called **conductors.**

Other materials where the electrons are bound within a rigid bonding structure are called **insulators.** Table 4.4.1 categorises some common materials under the headings conductors and insulators.

Conductors	Insulators
Silver	Plastics
Gold	Glass
Copper	Wood

Table 4.4.1

Resistors

In general we say that insulators prevent the movement of charge while conductors permit charges to pass. However if very high potential differences (p.ds) are applied across thin samples of glass for example there is a small amount of current. Even good conductors provide some opposition to the movement of charge. To describe the *opposition to the flow of charges* through a sample of material we often quote the **resistance** of the sample. **Resistance is measured in ohms (Ω).**

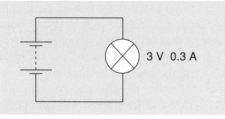

Figure 4.4.1

In figure 4.4.1 the lamp which is marked 3 V 0.3 A is connected to a 3 V battery. If the filament in the lamp were a perfect conductor with zero resistance the current from the battery would be much larger than 0.3 A. However, the filament of the lamp has a finite resistance which restricts current from the 3 V battery to 0.3 A.

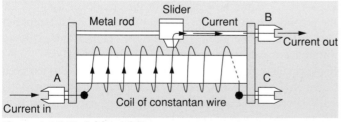

Figure 4.4.2 Variable resistor

Figure 4.4.2 shows an electrical component designed to allow the current in a circuit to be altered. The charge entering terminal A must pass through several turns of resistance wire before leaving through the sliding contact. When the sliding contact is near terminal A, the charge only has to pass through a short length of wire before leaving, so the resistance in the circuit is relatively low.

If the sliding contact is at terminal B the charge has to pass through more wire and with more resistance in the circuit, the current in the circuit is less. The component shown in figure 4.4.2 is called a **variable resistor.**

If the variable resistor is connected into a circuit with a 3 V battery and a lamp, as shown in figure 4.4.3, the current in the lamp alters as the sliding contact moves from one end to the other. Altering the current in this way alters the brightness of the lamp.

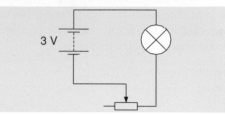

Figure 4.4.3

Changing resistances

We have seen how the resistance of the variable resistor depends on the length of the wire. For good conductors and moderate currents the resistance of a piece of wire, R, is directly proportional to its length, L.

$$R \propto L$$

This relationship allows us to compare the resistances of different lengths of wires. For comparisons to be valid the samples of wire must be made from the same material and have the same cross sectional area. For samples of equal length of the same material, the resistance of *thinner samples is greater than the resistance of thicker samples*. This is easily understood by remembering that current is a flow of charge. Less charge each second will be able to flow through narrower openings so the resistance of thin wires is greater than the resistance of thick wires. For samples of a specific material having the same length the resistance R of the sample is inversely proportional to the cross-sectional area, A.

$$R \propto \frac{1}{A}$$

It is not difficult to imagine that different materials will have different resistances. If you were to test the above relationship you would probably use nichrome resistance wire. Nichrome is an alloy of nickel and chromium. Strands of nichrome wire generally have a resistance of a few ohms per metre. The wires used to connect components in circuits are not made from nichrome. They are made of a material which has much less resistance. Connecting wires can generally be assumed to have no resistance.

Measuring resistance

EXPERIMENT Mathematically the resistance, R, of a conductor is defined as;

$$R = \frac{V}{I}$$

where V is the potential difference across a conductor in which there is a current of I amperes.

Resistors which are physically very small can have resistances of many thousands of ohms. The value of a resistor can be identified from the coloured bands marked on the body of the resistor.

The circuit shown in figure 4.4.4 can be used to investigate the behaviour of a resistor and verify the above relationships.

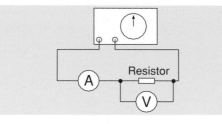

Figure 4.4.4

By altering the setting on the power supply the p.d. across the resistor can be changed and the current recorded for each setting. Typical results are shown in figure 4.4.5.

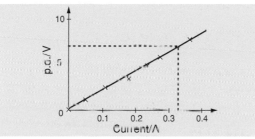

Figure 4.4.5

The graph of these results shows that for this resistor, the p.d., V, across it is directly proportional to the current, I, in it.

$$V \propto I$$

We can turn this proportionality into the following equation;

$$V = RI$$

Where the constant of proportionality, R, is the resistance. Rearranging the above equation gives,

$$R = \frac{V}{I}$$

We can also find the resistance from the graph by calculating the gradient;

$$Gradient = Resistance$$

$$Gradient = \frac{7 - 0}{0.325 - 0}$$

$$Resistance = 21.5\ \Omega$$

If the resistor originally tested is replaced by one with a higher resistance, a line could be plotted on the graph as in figure 4.4.6.

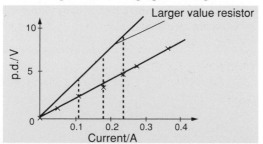

Figure 4.4.6

This line has a *steeper gradient* since *greater voltages* are needed to cause the same current to flow as in the 21.5 Ω resistor.

In some textbooks and questions this graph is drawn with the p.d. plotted on the x-axis and current on the y-axis. On such graphs the resistance will be given by $\frac{1}{gradient}$.

Other voltage current graphs

If the resistor in figure 4.4.4 is replaced with a different component such as a lamp or LED the shape of the p.d.-current graph may differ from that shown in figure 4.4.5. If a p.d.-current graph curves as shown in figure 4.4.7 the slope decreases as the current increases indicating that the resistance gets *lower* as the current increases. While the gradient indicates the *trend*, the actual resistance at any point on this curve is calculated from the p.d. and current values corresponding to that point on the line.

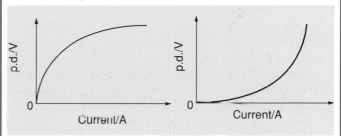

Figure 4.4.7   **Figure 4.4.8**

Figure 4.4.8 shows a situation in which the resistance *increases* as the current increases. If the temperature of a wire increases as more current passes, the resistance also increases as the increased vibrations of the atoms in the conductor impede the motion of the electrons. We would expect the p.d.-current graph for a filament lamp to be as shown in figure 4.4.8.

Multimeters

Figure 4.4.9 shows a multimeter capable of measuring resistance. The resistance function of these meters has circuitry which provides a very small, but constant, current in the resistor connected across the meter.

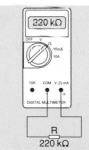

**Figure 4.4.9
Multimeter
and resistor**

In figure 4.4.9 the attached resistor has a value of 220 kΩ. The meter supplies a constant current of 1×10^{-6} A. The voltmeter within the multimeter will record a voltage across the resistor of;

$$V = IR = I \times 10^{-6} \times 220 \times 10^{3}$$
$$V = 0.220\ V$$

By moving the decimal point three places to the right the meter displays the resistance as 220 and the selector switch displays the units kΩ.

4.5 Consolidation questions

Worked example 1

Figures 4.5.1 and 4.5.2 show two diagrams. The circuit diagram shows a lamp connected in series with a 6 V battery.

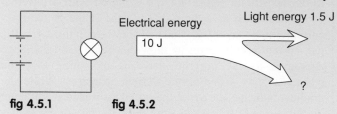

fig 4.5.1   **fig 4.5.2**

a) Redraw figure 4.5.1 showing how an ammeter should be connected to measure the current in the lamp.
b) With an ammeter correctly connected, the current in the lamp is 30 mA. Calculate the resistance of the lamp.
c) Figure 4.5.2 summarises the energy transformations in this circuit. From each 10 J of electrical energy supplied by the battery only 1.5 J are transferred to light energy by the filament of the bulb.
   (i) What has happened to the rest of the energy from the battery?
   (ii) Calculate the efficiency of this lamp.

Answers and comments

a) Ammeters measure current. Current is a flow of charge so the ammeter must be placed in series with the lamp. It does not matter whether the ammeter is placed before of after the lamp just as long as it is in series.

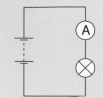

fig 4.5.3

b) We can calculate the resistance of the lamp when the p.d. across it is 6 V using Ohm's Law. The p.d. of 6 V causes a current of 30 mA.

$$R = \frac{V}{I} = \frac{6}{30 \times 10^{-3}}$$

$$R = 200 \ \Omega$$

c) The lamp operates because the current in the filament makes it heat up until it glows. The lamp therefore produces heat as well as light. In this case the heat is not a useful form of energy.

$$Efficiency = \frac{Useful \ energy \ output}{Total \ energy \ input} \times 100\%$$

$$Efficiency = \frac{1.5}{10} \times 100\%$$

$$Efficiency = 15\%$$

Worked example 2

An electron is accelerated in an oscilloscope by a potential difference of 750 V.

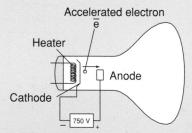

fig 4.5.4 Oscilloscope tube

a) Calculate the work done by the electric field in accelerating the electron.
b) How much kinetic energy does the electron gain as a result of the acceleration?
c) The electron is at rest before being accelerated. Calculate the velocity of the electron after it is accelerated. (The mass of an electron is 9.1×10^{-31} kg.)

Answers and comments

a) The electron in the electric field is acted on by a force. The work done (W) by this force in accelerating the electron is calculated using $W = Q \times V$ where Q is the charge on an electron and V the potential difference causing the acceleration;

$$W = Q \ V = 1.6 \times 10^{-19} \times 750$$
$$W = 1.2 \times 10^{-16} \ J$$

b) The work done by the electric field is equal to the energy gained by the electron, therefore;

kinetic energy gained by electron = 1.2×10^{-16} J

c) We can now calculate the velocity of the electron after it is accelerated using;

Work done = kinetic energy gained

$$1.2 \times 10^{-16} = \frac{1}{2} \ m \times v^2 = \frac{1}{2} \times 9.1 \times 10^{-31} \times v^2$$

$$v = \sqrt{\frac{2 \times 1.2 \times 10^{-16}}{9.1 \times 10^{-31}}}$$

$$v = 1.6 \times 10^7 \ m \ s^{-1}$$

Questions

1 A polythene rod can be charged negatively by rubbing it with a woollen cloth. A metal rod held in one hand and rubbed with a similar cloth cannot be charged.
   a) Explain why the polythene rod becomes negatively charged when rubbed with the cloth.
   b) Explain why the metal rod cannot be charged by rubbing in this way.

2 A charge of 6×10^{-6} C at point X is moved to point Y, against the electric field as shown in figure 4.5.5. An energy of 9 mJ is required.

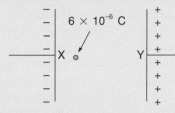

fig 4.5.5

   a) Is the 6×10^{-6} C of charge positively or negatively charged? Explain your answer.
   b) Calculate the potential difference between X and Y.
   c) The charge is held at plate Y while the p.d. of the supply is halved. The charge is then released. How much kinetic energy will it have when it arrives back at plate X?

3 A chemistry textbook states that in one mole of electrons there are 6×10^{23} electrons.
   a) Calculate the total charge of 1 mole of electrons.
   b) In a certain electrochemical process a current of 5 A is supplied to a solution for 20 hours.
      (i) Calculate the quantity of charge supplied to the solution.
      (ii) How many moles of electrons are supplied to the solution.

4 Figure 4.5.6 shows apparatus that can be used to determine the efficiency of a small electric motor.
   With the switch in position A the motor lifts the mass through a certain height. Changing the switch over to position B causes the mass to fall and the joulemeter calculates the energy produced as the mass falls.

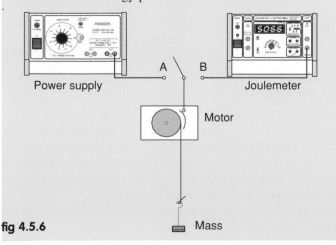

fig 4.5.6

In one test 25 J were required to raise the load whereas only 9 J were released. Calculate the efficiency of the system.

5 In an experiment to investigate the effect of an electric field, identical oil drops each with a charge of -4.8×10^{-10} C pass into the field at X and exit at Y.

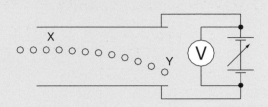

fig 4.5.7

A pupil considers the changes labelled I–IV;
I: Increasing the p.d. between the plates.
II: Using heavier oil drops.
III: Extending the plates at either end.
IV: Making the drops positively charged.
   a) Which of these changes could make the path XY horizontal? Explain your answer.
   b) Draw a diagram to show the vertical forces acting on a drop when the drops are travelling from X to Y.
   c) The drops have a mass of 2.5×10^{-6} kg. When the drops are travelling horizontally, calculate the force exerted on the drop by the electric field.

6 A particle accelerator increases the speed of alpha particles by accelerating them between a pair of parallel plates. The potential difference between the plates is 50 kV. The alpha particles have a charge of 3.2×10^{-19}.

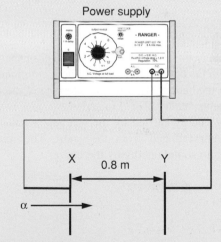

fig 4.5.8 Particle accelerator

   a) Draw a diagram to show the shape of the electric field between plates X and Y.
   b) Calculate the kinetic energy gained by the alpha particles as they travel between plates X and Y.
   c) On entering the field the alpha particles of mass 6.5×10^{-27} kg are travelling at 100 m s^{-1}. Calculate the speed of the alpha particles as they arrive at plate Y.

4.6 Series and parallel circuits

Introduction

The circuit of figure 4.6.1 is described as a lamp connected across a 6 V battery.

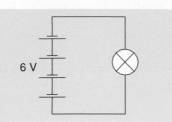

**Figure 4.6.1
Circuit with lamp**

If a buzzer is to be connected into the same circuit there are two ways in which it can be done. Figure 4.6.2 shows the buzzer connected in **series** with the lamp and battery while in figure 4.6.3 it is connected in **parallel** with the lamp.

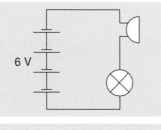

Figure 4.6.2 Circuit with lamp and buzzer in series

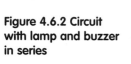

Figure 4.6.3 Circuit with lamp and buzzer in parallel

Series circuits

In the series circuit shown in figure 4.6.2 all of the 'energised charges' from the battery pass through both components before returning to the battery. This is characteristic of circuits where the components are connected in series.

Some of the energy carried by the charges is transferred into heat and light radiated from the lamp. The remainder is transferred into sound energy by the buzzer.

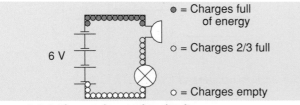

● = Charges full of energy

○ = Charges 2/3 full

○ = Charges empty

Figure 4.6.4 Charges in a series circuit

Since a 6 V battery is used there are 6 joules of energy available to every coulomb of charge leaving the battery. If 2 joules per coulomb are transferred to sound by the buzzer we can say that 4 joules per coulomb will be dissipated as heat and light by the lamp. Voltmeters connected across the buzzer and lamp will therefore read 2 V and 4 V respectively.

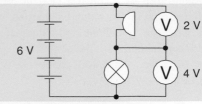

Figure 4.6.5 Voltages in a series circuit

This shows that for more than one component connected across a battery **the sum of the p.d.s across each of the components in series is equal to the p.d. across the supply.**

When introducing the idea of current as a flow of energised charges we stated that all of the 'energised charges' leaving one terminal of the battery returned to the other terminal having transferred their energy to the components in the circuit. Using this model it is easy to understand why there is no build up of charge at any one point in a circuit. Since there is no build up of charge we can say that the rate of charge flowing out of the battery is the same as the rate of flow of charge returning. **This means that the current at all points in a series circuit is the same.** In figure 4.6.6 all three ammeters will show the same reading.

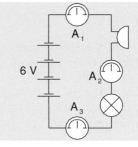

Figure 4.6.6 Current in a series circuit

Parallel circuits

In the parallel circuit of figure 4.6.7 some of the 'energised charges' leaving the battery pass through the buzzer and the rest pass through the lamp before returning to the battery. This is characteristic of circuits where the components are connected in parallel.

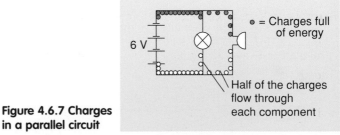

● = Charges full of energy

Half of the charges flow through each component

Figure 4.6.7 Charges in a parallel circuit

Since a 6 V battery is used, there are 6 joules of energy available to every coulomb of charge leaving the battery. For those charges passing through the lamp, all of the 6 joules of energy carried by each coulomb of charge is transferred to heat and light. Each coulomb of charge passing through the buzzer will dissipate 6 joules of energy as sound.

In a parallel circuit the p.d. across each component is the same.

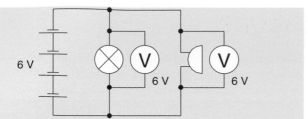

Figure 4.6.8 Voltages in a parallel circuit

Using the conduction model which assumes that charge does not build up at any point in a circuit, we can say that, **the current from the battery entering a parallel circuit is equal to the sum of the currents in each of the parallel branches.** For the ammeter readings in figure 4.6.9 we can say;

$$A_T = A_1 + A_2$$

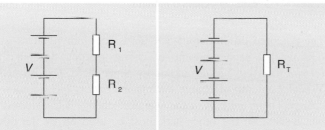

Figure 4.6.9 Current in a parallel circuit

Resistors in series

In figure 4.6.10 resistors R_1 and R_2 are joined in series to a power supply with an output voltage of V volts. In figure 4.6.11 another resistor, R_T, is connected to the same supply so that it has the same effect as the other two resistors together. The resistance of R_T is equal to the combined resistances of R_1 and R_2. The current in R_1 and R_2 must be the same as the current in R_T.

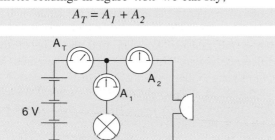

Figure 4.6.10 Resistors in series Figure 4.6.11 $R_T = R_1 + R_2$

Since R_1 and R_2 are in series, the p.d.s V_1 and V_2 across the series resistors must add up to the p.d. across the supply, V.

Therefore;

$$V = V_1 + V_2$$

Using Ohm's Law for each resistor we can say that;

$$V = IR_1 + IR_2 \qquad equation\ 1$$

For the circuit of figure 4.6.11 the p.d., V, across resistor R_T, also causes a current, I, so;

$$V = IR_T \qquad equation\ 2$$

By comparing equations 1 and 2;

$$IR_T = IR_1 + IR_2$$

Therefore;

$$R_T = R_1 + R_2$$

This analysis can be extended to work out the combined resistance of any number of resistors in series and can show that **the combined resistance of any number of resistors in series is the numerical sum of the individual resistances.**

Resistors in parallel

In figure 4.6.12 the resistors R_1 and R_2 are each connected across a power supply with an output voltage of V volts. In figure 4.6.13 another resistor, R_T, is connected across an identical supply so that it has the same effect as the two resistors in parallel. The resistance of R_T is equal to the combined resistances of R_1 and R_2.

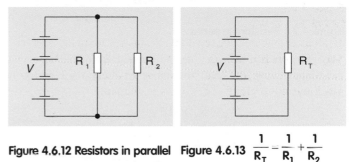

Figure 4.6.12 Resistors in parallel   Figure 4.6.13 $\dfrac{1}{R_T} = \dfrac{1}{R_1} + \dfrac{1}{R_2}$

Since R_1 and R_2 are in parallel and connected across the supply the p.d. across each is V volts. Similarly the p.d. across R_T is also V volts.

Because R_1 and R_2 are in parallel the currents I_1 and I_2 in the resistors must add up to the current, I, from the supply.

Therefore;

$$I = I_1 + I_2$$

Using Ohm's Law for each resistor we can say;

$$\frac{V}{R_T} = \frac{V}{R_1} + \frac{V}{R_2} \qquad equation\ 1$$

But since the voltage, V, is the same in each case;

$$\frac{1}{R_T} = \frac{1}{R_1} + \frac{1}{R_2} \qquad equation\ 2$$

For **two** resistors in parallel we can say that;

$$\frac{1}{R_T} = \frac{1}{R_1} + \frac{1}{R_2}$$

$$\frac{1}{R_T} = \frac{R_1 + R_2}{R_1 R_2}$$

$$R_T = \frac{R_1 R_2}{R_1 + R_2}$$

$$R_T = \frac{Product\ of\ resistances}{Sum\ of\ resistances}$$

This equation is a rearrangement of equation 2 above and makes it easier to calculate the effective resistance of two resistors in parallel. Both forms of this equation, when used correctly will give the same answer. You must remember that the second form only applies to cases where TWO resistors are joined in parallel.

4.7 Voltage dividers

Introduction

In figure 4.7.1 two 10 MΩ resistors are connected in series with a 6 V battery pack.

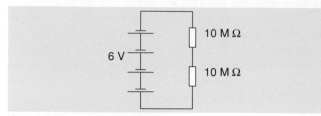

Figure 4.7.1 Resistors in series

The resistors are in series so the current from the battery is calculated using the total resistance of the circuit which is given by;

$$R_T = R_1 + R_2$$
$$R_T = 10\ M\Omega + 10\ M\Omega$$
$$R_T = 20\ M\Omega$$

The total resistance connected to the battery in figure 4.7.1 is 20 MΩ. The current from the battery can now be calculated using Ohm's Law;

$$I = \frac{V}{R}$$

$$I = \frac{6}{20 \times 10^6}$$

$$I = 3 \times 10^{-7}\ A$$

Figure 4.7.2 shows resistances of 2.2 kΩ and 4.7 kΩ connected in series with a 6 V battery. The total resistance of this combination is 6.9 kΩ.

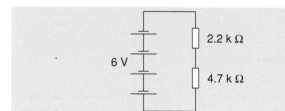

Figure 4.7.2 Resistors in series

We can again calculate the current from the battery using Ohm's Law as shown above. Normally when using Ohm's Law we express the potential difference in volts, the resistance in ohms and the current in amps. In the calculation shown below, the voltage is indeed in volts but the resistance is in kΩ, so the current calculated is in units of milliamps (mA), where 1 mA = 0.001 A.

$$I = \frac{V}{R}$$

$$I = \frac{6}{6.9}$$

$$I = 0.87\ mA$$

Both methods for calculating the current from a battery, if used correctly, will give the correct answer. The combinations of units can be summarised in the following table.

p.d.	Resistance	Current
V	Ω	A
V	kΩ	mA

Table 4.7.1

Currents in series resistors

In the circuit of figure 4.7.1 the 10 MΩ resistors are joined in series so the current at all points in the circuit is 0.3 µA.
A current in a resistor means that there is a potential difference across the resistor. We can find the size of the potential difference across one of the 10 MΩ resistors by applying Ohm's Law to that *individual* resistor.

$$V = IR$$
$$V = 0.3 \times 10^{-6} \times 10 \times 10^{6}$$
$$V = 3\ V$$

A voltmeter connected across a single 10 MΩ resistor as shown in figure 4.7.3 would show a reading of 3 V.

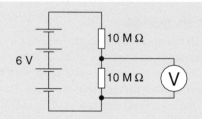

Figure 4.7.3 Voltage across a resistor

The potential difference of 3 V across the individual resistor tells us that each coulomb of charge passing through that resistor transfers 3 joules of energy. Since the battery is rated 6 V, each coulomb of charge leaving the battery has 6 joules of energy. Therefore half of the available energy is transferred to one resistor so the other half must be used up by the other resistor.
The total voltage available from the supply has been divided equally between the two identical resistors.
In the circuit shown in figure 4.7.2 the resistors connected in series with the battery have different values. However the current in each of these resistors is the same and was calculated earlier as 0.87 mA. We can find the potential difference across each resistor by applying Ohm's Law.

For the 2.2 kΩ resistor
$$V = I\,R$$
$$V = 0.87 \times 2.2$$
$$V = 1.91\ V$$
For the 4.7 kΩ resistor
$$V = I\,R$$
$$V = 0.87 \times 4.7$$
$$V = 4.09\ V$$

In this case the 6 joules of energy transferred by each coulomb of charge from the battery is not shared equally between the two resistors. *The larger resistor 'grabs' the bigger share of the available voltage.* Voltmeters connected across the resistors will read 1.91 V and 4.09V as shown in figure 4.7.4.

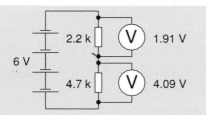

Figure 4.7.4 Voltage across two resistors in series

The supply voltage V_s is divided up by the resistors according to the following equations;

$$V_s = V_{2.2} + V_{4.7}$$

$$and \quad \frac{V_{2.2}}{V_{4.7}} = \frac{2.2}{4.7}$$

Replacing one of the 10 MΩ resistors with ten individual 1 MΩ resistors will keep the current from the battery at 0.3 µA. The p.d. across the remaining 10 MΩ resistor will stay at 3 V but the remaining 3 V will be shared equally among the ten 1 MΩ resistors. The p.d. across each of the 1 MΩ resistors will be 0.3 V.

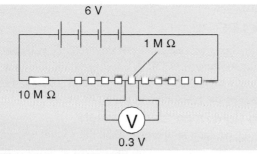

Figure 4.7.5 Voltage across resistors

Variable resistors and potentiometers

If the lead of the voltmeter in figure 4.7.6 marked X is connected to point A, the voltmeter will show a reading of 3 V. Connecting in turns to points B, C and D will increase the reading to 3.3 V, 3.6 V and 3.9 V.

Using the ten 1 MΩ resistors in figure 4.7.6 in place of a single 10 MΩ resistor allows the voltmeter reading to increase in steps of 0.3 V from 3 V up to 6 V.

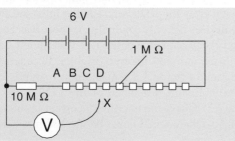

Figure 4.7.6 Using a potentiometer

Replacing the ten 1 MΩ resistors of figure 4.7.6 with a 10 MΩ potentiometer simplifies the circuit considerably (figure 4.7.7). The current from the battery is still 0.3 µA because the total resistance in the circuit is still 20 MΩ.

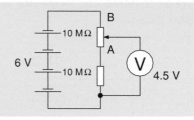

Figure 4.7.7 Using a potentiometer

Moving the sliding contact of the potentiometer between its limits allows you to tap off different fractions of the battery voltage. When the sliding contact of the potentiometer is at the mid point of its travel the voltmeter will indicate a reading of 4.5 V. Moving the sliding contact towards the end marked A will lower the reading shown on the voltmeter, while moving it towards B will increase the reading towards a maximum value of 6 V.

A note of caution

It would be easy to imagine that replacing the voltmeter of figure 4.7.7 with a lamp and moving the sliding contact would allow the p.d. across the lamp, and therefore the lamp's brightness to be altered smoothly. However if the circuit of figure 4.7.8 is set up the lamp will only light when the sliding contact is tight against position B. Move the contact even a little and the lamp will not light.

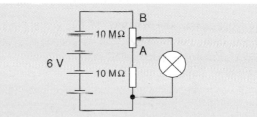

Figure 4.7.8

To change the brightness of the lamp up to maximum brightness using the potentiometer requires some sort of output driver to be connected to the bulb as in figure 4.7.9.

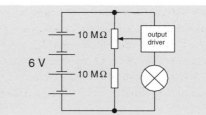

Figure 4.7.9

The voltage applied to the bulb is set by the potentiometer but no current is diverted from the voltage divider to the lamp. The current causing the lamp to light comes directly from the battery.

4.8 Resistors and voltage dividers in use

Introduction

The block diagram of figure 4.8.1 shows the main parts in a modern computer system.

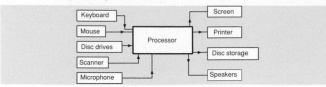

Figure 4.8.1 Systems in a computer

The processor is supplied with information by the keyboard, microphone, scanner or other input devices. The task requested by the user is performed by the processor and the outcome is shown on the screen or by one of the other output devices. One of the great advantages of complex computer systems is that they can perform many different tasks selected by the user. Other electronic systems can be dedicated to one particular task.

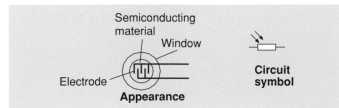

Figure 4.8.2 Automatic window opening

A gardener wants the window of his greenhouse, in figure 4.8.2, to open automatically on bright days. To solve this problem he requires a dedicated system capable of sensing the brightness of daylight, deciding if it is a bright day and then opening or closing the window.

Figure 4.8.3

Light dependent resistors

One of the simplest ways of sensing light level uses a light dependent resistor (LDR) as in figure 4.8.4.

Figure 4.8.4 LDR

The transparent window on the front of the LDR allows light to fall on a thin layer of semiconducting material. **When bright light falls on the material its conducting properties change so that the resistance between the electrodes decreases.**

When the LDR is covered so that **no light falls on the semiconducting material its resistance increases**. The actual value of its resistance depends on the light level and on the composition of the material in the LDR. You can use the apparatus shown in figure 4.8.5 to show that there is a significant change in the resistance when the LDR is covered.

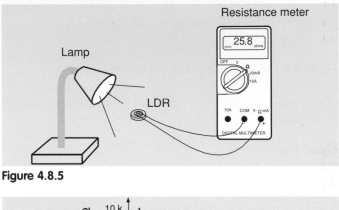

Figure 4.8.5

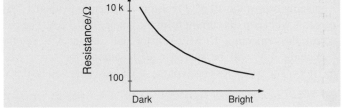

Figure 4.8.6

With the lamp off and the LDR covered, the resistance is about 10 kΩ but when the LDR is uncovered and the lamp is on, the resistance of the LDR falls to around 100 Ω. Figure 4.8.6 shows how the resistance of the LDR depends on light level.

Using Light Dependent Resistors

It would be easy to imagine that the circuit shown in figure 4.8.7 could be used to sound the buzzer at dawn each morning.

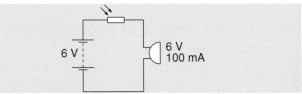

Figure 4.8.7

During the night the resistance of the LDR would be high and there would be no current in the buzzer so it would not sound. As the sun rises the LDR's resistance falls and the current in the circuit rises but only gradually. If there was enough current to sound the buzzer it would start very quietly and then get louder as the morning became brighter. In practice the buzzer in the circuit of figure 4.8.7 would never sound. In very bright light the minimum resistance of the LDR is around 100 Ω. With the 6 V battery the maximum current in this circuit would be 60 mA so there could never be the 100 mA needed to make the buzzer sound.

The circuits of figures 4.8.8 and 4.8.9 explain how an LDR, exposed to bright light, can be used to switch on a buzzer.

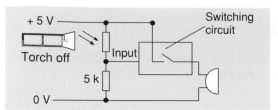

Figure 4.8.8

In the circuit of figure 4.8.8 the LDR is connected in series with a 5 kΩ resistor to form a voltage divider. In dark conditions the LDR has a much higher resistance than the 5 kΩ resistor. The voltage at the midpoint of the voltage divider is consequently low and insufficient to switch on the current to the buzzer.

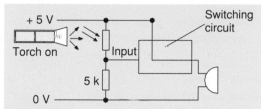

Figure 4.8.9

When the torch is switched on (figure 4.8.9) the resistance of the LDR is less than 5 kΩ so the voltage at the midpoint of the voltage divider rises to a level where it can switch on the current in the buzzer. The current in the buzzer comes directly from the supply but is controlled by the voltage at the input.

 The switching circuit of figure 4.8.8 switches on the current in the lamp when there is 1 V across the LDR. What is the resistance of the LDR?

If there is a potential difference of 1 V across the LDR the remaining 4 V, of the 5 V supply, must be across the 5 kΩ resistor. Therefore the current in the 5 kΩ resistor must be;

$$I = \frac{V}{R} = \frac{4}{5\,k\Omega}$$

$$I = 0.8\ mA$$

If all of this current passes through the LDR which has a p.d. of 1 V across it, then,

$$R_{LDR} = \frac{V}{I} = \frac{1}{0.8\,mA}$$

$$R_{LDR} = 1.25\ k\Omega$$

When a calibration graph is available for the LDR the light level needed to switch on the buzzer can be found.

The LDR is one device whose resistance depends upon a physical quantity – light intensity. There are many sensors which can monitor other physical properties such as temperature, force, atmospheric pressure, etc. All of these devices can be used to generate a voltage signal if they are made one of the two resistors in a voltage divider. A low output is generated when the LDR is illuminated and a high output when the LDR is covered.

Thermistors

Just as an LDR has a resistance which depends on illumination, a thermistor has a resistance which depends on its temperature. A voltage divider, containing a thermistor, can therefore be used to generate an output that is temperature dependent.

Many types of thermistor are available but the most common type has a resistance which *reduces as the temperature increases*. The resistances at 100 °C and 0 °C depend on the specific thermistor chosen but one common thermistor has the type of resistance-temperature variation shown in figure 4.8.10.

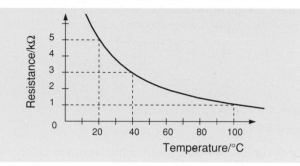

Figure 4.8.10

A simple thermostat

In the circuit of figure 4.8.11 the switching circuit switches on the heater when the p.d. across the thermistor rises to 1.0 V.

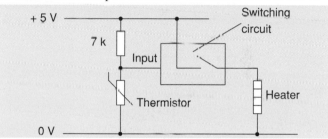

Figure 4.8.11

When the thermistor is warm its resistance is low and so the voltage at the midpoint of the potential divider in figure 4.8.11 is low. As the thermistor cools its resistance increases and the voltage at the input to the switching circuitry also increases. When this reaches 1.0 V the heater switches on and warms the surroundings.

When the heater switches on, the p.d. across the thermistor is 1.0 V. This means that the remaining 4 V of the 5 V supply are across the 7 kΩ resistor. The current in the 7 kΩ resistor must be;

$$I = \frac{V}{R} = \frac{4}{7\,k\Omega}$$

$$I = 0.57\ mA$$

If this is the current in the thermistor which has a p.d. of 1.0 V across it then;

$$R_{Thrm} = \frac{V}{I} = \frac{1.0}{0.57\,mA}$$

$$R_{Thrm} = 1.75\ k\Omega$$

The thermistor causes the heater to switch on when the resistance of the thermistor rises to 1.75 kΩ. If this thermostat uses the type of thermistor described by the graph of figure 4.8.10 we can see that it has a resistance of 1.75 kΩ at a temperature of 70 °C, so the heater turns on when the temperature of the thermistor falls below 70 °C.

4.9 Wheatstone bridges

Potential divider formula

When two resistors are connected in series with a battery the current in each of the resistors is the same and depends on the size of the supply voltage and the total circuit resistance.

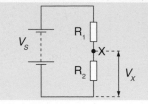

Figure 4.9.1

For the circuit of figure 4.9.1 the total resistance is $R_1 + R_2$ so the current I in each of the resistors is given by;

$$I = \frac{V_s}{R_1 + R_2}$$

If we assume that the negative terminal of the battery is 0 V the potential at the point marked X will be the same as the potential difference across resistor R_2. The potential difference across resistor R_2 can be found by applying Ohm's Law to R_2, individually. Therefore;

$$V_X = IR_2$$

But since we already have an expression for the current we can say;

$$I = \frac{V_s}{R_1 + R_2}$$

Therefore;

$$V_X = \frac{V_s}{R_1 + R_2} \times R_2$$

Rearranging;

$$V_X = \frac{R_2}{R_1 + R_2} \times V_s$$

This expression describes the voltage at the output of a potential divider in terms of the resistances and the supply voltage. In the circuit of figure 4.9.2 the same supply voltage V_s is applied across resistors R_3 and R_4.

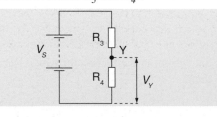

Figure 4.9.2

By an analysis similar to that outlined above it is possible to show that the output voltage, V_Y, in this circuit is given by;

$$V_Y = \frac{R_4}{R_3 + R_4} \times V_s$$

In the circuit of figure 4.9.3 *both* of the potential dividers already considered are connected to the *same* power supply.

Since they are connected in parallel the p.d. across each branch is the same.

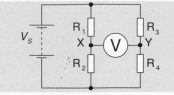

Figure 4.9.3

When the reading on the voltmeter is 0 V we can say that there is no potential difference between points X and Y, therefore;

$$V_X = V_Y$$

Combining the expressions for V_X and V_Y which we have already derived we can state that;

$$V_X = V_Y$$

Therefore;

$$\frac{R_2}{R_1 + R_2} \times V_s = \frac{R_4}{R_3 + R_4} \times V_s$$

$$\frac{R_2}{R_1 + R_2} = \frac{R_4}{R_3 + R_4}$$

Cross multiplying;

$$R_1R_4 + R_2R_4 = R_2R_3 + R_4R_2$$

Simplifying;

$$R_1R_4 = R_2R_3$$

Rearranging;

$$\frac{R_1}{R_2} = \frac{R_3}{R_4}$$

The circuit of figure 4.9.3 is known as a **Wheatstone bridge** circuit. **If the voltages at the outputs of the potential dividers are the same the bridge circuit is said to be** *balanced*.

Therefore the mathematical relationship;

$$\frac{R_1}{R_2} = \frac{R_3}{R_4}$$

is described as the **balance condition**.

Wheatstone bridges

The resistor network known as the Wheatstone bridge was devised by Sir Charles Wheatstone around 1843 to allow unknown resistances to be measured accurately. If the resistances of R_3 and R_4 are known accurately the value of the unknown resistor R_1 can be found by altering R_2 until the bridge is balanced. The value of the unknown resistance R_1 can then be calculated. With this method R_2 has to be altered over a range of accurately known values. There are a number of ways to do this but one of the easiest is to alter the dial settings on the type of resistance substitution box shown in figure 4.9.4.

Figure 4.9.4 Resistance substitution box

Figure 4.9.5, like figure 4.9.3, has four resistors arranged in two parallel branches, each of two resistors. The voltmeter is connected between the mid points of the pairs of resistors in each branch. Altering the way the bridge circuit is drawn, like altering the supply voltage does not change the resistance needed to balance the bridge.

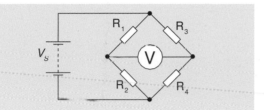

Figure 4.9.5

 In the circuit shown in figure 4.9.6 a temperature dependent resistor called a **thermistor** is placed in a beaker of boiling water.

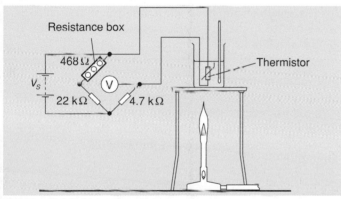

Figure 4.9.6 Using a thermistor

The resistance substitution box is altered until the reading on the voltmeter is 0 V. The resistances for the balanced bridge circuit are as shown in the diagram. Calculate the resistance of the thermistor.

A For balance;

$$\frac{R_1}{R_2} = \frac{R_3}{R_4}$$

$$\frac{0.468}{22} = \frac{R_3}{4.7}$$

$$R_3 = \frac{4.7 \times 0.468}{22}$$

$$R_3 = 0.1 \ k\Omega$$

Please note that in this example since we have converted all of the resistance values into kΩ, the final answer is also in kΩ.

Out-of-balance bridges

For the example in figure 4.9.6 the bridge is balanced when the resistance of the thermistor is 100 Ω. When the thermistor cools its resistance changes so the bridge becomes out of balance.

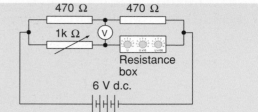

Figure 4.9.7

The circuit shown in figure 4.9.7 allows you to investigate how the out-of-balance voltage, ΔV, depends upon how far, ΔR, the resistance is away from its value for the balance condition. With the resistance box set at 470 Ω, the variable resistor can be altered until the bridge is balanced. The resistance box can then be set to different values between 270 Ω and 670 Ω and the out-of-balance voltages, ΔV, noted. The graph of ΔV verses ΔR would be as shown in figure 4.9.8.

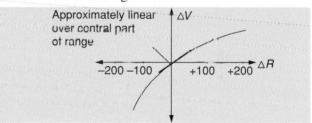

Figure 4.9.8

Advanced circuit theory predicts that **for small changes of resistance** (around the resistance needed to balance the bridge) ΔV **is directly proportional to** ΔR. This can be confirmed by taking readings for out-of-balance resistances between 420 Ω and 520 Ω.

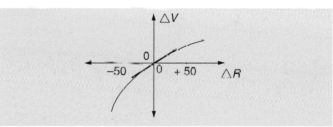

Figure 4.9.9

The relationship ΔV is proportional to ΔR, is true for resistance changes within approximately 10 % of the resistance needed to balance the bridge. ΔV can have both positive and negative values depending on whether the variable resistance is above or below the resistance required to balance the bridge.

Out-of-balance bridges have historically been used in instrumentation where thermistors, light dependent resistors or strain gauges are used in the bridge. When the monitored property changes by a small amount, the magnitude of the change can be indicated by the out-of-balance voltage. Whether the resistance of the sensor is increasing or decreasing can also be determined from the sign of the out-of-balance voltage.

4.10 Consolidation questions

Worked example 1

Figure 4.10.1 shows two different combinations of resistors.

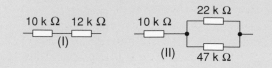

fig 4.10.1

a) Calculate the effective resistance for combinations I and II.

b) Each combination is joined in turn to a 6 V battery. Which combination will cause the least current from the battery? Explain your answer.

Answers and comments

a) In combination I the resistors are in series so we can calculate the resistance using;

$$R_T = R_1 + R_2$$
$$R_T = 12 + 10$$
$$R_T = 22 \ k\Omega$$

In combination II the 10 kΩ resistor is in series with the 22 kΩ and 47 kΩ resistors which are in parallel with each other. We will first find the resistance of the two resistors in parallel.

$$\frac{1}{R_T} = \frac{1}{R_1} + \frac{1}{R_2}$$

$$\frac{1}{R_T} = \frac{1}{22} + \frac{1}{47}$$

$$R_T = \frac{22 \times 47}{22 + 47}$$
$$R_T = 15 \ k\Omega$$

Now we can find the effective resistance of combination II by adding on the series resistor.

Effective resistance = 15 + 10 = 25 kΩ.

b) The smallest current is when the largest resistance is connected to the 6 V battery. Therefore combination II will result in the lowest current.

Worked example 2

The maximum voltage allowed across the LED shown in figure 4.10.2 is stated as 1.6 V. It is also stated that the current in the LED must not exceed 8.5 mA.

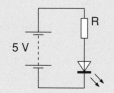

fig 4.10.2

a) Calculate the p.d. across the resistor when the p.d. across the LED is 1.6 V.

b) When the p.d. across the LED is 1.6 V the current in the LED is 8.5 mA. What is the current in the resistor? Explain your answer.

c) Calculate the value of the resistor which will keep the current in the LED below the stated maximum.

d) Calculate the power dissipated in the resistor.

Answers and comments

a) The LED and resistor are connected in series and joined to a 5 V battery. The total voltage across both components is therefore 5 V and since the p.d. across the LED is 1.6 V the p.d. across the resistor, V_R, will be given by;

$$V_R = 5 - 1.6$$
$$V_R = 3.4 \ V$$

b) Once again we must use the fact that the resistor and LED are in series. Since they are in series the current in both will be the same. So the current in the resistor is also 8.5 mA.

c) We now know both the current in the resistor and the value of the p.d. across the resistor so we can calculate its resistance using;

$$R = \frac{V}{I}$$

$$R = \frac{3.4}{8.5}$$

$$R = 0.4 \ k\Omega$$
$$R = 400 \ \Omega$$

d) Once again we can use the fact that we know the current in the resistor and the p.d. across the resistor to calculate the power dissipated using;

$$P = V I$$
$$P = 3.4 \times 8.5$$
$$P = 28.9 \ mW$$

Worked example 3

The circuit of figure 4.10.3 shows a thermistor and a resistor connected in series with a 6 V battery.

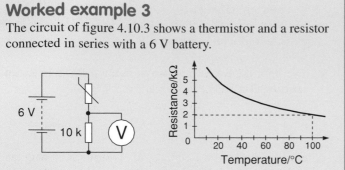

fig 4.10.3　　　　　　　　**fig 4.10.4**

Figure 4.10.4 shows how the resistance of this thermistor changes with temperature.
a) Describe how the voltmeter reading will change as the temperature of the air surrounding the thermistor rises.
b) What is the resistance of the thermistor when it is placed in a beaker of boiling water? Explain your answer.
c) Calculate the voltage shown on the voltmeter when the thermistor is placed in boiling water.

Answers and comments

a) As the thermistor warms its resistance decreases. This means that the upper resistance in the voltage divider is reducing. A progressively larger fraction of the supply voltage will therefore be across the lower resistor. So as the thermistor warms we would expect the reading shown on the voltmeter to increase.
b) Figure 4.10.4 shows the calibration curve for the thermistor. From this graph we can see that the resistance of the thermistor is approximately 2 kΩ at 100 °C.
c) We can calculate the voltage at the point midway between the resistors in a potential divider using;

$$V_{MP} = \frac{R_B}{R_B + R_T} \times V_s$$

$$V_{MP} = \frac{10}{10+2} \times 6$$

$$V_{MP} = 5\ V$$

Worked example 4

a) Calculate the reading that will be shown on the voltmeter in figure 4.10.5.

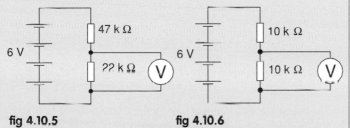

fig 4.10.5　　　　　　　　**fig 4.10.6**

b) Calculate the reading on the voltmeter in figure 4.10.6.
c) The four resistors are now joined to the battery as shown in figure 4.10.7. Calculate the reading on the voltmeter in this circuit.
d) Resistor P is replaced by another resistor so that the voltmeter shows a reading of 0 V. Calculate the resistance of this new resistor.

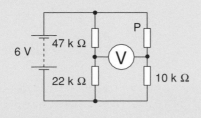

fig 4.10.7

Answers and comments

a) The circuit of figure 4.10.5 is a potential divider. The fraction of the 6 V across the 22 kΩ resistor can be calculated using;

$$V = \frac{R_B}{R_B + R_T} \times V_S = \frac{22}{22+47} \times 6$$

$$V = 1.91\ V$$

b) In the circuit of figure 4.10.6 both of the resistors in the potential divider have the same value. This means that the p.d. across each resistor is equal to half of the supply voltage. The voltmeter therefore shows 3 V.
c) We can use the values already calculated to determine the potential difference across the voltmeter. One side of the voltmeter is at a potential of 1.91 V while the other side is at 3 V. The potential difference is 3 − 1.91 = 1.09 V.
d) We can calculate the resistance needed to balance the bridge using;

$$\frac{R_1}{R_2} = \frac{R_3}{R_4} \qquad \frac{47}{22} = \frac{R_3}{10}$$

$$R_3 = \frac{47 \times 10}{22}$$

$$R_3 = 21.4\ k\Omega$$

A resistance of 21.4 kΩ will cause the reading on the voltmeter to reduce to 0 V.

Questions

1 In the circuit shown in figure 4.10.8 the reading on ammeter A_1 is 0.2 A.

a) What is the reading on ammeter A_2?

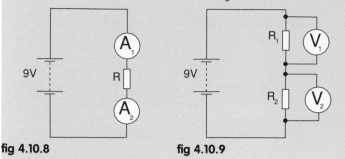

fig 4.10.8   **fig 4.10.9**

b) In the circuit of figure 4.10.9 two different resistors are connected to the 9 V battery. The reading on voltmeter V_1 is 3.5 V. What is the reading on voltmeter V_2?

c) When an ammeter is added to the circuit of figure 4.10.9 it indicates a current of 0.2 A.
  (i) What is the resistance of R_1?
  (ii) Write down a relationship connecting the resistor values R, R_1 and R_2. Explain your answer.
  (iii) Calculate numerical values for R and R_2 to check your answer to part (c) (ii).

2 In the circuit of figure 4.10.10 resistors with values 2.2 kΩ and 4.7 kΩ are connected in parallel with a 5 V power supply.

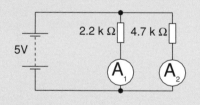

fig 4.10.10

a) What is the potential difference across each of the resistors?

b) Calculate the current in the 2.2 kΩ resistor.

c) Calculate the current in the 4.7 kΩ resistor.

d) Show that the total current from the supply is 3.34 mA.

e) Calculate the single value of resistor which will take a current of 3.34 mA from this 5 V supply.

3 While investigating resistor combinations a pupil is given resistors labelled A, B and C with values 10 kΩ, 22 kΩ and 39 kΩ respectively.

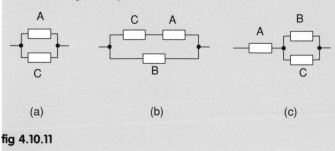

(a)   (b)   (c)

fig 4.10.11

a) Calculate the effective resistances of the combinations shown in figure 4.10.11.

b) What is the smallest possible resistance that can be achieved with resistors A, B and C?

4 The light dependent resistor in figure 4.10.12 is joined in series with a 10 kΩ resistor and a 3 V battery.

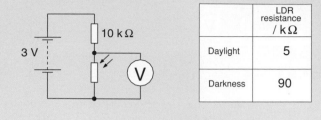

fig 4.10.12   **fig 4.10.13**

	LDR resistance / kΩ
Daylight	5
Darkness	90

The resistance of the LDR in daylight and darkness is as shown in figure 4.10.13.

a) Show by calculation that in daylight the voltmeter will show a reading of 1 V.

b) Calculate the reading on the voltmeter when the LDR is in darkness.

c) Explain how the components in figure 4.10.12 could be used to give a reading of 2 V on the voltmeter.

5 In figure 4.10.14 a lamp rated 1.25 V and 250 mA is to be connected to a 5 V battery.

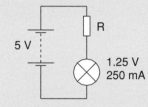

fig 4.10.14

The lamp lights at normal brightness when there is a p.d. of 1.25 V across it and when the current in it is 250 mA.

a) What is the p.d. across the resistor when the lamp is at normal brightness? Explain your answer.

b) Calculate the resistance of R which will cause the lamp to light at normal brightness.

6 In the circuit of figure 4.10.15 a red and a green LED are connected in series with a resistor and a 6 V battery.

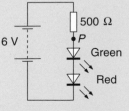

fig 4.10.15

Each of the LEDs is designed to operate at normal brightness when the p.d. across it is 1.75 V. The current in the LED is then 5 mA.

a) When the LEDs are lit at normal brightness what is the voltage at the point labelled P? Explain your answer.

b) Calculate the current in the resistor when the LEDs are lit at normal brightness.

c) Calculate the current in the resistor when the green LED is accidentally short circuited.

7 A pupil studying potential dividers sets up the circuit shown in figure 4.10.16.

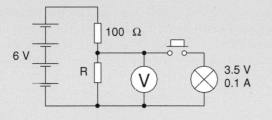

fig 4.10.16

a) When the switch is not pressed the voltmeter shows a reading of 3.5 V. What is the p.d. across the 100 Ω resistor?

b) Calculate the current in the 100 Ω resistor when the switch is open.

c) Calculate the value of resistor R.

d) The pupil now closes the switch.

   (i) Explain why the reading shown on the voltmeter reduces.

   (ii) Calculate the reading shown on the voltmeter when the switch is closed.

8 A pupil sets up the Wheatstone bridge circuit shown in figure 4.10.17.

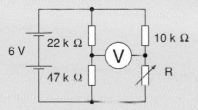

fig 4.10.17

The variable resistor is altered until the bridge balances.

a) Explain what is meant by the term balanced Wheatstone bridge.

b) Write down an expression connecting the values of the resistors in the bridge when the bridge is balanced.

c) Calculate the value of the variable resistor when the bridge is balanced.

d) Explain the effect on the balanced bridge of replacing the 6 V battery with a 9 V battery.

e) When the bridge is balanced the voltmeter is replaced by a very sensitive ammeter of the type shown in figure 4.10.18. What will be the value of the reading shown on this ammeter? Explain your answer.

fig 4.10.18

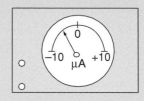

Worked example

A pupil investigating voltage dividers sets up the circuit shown in figure 4.11.1.

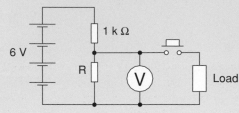

fig 4.11.1

When the switch is pressed the voltmeter shows a reading of 2V.

a) What is the p.d. across the 1 kΩ resistor when the switch is closed?

b) What is the *effective* resistance of the load resistor and the resistor labelled R?

c) The load resistor has a value of 5 kΩ. Calculate the value of resistor R.

Answers and comments

a) The voltmeter shows that the potential difference across the lower part of the potential divider is 2 V. Since the supply voltage is 6 V we can say that the potential difference across the 1 kΩ resistor is given by;

$$p.d.\ across\ 1\ k\Omega\ resistor = 6 - 2 = 4\ V$$

b) The p.d. of 4 V across the top 1 kΩ resistor will cause a current of 4 mA in the 1 kΩ resistor. The current in the lower effective resistor is also 4 mA. Since the p.d. across the effective resistor is 2 V we can calculate its resistance using;

$$R_{eff} = \frac{V}{I} = \frac{2}{4} = 0.5\ k\Omega$$

c) The load and R are in parallel and have an effective resistance of 0.5 kΩ while the load has a resistance of 5 kΩ. We can find R using;

$$\frac{1}{R_{eff}} = \frac{1}{R} + \frac{1}{R_{load}} \qquad R_{eff} = \frac{RR_{load}}{R + R_{load}}$$

$$0.5 = \frac{R \times 5}{R + 5}$$

$$R = 0.56\ k\Omega$$

Past paper question

A Wheatstone bridge is used to monitor the temperature of gas in a pipe. A length of platinum resistance wire forms one part

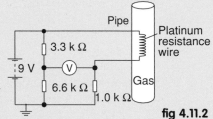

fig 4.11.2

of the Wheatstone bridge circuit. The wire is inserted into a pipe containing the gas as shown. (The 9 V supply has negligible internal resistance.)

a) (i) The bridge is initially balanced. What is the reading on the voltmeter?

(ii) Calculate the resistance of the platinum wire.

b) The graph of figure 4.11.3 shows how the resistance of the platinum wire varies with temperature.

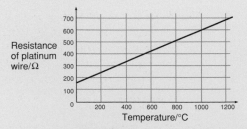

fig 4.11.3

The temperature of the gas and the platinum wire is changed to 600 °C. The Wheatstone bridge is now out of balance.

(i) What is the resistance of the platinum wire at 600 °C?

(ii) Calculate the reading on the voltmeter. **SQA 1996**

Answers and comments

a) (i) When the bridge is balanced the reading shown by the voltmeter is 0 V.

(ii) Since the bridge is balanced we can use the balanced bridge equation;

$$\frac{R_1}{R_2} = \frac{R_3}{R_4} \qquad \frac{3.3}{6.6} = \frac{R_3}{1}$$

$$R_3 = \frac{3.3}{6.6} \times 1 = 0.5\ k\Omega$$

b) (i) From the graph we can deduce that the resistance of the wire at 600 °C is 410 Ω.

(ii) The potential on the other side of the voltmeter created by the 3.3 kΩ and 6.6 kΩ resistors is calculated from;

$$V = \frac{R_2}{R_1 + R_2} \times V_S = \frac{6.6}{3.3 + 6.6} \times 9$$

$$V = 6.0\ V$$

With one side of the voltmeter at a potential of 6 V and the other side at a potential of 6.38 V the potential difference indicated by the voltmeter is 6.38 − 6 = 0.38 V

Voltmeter reading = 0.38 V.

Questions

1 Resistors are connected to a 12 V power supply as shown in figure 4.11.4. S_1 and S_2 represent switches.

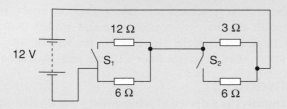

fig 4.11.4

a) Calculate the total resistance of the circuit when both switches S_1 and S_2 are closed.
b) Calculate the total current supplied by the power supply when both switches S_1 and S_2 are closed.
c) Calculate the current in the 3 Ω resistor when switch S_1 is closed and S_2 is open.
d) Calculate the current supplied by the power supply when both switches are open.

2 In the circuit of figure 4.11.5 a 2 V cell is in a circuit as shown.

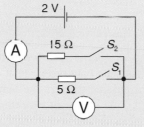

fig 4.11.5

a) Calculate the readings on the ammeter and voltmeter when switch S_1 is closed and S_2 is open.
b) Calculate the readings shown on the ammeter and voltmeter when S_1 and S_2 are both closed.
c) What readings will be shown on the meters when both switches are open? Explain your answer.

3 A pupil sets up a Wheatstone bridge circuit which includes a resistance substitution box which can alter the resistance by small amounts.

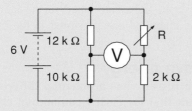

fig 4.11.6

a) Show that the bridge is balanced when the resistance of R is 2400 Ω.
b) When the value of R is increased by 10 Ω the voltmeter shows a reading of –3 mV. Sketch a graph to show how you would expect the reading on the voltmeter to vary as the resistance of R is varied between 2.3 kΩ and 2.5 kΩ.

4 A student investigating electronic components sets up the test circuit shown in figure 4.11.7. She is given three components labelled A, B and C. A and B are resistors while C is a light bulb.

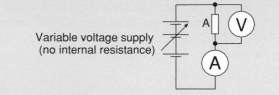

fig 4.11.7

When testing the components she adjusts the power supply and records pairs of ammeter and voltmeter readings.
She repeats this procedure with each component and then plots the results on a single set of graph axis. Her results are shown in figure 4.11.8.

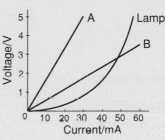

fig 4.11.8

a) (i) What is the current in the lamp when the voltage across it is 3 V?
   (ii) What is the current in resistor A when the voltage across it is 3 V?
   (iii) Which resistor A or B has the larger resistance? Justify your answer.

b) The student is now asked to use her graphs to answer questions about the circuit of figure 4.11.9 where resistors A and B are connected in parallel with the power supply.

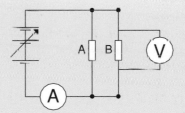

fig 4.11.9

(i) When the output voltage from the power supply is 3 V, what is the reading on the ammeter?
(ii) Using your answer to the previous part calculate the effective resistance of A and B in parallel.
(iii) What is the reading on the ammeter when resistor B is replaced by the lamp and the supply voltage left at 3 V?

c) Resistor A is removed from the circuit and resistor B and the lamp are now connected in series with the ammeter and the power supply.
What is the voltage of the power supply when the reading on the ammeter is 50 mA? Explain your answer.

4.12 Energy, work and power

Introduction

There are many different forms of energy; light, sound, chemical energy and kinetic energy. Heat can raise the temperature of matter while fuels are a source of chemical energy.

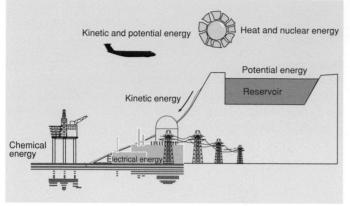

Figure 4.12.1 Forms of energy

A plane moves because of the energy produced when aviation fuel is burned. Chemical energy is transferred into kinetic, potential and some sound energy. Power stations burning fossil fuels produce electrical energy which can then be transferred into other types of energy by devices such as heaters, lamps and loudspeakers. The electrical energy generated is used to make charges flow around a circuit. **In general, energy can be thought of as the capability to produce movement or some sort of other useful activity**.

Electrical energy

In the circuit of figure 4.12.2 the electrical energy needed to make the charges flow through the lamp and buzzer comes from the chemical energy stored in the battery.

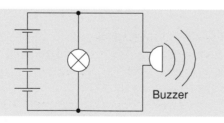

Figure 4.12.2

The energy transfers taking place in the lamp and buzzer can be summarised as shown in the figure 4.12.3.

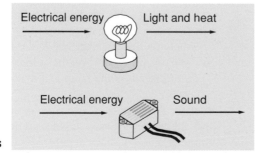

Figure 4.12.3
Energy transfers

The Law of Conservation of Energy is one of a number of conservation laws and states that **energy cannot be created or destroyed; it can only be transferred from one form to another**. First proposed by James Prescott Joule in the nineteenth century, it has been extended slightly following work by Albert Einstein. However, for the type of conditions met in everyday situations Joule's Law is perfectly adequate.

Transducers

A transducer is a device which transfers one kind of energy into another. The windmills on a wind farm transfer wind energy into electrical energy, while solar powered heating panels produce heat energy from the infrared radiation from the sun. In an electric heater the resistance wire in the element transfers electrical energy into heat energy.

The lamp and buzzer shown in figure 4.12.3 are output transducers commonly used in electrical and electronic circuits. If the lamp is connected to a 6 V battery, 6 J of energy are supplied to each coulomb of charge leaving the battery.

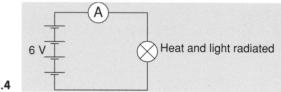

Figure 4.12.4

The filament lamp of figure 4.12.4 will transfer the 6 J of chemical energy from the battery into heat and light for every coulomb of charge passing through the lamp. The filament in the lamp is a small coiled piece of very thin wire. When there is an electric current in the filament it heats up and radiates heat and light.

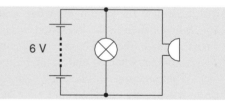

Figure 4.12.5

When a buzzer is connected in parallel with the lamp as shown in figure 4.12.5 it too will have a p.d. of 6 V across its terminals. An additional 6 J of sound will be heard for every coulomb of charge passing through the buzzer. The combined production of heat, light and sound in the lamp and buzzer means that the chemical energy stored in the battery will be used more quickly.

Quantitative energy

Stating that a resistor has a potential difference of 1 volt across it means that for each 1 coulomb of charge passing through the resistor, 1 joule of energy is given out. This allows us to say;

$$Voltage = \frac{Energy}{Charge}$$

Rearranging;

$$Energy = Voltage \times Charge$$
$$E = QV$$

The energy flowing from an electrical supply each second is called the **power**. Therefore;

> *Energy per second = Voltage × Charge per second*
> *Power = Voltage × Charge per second*

This relationship can be further simplified by realising that in previous work we have stated that the current is the charge flowing per second.

> *Power = Voltage × Charge per second*
> *Power = Voltage × Current*

This mathematical relationship links the potential difference across a device, the current in it and the electrical energy produced each second (the power). In symbols;

$$P = VI$$

When current is expressed in amps and the potential difference in volts, the power will be calculated in watts; where 1 W equals 1 J per second.

Q When the lamp shown in figure 4.11.4 has a p.d. of 6 V across it the current from the battery is 0.3 A. How much heat and light is produced each second?

A Using;

$$P = VI$$
$$P = 6 \times 0.3$$
$$P = 1.8\ W$$

This tells us that the lamp is taking 1.8 J of energy each second from the battery.

How long it will take for the lamp to extract 1000 J from the battery?

$$Power = \frac{Energy}{Time}$$

$$P = \frac{E}{t}$$

Rearranging;

$$E = Pt$$

Substituting;

$$1000 = 1.8 \times t$$

Therefore;

$$t = \frac{1000}{1.8}$$

$$t = 556\ s$$

It will take the lamp 556 seconds to extract 1000 J of energy from the battery.

Power and Ohm's Law

In the previous example it was easy to calculate the power dissipated (given out) by the lamp simply by substituting the current and voltage values into the formula $P = VI$. In figure 4.12.6 a 6 V battery produces a current in a 100 Ω resistor.

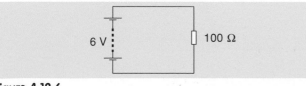

Figure 4.12.6

To calculate the energy dissipated by the resistor in this situation we must first calculate the current. From Ohm's Law;

$$I = \frac{V}{R} \qquad I = \frac{6}{100} \qquad I = 0.06\ A$$

Now we can use $P = VI$ to calculate the power;

$$P = VI \quad P = 6 \times 0.06 \quad P = 0.36\ W$$

In doing this calculation we have combined the formula for electrical power with a value generated from an Ohm's Law calculation. When we combine $P = VI$ with Ohm's Law directly we can generate two other equations equivalent to $P = VI$.

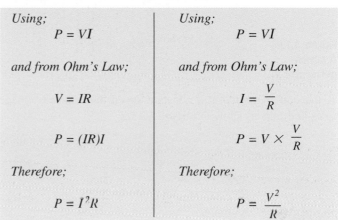

With this analysis we have shown that;

$$P = VI = I^2R = \frac{V^2}{R}$$

The most appropriate one to use in a particular question depends on the data given in the question.

For the circuit of figure 4.12.6 we could have calculated the power dissipated by the resistor using;

$$P = \frac{V^2}{R} \qquad P = \frac{6^2}{100} \qquad P = 0.36\ W$$

Power and fuses

A fuse is a thin piece of wire designed to melt when the current in it exceeds the maximum current that the connecting wires in a circuit can safely carry.

Q What is the current in the fuse when a 2 kW heater is connected to the 230 V domestic supply?

A

$$P = VI$$
$$2000 = 230 \times I$$
$$I = 8.7\ A$$

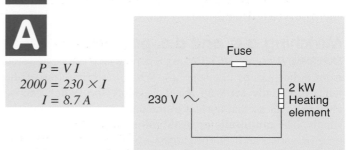

Figure 4.12.7

The fuse must be capable of carrying a current of 8.7 A without melting. A fuse rated 13 A would be suitable.

4.13 Alternating currents

Generators

In the simple generator, a rectangular coil between the poles of a magnet is rotated by an external source of energy.

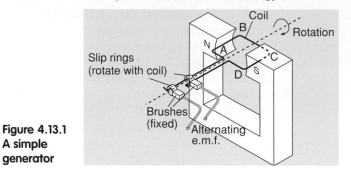

**Figure 4.13.1
A simple
generator**

If the coil is rotated at a constant speed the induced voltage rises and falls as shown in figure 4.13.2.

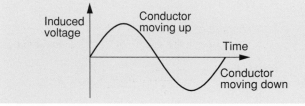

Figure 4.13.2

This is called an **alternating voltage**. If the coil is connected to a resistor, the alternating voltage causes an alternating current in the resistor. Alternating currents reverse their direction at regular intervals. The number of times each second that an alternating current reverses its direction is called its **frequency**. In the UK the mains a.c. supply has a frequency of 50 Hz but in the USA mains alternating current has a frequency of 60 Hz.

Comparing a.c. and d.c. supplies.

When a 6 V battery of negligble internal resistance is connected across a 6 Ω resistor, there is a steady direct current of 1 A in the resistor, transferring electrical energy into heat. The current supplied to a 6 Ω resistor by an alternating supply with a peak voltage of 6 V varies from zero up to a peak of 1 A. For most of the time the current supplied by the a.c. supply is less than 1 A, so the 6 V a.c. supply does not provide as much energy to the resistor as the 6 V d.c. supply.

Matching a.c. and d.c. powers

EXPERIMENT Figure 4.13.3 shows a lamp which can be powered from either a.c. or d.c. supplies. When the brightness of the lamp is initially set with the d.c. supply the switch can be changed over so that current in the lamp is now from the a.c. supply. The **amplitude** or **peak value** of the a.c. voltage can then be varied until the brightness of the lamp is the same as that produced with the d.c. supply. Once the brightness of lamp appears to be the same from the a.c. and d.c. supplies, a light intensity meter could be used to make sure that the brightnesses are exactly the same.

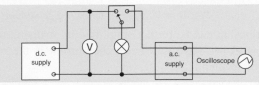

Figure 4.13.3

When the brightness of the lamp is the same using each supply, the power being delivered from each supply is the same. Using an oscilloscope you will be able to show that the peak value of the a.c. signal is greater than the selected d.c. voltage, even though both supplies are delivering the same amount of energy to the lamp.

Peak and effective (or r.m.s.) values

If the d.c. supply voltage in the experiment shown in figure 4.13.3 is set to 6 V, the a.c. supply giving the same energy will have a peak value of 8.5 V. Therefore the effective value of an a.c. supply with an 8.5 V peak is 6 V. *In general we can say that the effective (or quoted) value of an a.c. signal is less than its peak.*

The domestic mains supply is quoted as having a value of 230 V. This means that a heater plugged into the mains will dissipate the same power as one connected to a 230 V d.c. supply. The effective value of an alternating supply is called the **r.m.s. (root mean square)** value. Mathematically the r.m.s. value is related to the peak value by;

$$V_{peak} = \sqrt{2} \times V_{rms} \qquad I_{peak} = \sqrt{2} \times I_{rms}$$

The domestic mains quoted as supplying 230 V will have a peak voltage of approximately 325 V.

A.C. signals and CRO traces

Not only can we use the trace on an oscilloscope to measure the peak voltage of an a.c. supply, but we can also use it to determine the frequency of the supply.

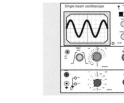 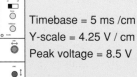

Timebase = 5 ms /cm
Y-scale = 4.25 V / cm
Peak voltage = 8.5 V

**Figure 4.13.4
Oscilloscope**

The timebase setting on the oscilloscope of figure 4.13.4 is set to 5 ms/cm. This means that the dot on the oscilloscope takes 5 milliseconds to travel each centimetre across the screen. Each cycle is completed in 4 cm so the total time for 1 wave is 20 milliseconds. This is called the period. We can calculate the frequency using;

$$Frequency = \frac{1}{Period} = \frac{1}{0.020}$$

$$Frequency = 50 \ Hz$$

4.14 Consolidation questions

Worked example

In the model power line shown in figure 4.14.1 a 12 V a.c. supply with negligible internal resistance is connected by long wires of total resistance 4 Ω to a lamp which has a resistance of 6 Ω.

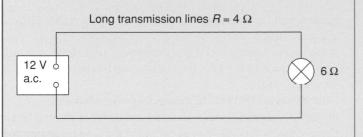

fig 4.14.1

Calculate:

a) The current in the wires
b) The power loss in the wires
c) The voltage across the wires
d) The voltage available to the lamp
e) The power converted in the lamp
f) The percentage of the input power that is dissipated in the lamp.

Answers and comments

a) The resistance of the power lines and lamp gives a total load resistance of 4 Ω + 6 Ω. We can calculate the current from;

$$I = \frac{V}{R} = \frac{12}{10} = 1.2\ A$$

b) The power loss in the wires can be calculated from $P = I^2R$ where R is the resistance of the wires.

$$P = I^2R = 1.2^2 \times 4 = 5.76\ W$$

c) The voltage across the wires can be calculated from $V = IR$ where R is the resistance of the wires;

$$V = IR = 1.2 \times 4 = 4.8\ V$$

d) The voltage available for the lamp is the difference between the voltage across the wires and the supply voltage.

$$Voltage\ across\ lamp = 12 - 4.8 = 7.2\ V$$

e) The power converted in the lamp can be calculated from;

$$P = \frac{V^2}{R} = \frac{7.2^2}{6} = 8.64\ W$$

f) The input power can be calculated from $P = VI$

$$P = VI = 12 \times 1.2 = 14.4\ W$$

$$Efficiency = \frac{Useful\ power\ output}{Total\ input\ power} \times 100\%$$

$$Efficiency = \frac{8.64}{14.4} \times 100\% = 60\%$$

60% of the input power is dissipated in the lamp.

Questions

1 An alternating voltage with a peak of 7 V is connected in series with a 100 Ω resistor and an ammeter.
   a) Calculate the r.m.s. current in the circuit.
   b) Calculate the power dissipated in the resistor.
   c) Another 100 Ω resistor is now connected in parallel with the first one. Calculate the current in the circuit.
   d) The power supply is replaced with a battery and the ammeter is switched to measure direct current. Calculate the value of voltage required for the d.c. power supply to dissipate 245 mW of power in the 100 Ω resistor.

2 An a.c. signal with a frequency of 80 Hz and a peak voltage of 10 V is connected in a circuit with a 100 Ω resistor.
   a) What d.c. voltage would produce the same heating effect in the 100 Ω resistor?

   b) The a.c. frequency is now doubled. What d.c. voltage is now required to produce the same heating effect in the 100 Ω resistor? Justify your answer.

3 A pupil connects an alternating voltage with a frequency of 2 Hz to the circuit shown in figure 4.14.2 and notices that the LED appears to flash roughly twice per second.

fig 4.14.2

   a) Explain why the LED appears to flash.
   b) The alternating voltage has a peak value of 5 V and the LED lights when the voltage across it is 1.6 V. Calculate the peak current in the LED.

4.15 E.m.f. and internal resistance

Introduction

When a 60 Ω resistor is connected across a 6 V supply the current from the battery can be calculated from Ohm's Law.

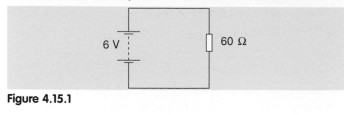

Figure 4.15.1

$$I = \frac{V}{R} = \frac{6}{60}$$

$$I = 0.1\ A$$

If a second 60 Ω resistor is connected in parallel with the first the combined resistance is given by;

$$\frac{1}{R_T} = \frac{1}{R_1} + \frac{1}{R_2}$$

$$\frac{1}{R_T} = \frac{R_1 + R_2}{R_1 R_2}$$

$$R_T = \frac{R_1 R_2}{R_1 + R_2} = \frac{60 \times 60}{60 + 60}$$

$$R_T = 30\ \Omega$$

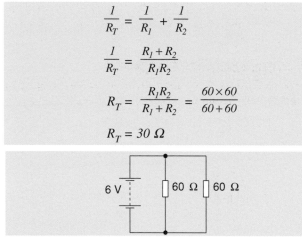

Figure 4.15.2

By joining the two resistors in parallel the total load resistance across the 6 V supply has halved. Thus we would expect the current from the supply to double.

If a third resistor is joined in parallel with the other two, the total load resistance across the 6 V supply falls to 20 Ω and we would expect the circuit current to rise to 0.3 A.

Loading supplies

The simple circuit theory used above predicts that if the load resistance connected to a voltage supply is reduced in turn from R to $R/2$, $R/4$ and $R/8$, the current would increase in turn from I to $2I$, $4I$ and $8I$. If taken to its logical conclusion this simple theory would predict that a very very small load resistance connected to a battery would draw a very large, almost infinite current.

When the resistance of the variable resistor in figure 4.15.3a is reduced the ammeter shows that the current from the battery increases. When the variable resistor is at its minimum value, or if it is short circuited with a piece of conducting wire (figure 4.15.3b), the ammeter will register a current of under 10 A. By bypassing the variable resistor we have short circuited the

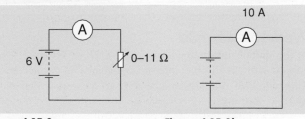

Figure 4.15.3a   **Figure 4.15.3b**

battery, and the ammeter is showing the short circuit current (figure 4.15.3b).

If the supply voltage is 6 V and the current 10 A there must still be a resistance of 0.6 Ω in the circuit even though the battery is short circuited.

Only the ammeter and the connecting wires are connected to the battery externally and their total resistance is about 0.01 Ω. Clearly this is too small to account for the resistance limiting the short circuit current to 10 A.

To explain why the current is limited we must introduce the concept of **internal resistance.** In this model we picture *each cell in the battery as having a small resistance* through which the charges must pass before they enter the external circuit.

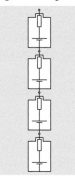   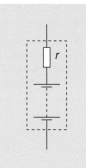

Figure 4.15.4   **Figure 4.15.5**

Figure 4.15.4 shows how we must picture a battery using this model for internal resistance. Each cell is a source of electrical energy with an internal resistance. In a circuit diagram we draw a battery with internal resistance as shown in figure 4.15.5 where 'r' represents the entire internal resistance of the battery. If the battery of figure 4.15.5 is short circuited, the internal resistance limits the current that the battery can supply.

Internal resistance and energy

In figure 4.15.6 a load resistor is connected across a battery with internal resistance. The internal resistance is effectively in series with the load resistor. The load resistor could be an LED, lamp, buzzer or other transducer requiring current from the supply.

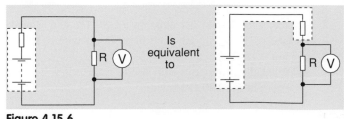

Figure 4.15.6

The voltmeter connected across the load resistor in figure 4.15.6 is effectively measuring the voltage available across the terminals of the supply. This is called the **terminal p.d.** (V_{tpd}) and is the voltage that the battery supplies to circuitry connected across its terminals.

If we imagine a situation where the load resistance, R, exactly equals the internal resistance, r, we can determine the effect that the internal resistance will have on the operation of the circuit. As for any circuit which has two equal resistances in series, the voltage across each resistor will be equal to half the supply voltage. Therefore the terminal p.d. is only half of the total supply voltage. The other half of the supply voltage will be across the internal resistance and is 'lost' to the external circuit. For this reason the voltage across the internal resistance is called V_{lost}. Simple circuit theory allows us to say that;

$$Total\ supply\ voltage = V_{tpd} + V_{lost}$$

The total supply voltage is called the e.m.f. (electromotive force, E); therefore

$$E = V_{tpd} + V_{lost}$$

E.m.f. and energy

Since voltage is defined as the electrical energy available per coulomb of charge, the terminal p.d. can be thought of as the energy available to each coulomb of charge passing through the circuitry connected to the supply. V_{lost} is the electrical energy per coulomb 'lost' across the internal resistance. **The e.m.f. is the total electrical energy supplied to each coulomb of charge which passes through the source.**

Any power supply can be thought of as a source of e.m.f., E, with an internal resistance, r.

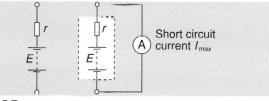

Figure 4.15.7

If the battery shown in figure 4.15.7 is short circuited the p.d. across the internal resistance will be equal to the e.m.f. The short circuit current I_{max} is given by;

$$I_{max} = \frac{E}{r}$$

Different terminal p.d.s

In an earlier example, we considered the situation where the load resistance and the internal resistance were equal. In the circuit

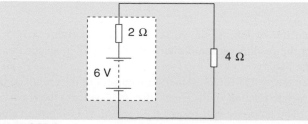

Figure 4.15.8

shown in figure 4.15.8 a supply with an e.m.f. of 6 V and internal resistance of 2 Ω is connected across a 4 Ω load resistor.
The current from the battery depends on the total resistance in the circuit.

$$R_T = R + r$$
$$R_T = 4 + 2$$
$$R_T = 6\ \Omega$$

Therefore the current in the circuit is given by;

$$I = \frac{E}{R_T} = \frac{6}{6}$$
$$I = 1\ A$$

There is a current of 1 A in both the internal and the load resistors. We can calculate the voltage across the load resistor, V_{tpd}, using;

$$V_{tpd} = I\ R$$
$$V_{tpd} = 1 \times 4$$
$$V_{tpd} = 4\ V$$

Only 4 V of the 6 V available from the battery are available to the load resistor, the other 2 V are 'lost' due to the internal resistance of the supply.

Combining electrical supplies

The circuit in figure 4.15.9 shows two of the batteries used in the previous example connected across a 4 Ω load resistor.

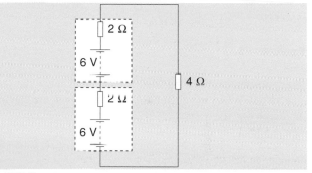

Figure 4.15.9

The total e.m.f. in this circuit is 12 V and the total internal resistance is 4 Ω. Using an analysis similar to that shown in the previous section it is possible to show that the current in the 4 Ω resistor is 1.5 A. The current of 1.5 A in each 2 Ω resistor means that there must be a p.d. of 3 V across each 2 Ω resistor. **This example has shown that the sum of the e.m.f.s in the circuit (6 V + 6 V) equals the sum of the p.d. across the resistors (3 V + 3 V + 6 V).** This must be the case since the energy dissipated per coulomb in the resistors must equal the energy per coulomb supplied by the battery.

4.16 Measuring internal resistance

Introduction

Any supply of electrical energy is equivalent to a source of e.m.f. with an internal resistance as shown in figure 4.16.1.

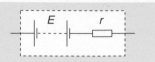

Figure 4.16.1

When a load resistor is connected across a supply the electrical energy available to each coulomb of charge is shared; some is dissipated in the load resistance and some is 'wasted' in the internal resistance. This leads to the equation;

$$E = V_{tpd} + V_{lost} \qquad \text{equation 1}$$

E.m.f.s and short circuits

When the load resistor is connected to a battery it is in series with the internal resistance so the current in each resistor is the same.

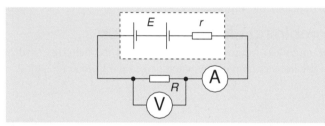

Figure 4.16.2

The voltmeter across the load resistor will show the terminal p.d. (V_{tpd}). When the current in the circuit is, I, we can express V_{tpd} and V_{lost} as;

$$V_{tpd} = I R \qquad V_{lost} = Ir$$

Therefore;

$$E = I R + I r \qquad \text{equation 2}$$

Rearranging this equation gives;

$$I = \frac{E}{R + r}$$

When the resistance, R, of circuitry attached across a battery is reduced to 0 Ω, the battery is short circuited. With $R = 0$ the short circuit current if given by;

$$I = \frac{E}{O + r}$$

$$I = \frac{E}{r}$$

For a short circuit we have assumed that the load resistance is 0 Ω. This assumption implies that the terminal p.d. for a short circuit is 0 V. In practice a short circuit occurs when two or more terminals of a source of electrical energy are connected through a path of low resistance.

Measuring e.m.f.

Equation 3 below is derived from equations 1 and 2.

$$E = V_{tpd} + I r \qquad \text{equation 3}$$

When there is no current from the supply I = 0 so;

$$E = V_{tpd}$$

From this equation we can state that the e.m.f. of a supply is equal to the terminal p.d. when no current flows. Sometimes this is stated as the e.m.f. equals the terminal p.d. on 'open circuit'. In figure 4.16.3 the battery is connected to a digital voltmeter. The input resistance of this voltmeter is very high (10 MΩ). There is almost no current taken from the battery. The digital voltmeter is showing the value of the e.m.f. of the battery.

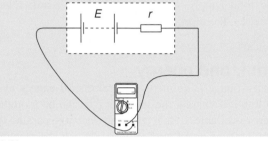

Figure 4.16.3

If a moving coil voltmeter were connected to the same battery you might imagine that it would show the same value for the e.m.f.

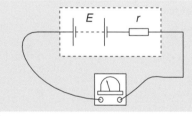

Figure 4.16.4

Although it might appear that there is no load resistance in this circuit, the meter itself requires a small quantity of current to move the needle. There is a current in the circuit so there will be some 'lost volts' across the internal resistance. The reading on the analogue voltmeter is therefore less than the e.m.f. of the battery.

Finding internal resistance

EXPERIMENT Equation 3 can be rearranged to give;

$$E = V_{tpd} + I r$$

$$r = \frac{E - V_{tpd}}{I}$$

When the switch, S, shown in the circuit of figure 4.16.5 is open the reading on the digital voltmeter will equal the e.m.f. of the battery.

When the switch is closed there is a current in the circuit so the voltmeter reading will equal the p.d. across the resistor, V_{tpd}. The ammeter will record the current, I. Using the values for E, V_{tpd}, and I the internal resistance, r, can be calculated.

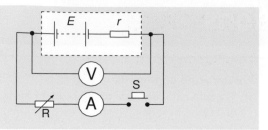

Figure 4.16.5

Setting the variable resistor to another value of resistance alters the current in the circuit. The terminal p.d. will change but the e.m.f. stays the same. The internal resistance can again be calculated. This can be repeated and a mean value for r determined.

Terminal p.d. and current

The current in the circuit of figure 4.16.6 can be varied by altering the variable resistor R.

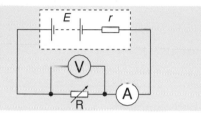

Figure 4.16.6

As the current increases, the value of V_{lost} increases so the terminal p.d. given by the equation below, reduces;

$$V_{tpd} = E - V_{lost}$$

When the variable resistor in figure 4.16.6 is set to different values and the V_{tpd} and current values are recorded, the resulting graph will be as shown in figure 4.16.7.

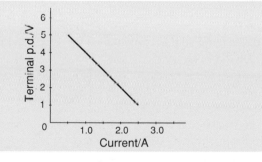

Figure 4.16.7

Knowing that the e.m.f. is equal to the terminal p.d. when there is no current, we can extend the line to find the e.m.f.

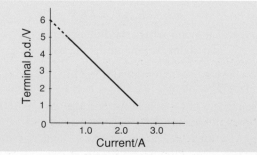

Figure 4.16.8

Since this line cuts the y-axis at 6 V we can say that the e.m.f. of the battery is 6 V. To find the internal resistance we can use this value of e.m.f. and take any pair of terminal voltage and current readings from the graph.

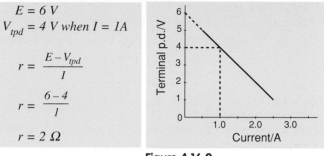

$$E = 6 V$$
$$V_{tpd} = 4 V \text{ when } I = 1A$$

$$r = \frac{E - V_{tpd}}{I}$$

$$r = \frac{6 - 4}{1}$$

$$r = 2\ \Omega$$

Figure 4.16.9

It is also possible to find the short circuit current from the graph of figure 4.16.7 by extending the line to the point where it cuts the axis for current. This gives the maximum current that can be drawn from the battery.

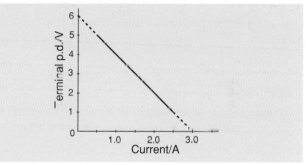

Figure 4.16.10

In this example the short circuit current is 3 A. We already know that the short circuit is given by;

$$I = \frac{E}{r}$$

The internal resistance can be confirmed from;

$$I = \frac{E}{r}$$

$$r = \frac{E}{I}$$

$$r = \frac{6}{3}$$

$$r = 2\ \Omega$$

From the graph of figure 4.16.10 we can see that the e.m.f., E, is the intercept on the y-axis while the short circuit current, I_{max}, is the intercept on the x-axis. Since the internal resistance is given by;

$$r = \frac{E}{I_{max}}$$

we can see that the gradient of the graph in figure 4.16.10 is numerically the same as the internal resistance.

4.17 Consolidation questions

Worked example 1

In the circuit of figure 4.17.1 a cell with an e.m.f. of 1.5 V and having internal resistance of 0.1 Ω is connected in series with a 2 Ω resistor.

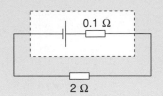

0.1 Ω

2 Ω

fig 4.17.1

a) Calculate the total resistance and current in the circuit.
b) Calculate the 'voltage lost' due to the internal resistance of the battery.
c) Calculate the p.d. across the 2 Ω resistor.

Answers and comments

a) The internal resistance of the cell and the load resistance of 2 Ω are in series. The total resistance in the circuit, R_T, is given by;

$$R_T = 0.1 + 2 \qquad R_T = 2.1\ \Omega$$
$$I = \frac{E}{R_T} = \frac{1.5}{2.1} = 0.71\ A$$

b) We can calculate the voltage lost across the internal resistance from

$$V_{lost} = I\,r = 0.71 \times 0.1$$
$$V_{lost} = 0.07\ V$$

c) The p.d. across the load is the difference between the e.m.f. and the lost volts.

$$E = V_{tpd} + V_{lost}$$
$$V_{tpd} = E - V_{lost} = 1.5 - 0.07$$
$$V_{tpd} = 1.43\ V$$

Worked example 2

Figure 4.17.2 shows the type of apparatus that could be used in an experiment to determine the internal resistance of a battery.

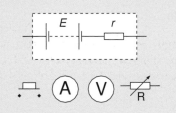

fig 4.17.2

a) Draw the circuit diagram required to allow you to take the readings necessary to determine the internal resistance of the battery.
b) In one experiment the terminal p.d. and current are measured and recorded for a number of different values of resistor connected across the battery. The results are recorded in the table 4.17.1

V/volts	0.3	0.5	0.7	0.9	1.1	1.3	1.5
I/amps	0.75	0.68	0.55	0.45	0.35	0.25	0.15

Table 4.17.1

c) Draw a graph of V versus current for the readings shown in the table. From your graph determine;
(i) the e.m.f. of the battery
(ii) the maximum current that the battery can supply
(iii) the internal resistance of the battery.

Answers and comments

a) The ammeter and variable resistor should be connected in series with the battery. The voltmeter needs to be connected across the variable resistor. The circuit is shown in figure 4.17.3.

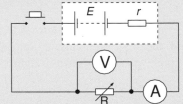

fig 4.17.3

b) The terminal p.d. is plotted on the y-axis and the current on the x-axis.

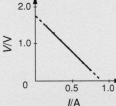

fig 4.17.4

c) (i) The terminal p.d. when the circuit current is zero gives the e.m.f. of the battery. This is found from the point where the line cuts the y-axis.

$$E.m.f. = 1.8\ V$$

(ii) The point where the line cuts the x-axis shows the maximum current which the battery can supply.

$$I_{max} = 0.9\ A$$

(iii) The internal resistance, r, is found from

$$r = \frac{E}{I_{max}} = \frac{1.8}{0.9}$$
$$r = 2\ \Omega$$

Questions

1 A student making notes on internal resistance writes down the following statements.

I The e.m.f. of a supply is the electrical energy supplied to each coulomb of charge passing through the supply.

II An electrical supply can be represented as a source of e.m.f. with a resistor in series.

III The e.m.f. of a supply is equal to the p.d. across the terminals of the source when there is no current in the circuit.

Which of these statements is/are correct?

   A I only
   B II only
   C I and II only
   D II and III only
   E I, II and III.

2 In figure 4.17.5 a 12 V battery having an internal resistance of 1 Ω is connected across a 2 Ω resistor.

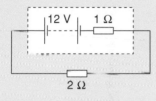

fig 4.17.5

   a) Calculate the current in the circuit.
   b) Calculate the terminal p.d. of the battery.
   c) Calculate the power dissipated in the load resistor.
   d) Calculate the current in the circuit when the battery is short circuited.

3 Figure 4.17.6 shows a power supply with an e.m.f. of 12 V and an internal resistance of 3 Ω.

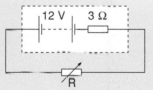

fig 4.17.6

   a) Calculate the current from the battery when the variable load resistance is set at 0 Ω, 1 Ω, 2 Ω, 3 Ω, 4 Ω, 5 Ω and 6 Ω. Enter these values in a table such as 4.17.2.

R/ohms	0	1	2	3	4	5	6
I/amps							
V/volts							

Table 4.17.2

   b) Calculate the terminal p.d. for each of the resistance values shown in the table. Add these values to your table.
   c) Calculate the power dissipated for each of the load resistances.
   d) Plot a graph to show the variation of power dissipated with load resistance.

4 In the circuit of figure 4.17.7 two 10 Ω resistors are connected across a 12 V power supply which has an internal resistance of 1 Ω.

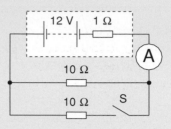

fig 4.17.7

   a) Calculate the current in the ammeter when the switch is open.
   b) Calculate the power dissipated in the 10 Ω resistor when the switch is open.
   c) Calculate the current in the ammeter when the switch is closed.
   d) What is the voltage 'lost' across the internal resistance when the switch is closed? Explain your answer.
   e) What is the voltage across each of the 10 Ω resistors when the switch is closed? Explain your answer.
   f) Calculate the total power dissipated in the load resistors when the switch is closed.

5 Figure 4.17.8 shows a lamp connected to a 6 V battery which has negligible internal resistance.

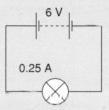

fig 4.17.8

The current in the lamp is 0.25 A.

   a) Calculate the resistance of the lamp.
   b) Calculate the electrical energy converted to heat and light by the lamp in 10 seconds.
   c) A second identical lamp is now connected in parallel with the first.
      (i) Calculate the quantity of charge flowing from the battery in 10 s.
      (ii) Calculate the quantity of energy taken from the 6 V battery in 10 s.

6 The resistance, R, of a resistor and the power, P, dissipated in that resistor can each be expressed in terms of the current, I, in the resistor and the voltage, V, across the resistor.

   a) Write down the relationship connecting R, V and I.
   b) Write down the relationship connecting P, V and I.
   c) Show that the power dissipated in a resistor can be expressed as;

$$P = I^2R \quad or \quad P = \frac{V^2}{R}$$

Past paper question 1

A heater of resistance 0.32 Ω is connected to a power supply of e.m.f. 2.0 V and internal resistance r as shown in figure 4.18.1.

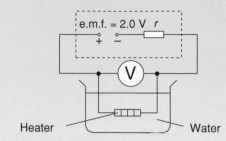

fig 4.18.1

a) State what is meant by the term *electromotive force*.

b) The power output of the heater is 8 W.
   Calculate;
   (i)   the current in the heater
   (ii) the reading on the voltmeter
   (iii) the internal resistance of the power supply.

c) Another identical heater is now placed in the water and connected in parallel with the original heater. The rest of the circuit is unaltered. How does this affect the rate at which heat is supplied to the water? Justify your answer by calculation. **SQA 1993**

Answers and comments

a) The electromotive force of the power supply is stated as 2 V. This means that the power supply provides 2 joules of electrical energy to each coulomb of charge leaving the power supply.

b) (i)   We are told that the power output of the heater is 8 W and that the heater has a resistance of 0.32 Ω. We can calculate the current in the heater using;
$$P = I^2 R$$
$$I^2 = \frac{P}{R} = \frac{8}{0.32}$$
$$I = 5\ A$$

   (ii) The reading on the voltmeter can be calculated from;
$$V = IR = 5 \times 0.32$$
$$V = 1.6\ V$$

   (iii) We can find the voltage lost across the internal resistance from;
$$V_{lost} = E - V_{tpd} = 2 - 1.6$$
$$V_{lost} = 0.4\ V$$

   From this we can calculate the internal resistance
$$V_{lost} = I\,r = 5 \times r$$
$$r = 0.4 / 5 = 0.08\ \Omega$$

c) When a second heater with a resistance of 0.32 Ω is connected in parallel with the other their total resistance is 0.16 Ω. The total resistance in the circuit is now 0.16 + 0.08 = 0.24 Ω.
The current in the circuit is now given by;
$$I = \frac{2.0}{0.24}$$

The power output of the heaters is now given by $P = I^2 R$
$$P = \left(\frac{2.0}{0.24}\right)^2 \times 0.16 = 11.1\ W$$

The power output with 2 heaters is greater than with one.

Past paper question 2

A rechargeable cell is rated at 0.5 A h (ampere hour). This means that, for example, it can supply a current of 0.5 amps for 1 hour. The cell then requires to be recharged.

a) What charge, in coulombs, is available from a fully charged cell?

b) A fully charged cell is connected to a load resistor and left until it needs recharging. During this time the p.d. across the terminals of the cell remains constant at 1.2 V. Calculate the electrical energy supplied to the load resistor in this case. **SQA 1994**

Answers and comments

a) The charge flowing from a power supply or battery can be found using $Q = It$ (where t is in seconds). In this case a steady current of 1 amp can flow for 0.5 hours
$$Q = It$$
$$Q = 0.5 \times 60 \times 60$$
$$Q = 1800\ C$$

b) The volt is defined from
$$V = \frac{W}{Q}$$
$$W = Q\,V$$
$$W = 1800 \times 1.2$$
$$W = 2160\ J$$

Questions

1 The work done in moving 1 coulomb of electrical charge from point P on a wire to another point Q on the same wire is 2.2 J.
   a) What is the potential difference between points P and Q? Explain your answer.
   b) Calculate the work done in transferring 250 C of charge from P to Q.
   c) 300 C of charge flow from P to Q as a steady current of 2.5 A.
      (i) Calculate the time taken for the 300 C of charge to flow from P to Q.
      (ii) Calculate the power dissipated as the 300 C of charge moves from P to Q.

2 When a car battery with an e.m.f. of 12 V is connected to the car's electric starter motor the terminal voltage falls to 10.8 V.

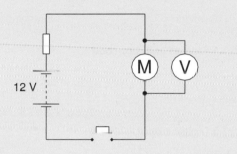

fig 4.18.2

   The motor and the connecting wires have a combined resistance of 0.06 Ω.
   a) Calculate the voltage lost across the internal resistance.
   b) Calculate the current required to start the motor.
   c) Calculate the internal resistance of the battery.

3 A battery with an e.m.f. of 100 V and internal resistance 8 Ω is connected to a 140 Ω resistor by two leads each of resistance 6 Ω.
   a) Calculate the current in the resistor.
   b) Calculate the power dissipated in the resistor.
   c) Calculate the voltage across the load resistor.

4 A battery with an e.m.f. of 6 V and internal resistance 1.0 Ω is connected into the circuit as shown in figure 4.18.3.

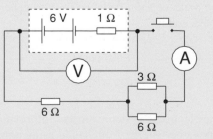

fig 4.18.3

   a) When the switch is open what is the reading on the digital voltmeter? Explain your answer.
   b) Calculate the current in the 3 Ω resistor when the switch is closed.

5 A string of Christmas tree lights consists of 20 identical 12 V lamps connected in series across a 240 V power supply of negligible internal resistance. The power consumption of the whole string of lights is 24 W.
   a) Calculate the resistance of each lamp.
   b) Two lamps become short circuited. Calculate the power now dissipated by the string of lights.

6 In the circuit of figure 4.18.4 a battery of e.m.f. E and internal resistance r is connected across a variable resistor as shown.

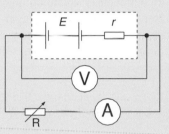

fig 4.18.4

   Current and voltage readings are taken for a series of different settings of the variable resistor. These results are shown in table 4.18.1.

V/volts	3	1.5	3.75	4.5	5.25
I/amps	2	3	1.5	1	0.5

Table 4.18.1

   a) Analysis of the circuit shows that;
   $$E = V_{tpd} + V_{lost} \qquad \text{equation 1}$$
   Show that this equation can be rearranged to give;
   $$V_{tpd} = (-r)I + E$$

   b) Rearrange equation 1 above to show that;
   $$R = E\left(\frac{1}{I}\right) - r$$

   c) Plot a suitable graph using this equation and the data in table 4.18.1 to determine the e.m.f. and internal resistance of the battery.

5.1 Preparing for analogue electronics

Introduction

There can be little doubt that electronics has been the fastest growing area of science for many years. Scotland, famed in the early 1900s for the quality of its shipbuilding, is now an established force in the European electronics industry. The speed that information can be gathered, stored, altered and transmitted by electronic systems is increasing all the time, making possible ever more complex applications.

Digital or analogue?

In a conventional tape recorder a tape head tracking a cassette tape generates a continuously changing analogue voltage. In contrast the laser head in a CD player gathers digital information from the surface of a spinning disc. This digital information is transmitted to the electronics in the CD player.

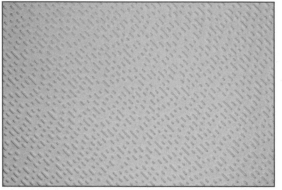

Figure 5.1.1 Magnified pits and troughs of a CD

Figure 5.1.3 shows part of a trace which might be obtained if a light sensor such as that shown in Figure 5.1.2 is used to monitor the light level over a 24 hour period.

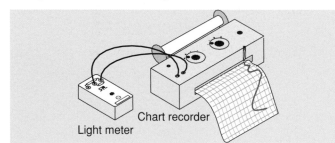

Figure 5.1.2 Light sensing equipment

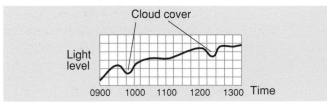

Figure 5.1.3 Light levels over 24 hours

The trace in figure 5.1.3 shows that the light level varies continuously so that **the output voltage generated by the light sensing circuitry can have any value between the displayed maximum and minimum**. This characteristic is typical of an **analogue voltage**.

When the bar code shown in figure 5.1.4 is swiped across a scanner the signal reflected back to the detectors is made up from a pattern of bright and dark lines. Different parts of the pattern are *either bright or dark*, there are no grey areas. The detectors collecting the reflected light are sensing only one of two light levels; bright or dark. **The voltage output from these detectors is therefore digital** (see figure 5.1.5). Items are identified by the digital information stored in their bar codes.

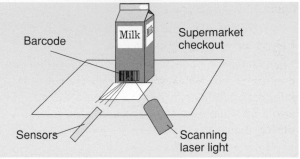

Figure 5.1.4 Product with bar code

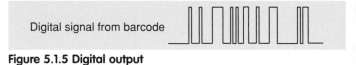

Figure 5.1.5 Digital output

Although analogue and digital signals are quite different both can be stored, altered and transmitted by electronic systems.

Gathering information

When light falls on the light dependent resistor (LDR) of figure 5.1.6 its resistance decreases. The LDR is one example of a photosensitive device which gathers information about light.

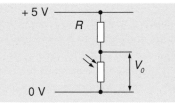

Figure 5.1.6 LDR

The solar cell is another light sensitive device made from silicon which converts some of the light it receives into electrical energy. Solar cells can also be called **photovoltaic cells**. Large areas of solar cells are used to generate electrical power for spacecraft.

Figure 5.1.7

While some electronic systems process information carried by light, others gather energy from sound using a microphone. A **microphone** contains a small magnet and a coil of wire arranged so that the vibrations in the sound wave move the coil relative to the magnet, thereby generating an induced voltage with the same frequency as the sound itself.

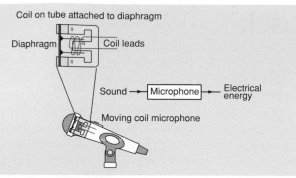

Figure 5.1.8 A microphone

Light emitting diodes

Microphones, solar cells and thermocouples are commonly used as input devices for electronic systems. They collect information for processing. The light emitting diode (LED) is also a commonly used component, but is used as an **output device**. It indicates the outcome of a process.

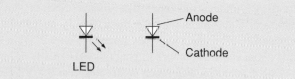

Figure 5.1.9 LED

When the anode of the LED is connected to the positive of a battery and the cathode is connected to the negative of the battery, the LED will emit light. If the anode of an LED is connected to the negative terminal of the power supply there will be no current. The polarity *must* be correct.

LEDs are made out of a very small treated silicon crystal embedded in a plastic casing. In section 8.12 you will learn why they emit light of a certain wavelength.

When connected to the correct voltage, LEDs operate with much smaller currents than filament lamps and do not get hot even when operating for long periods. Low power consumption and long operational lifetimes make LEDs ideal indicators for portable electronic appliances.

Figure 5.1.10 shows an LED connected to a 5 V power supply. The current in the LED is set by the series resistor. LEDs differ in the current and voltage that they require for optimum operation but, in general, LEDs operate with a potential difference of about 2 V between their anode and cathode.

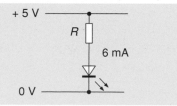

Figure 5.1.10

When the voltage across the LED is 2 V the remaining 3 V (5 – 2 V) must be across the resistor. The resistance of this resistor is determined by this voltage and the current requirements of the LED. When the current in the LED is 6 mA the value of the resistance can be found from;

$$R = \frac{V}{I} = \frac{3}{6}$$
$$R = 0.5 \; k\Omega$$
$$R = 500 \; \Omega$$

(Again we should note that since the voltage in the above example is quoted in volts and the current value given in milliamps then the resistance is calculated in units of kilohms.) The resistor connected in series with the LED is called the **protective** or **current limiting resistor**. Connecting an LED directly to a 5 V power supply will cause too much current in the LED, and may damage it.

A little further…

LED arrays

LEDs are now so common in modern circuitry that manufacturers' catalogues often feature over 100 varieties

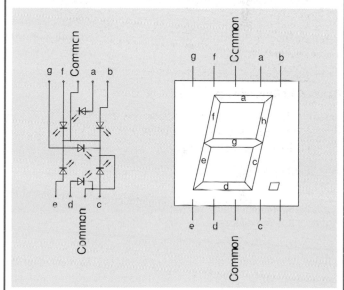

Figure 5.1.11 LED array

The arrangement of 7 rectangularly shaped LEDs shown in figure 5.1.11 is called a 7-segment display. By switching on specific combinations of LEDs it can display all the numbers from 0 to 9 as well as a number of letters and other symbols.

5.2 Transistors and feedback

Introduction

There can be little doubt that the invention of the transistor in 1947 by Bardeen and Brattain was one of the major technological advances of the twentieth century. Before its invention, engineers used thermionic valves to perform some of the functions now done by transistors.

Figure 5.2.1 A thermionic valve

Transistors are smaller, more robust and cheaper to produce than valves so they can be used in a larger variety of applications. The equivalent of many millions of transistors can be integrated into a piece of silicon the size of a pin head.

Types of transistor

Simply looking at a catalogue from any supplier of electronic components will show the wide selection of different transistors now available.

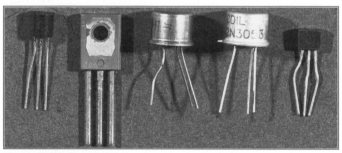

Figure 5.2.2 Transistors

Figure 5.2.3 shows the circuit symbol for an **npn bipolar transistor**. *Bipolar* describes its internal operation and *npn* describes how the slices of n-type and p-type silicon are assembled.

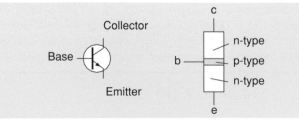

Figure 5.2.3 An npn bipolar transistor

The connections to the npn transistor are called the **collector**, **base** and **emitter**. The circuit symbol for an npn transistor has the arrow pointing from the base towards the emitter. When the transistor is switched on this arrow indicates the direction of conventional current in the circuit.

Figure 5.2.4 shows another variety of transistor; the **n-channel MOSFET** (pronounced mos-fet). This is not bipolar but is one of a family of **field effect transistors**.

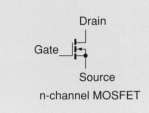

Figure 5.2.4 MOSFET transistor

What do transistors do?

The internal operation of even the simplest transistor can be difficult to understand but, like many modern devices, scientists and engineers can use them simply by considering *what* they do rather than *how* they operate. You will learn more details about semiconductors in chapters **8.10–8.14**.

Applying a potential difference of greater than 0.7 V between the base and emitter terminals of the npn transistor permits a current to flow from the collector terminal to the emitter. If the base-emitter voltage is below 0.7 V there is no current between the collector and emitter and the transistor is switched off.

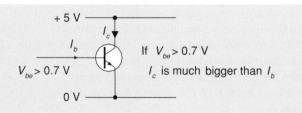

Figure 5.2.5

The base voltage can be used to switch on the collector current. The special feature of a transistor is that the current in the collector-emitter circuit is much larger than the current in the base-emitter circuit. As shown in figure 5.2.6, a very small current supplied to the base can switch on a large current through the lamp.

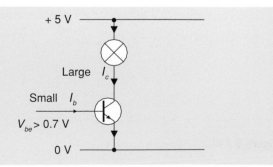

Figure 5.2.6

The transistor is acting as a current amplifier. The current in the lamp is many times bigger than the current in the base of the transistor.

Transistor switch

When the LDR in figure 5.2.7 is covered, its resistance increases and the signal at the midpoint of the voltage divider rises to nearly 5 V.

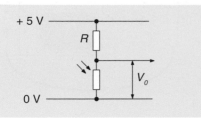

Figure 5.2.7

It would be wrong, however, to imagine that connecting a lamp to the midpoint and covering the LDR would make the lamp light. For this circuit to work the large current required to light the lamp would have to come from the midpoint of the voltage divider. If most of the current in the resistor is taken from the midpoint of the voltage divider the current in the LDR is small. This small current cannot make the voltage at the midpoint of the voltage divider rise when the LDR is covered.

With the circuit of figure 5.2.8 the midpoint of the voltage divider is connected to the base of an npn transistor.

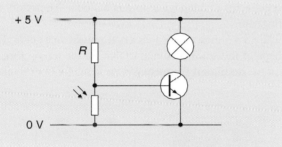

Figure 5.2.8 Transistor switch

All of the current needed to light the lamp comes *directly from the power supply*. Only the current needed to switch on the transistor is taken from the voltage divider. This current is small so the voltage at the midpoint of the voltage divider rises when the LDR is covered.

Temperature control system

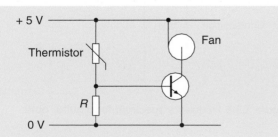

Figure 5.2.9 Temperature control circuit

As the room in which a temperature control system is placed warms up, the resistance of the thermistor shown in figure 5.2.9 decreases and the potential at the midpoint of the voltage divider

rises. When the voltage at the base of the transistor rises above 0.7 V, current flows from the battery through the fan. After a few moments the stream of air generated by the fan will cool the thermistor, its resistance will rise, the voltage at the midpoint of the voltage divider will fall below 0.7 V and the transistor will switch off the current to the fan. Without the stream of cooler air, the room will begin to warm and eventually the fan will switch on again.

Feedback

With the temperature control system shown in figure 5.2.9, the fan cools the thermistor and eventually the fan is switched off. The output of the system has an effect on its input. **This is called feedback**. In this example, the feedback tries to stabilise the room at a certain temperature. Stabilising feedback is called **negative feedback**.

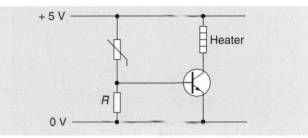

Figure 5.2.10. Heater

In the circuit of figure 5.2.10 the cooling fan is replaced by a heater. In the cool of the morning the resistance of the thermistor is high so the transistor is switched off. When the room warms naturally to the point where the transistor switches on, the heater turns on and warms the room further, forcing the input further away from stable conditions. **Destabilising feedback is called positive feedback**.

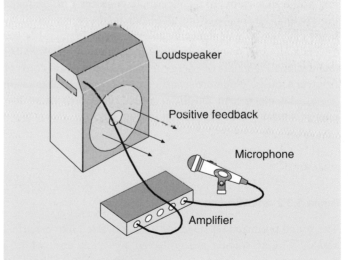

Figure 5.2.11 Positive feedback

The most common example of positive feedback is in amplifier systems where the sound from the speakers is picked up by the microphone, further amplified and re-emitted by the loudspeaker until the amplifier produces a high pitched squeal. Depending on your musical taste, positive feedback in sound amplification systems can be undesirable!

5.3 Amplifiers

Introduction

Pop concerts can attract many thousands or indeed tens of thousands of spectators, all eager to hear the music of their favourite group. Many of these fans will be so far from the stage that their idols will appear only as small dots in the distance. Despite their distance from the stage, all in the audience expect the sound produced by the guitars, keyboards, drums and vocals to be as clear as if it were played on the stereo at home or in the car.

Figure 5.3.1 At a pop concert

Modern sound systems require very sophisticated electronics to provide tone balance, spectrum equalisation and volume control but the basic building block of any public address system, relaying the sound produced on stage to an audience, is the **amplifier**.

What is an amplifier?

A retailer of electronic goods sells an amplifier as one box in a stacking hi-fi system. To the electronic engineer who designed the circuitry in the box, the amplifier might represent many hours of complex calculations during its development. For our understanding of what an amplifier does, we must remember that the small amounts of energy from a singer's vocal cords or from an acoustic guitar are insufficient to fill a large arena with sound. **An electronic amplifier must 'listen' to these low energy signals and output high energy copies**.

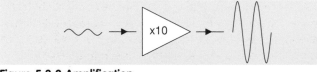

Figure 5.3.2 Amplification

The sound produced on stage must be converted into electrical signals. Guitar sounds are collected by 'pick ups' while a microphone turns the sound from a singer's voice into an electrical signal. The amplitude of both signals must be increased before they are relayed to the audience by the loudspeakers.

Amplifier inputs

A very pure note from a tuning fork, played near a microphone, will produce an electrical signal similar to that shown in figure 5.3.3.

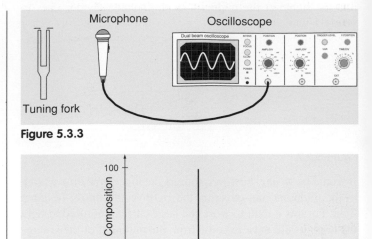

Figure 5.3.3

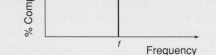

Figure 5.3.4 Pure note

This is called a 'pure' note because it has a single frequency. Figure 5.3.4 shows that 100% of this signal has one particular frequency. A diagram like figure 5.3.4 which shows the frequencies making up a note is called a **frequency spectrum diagram**.

Figure 5.3.5 shows a trace for a note played on a guitar. The general shape of this trace is similar to that for the tuning fork indicating that the **dominant frequency** is the same for both sounds.

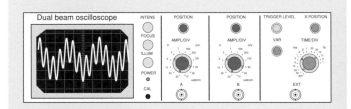

Figure 5.3.5

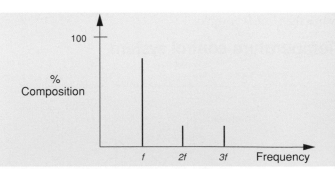

Figure 5.3.6 Frequency spectrum for a guitar note

The dominant frequency of the guitar note is accompanied by other frequencies giving the note '**richness**' or '**quality**'. The frequency spectrum for the guitar note could be represented as shown in figure 5.3.6. Other instruments, or even different guitars, playing this *same note* would all produce the *same dominant frequency* but each would *differ in the number and amplitude of the accompanying frequencies*.

An ideal amplifier

If the output from an amplifier is to sound identical to the input, the frequency spectrum of the output must be the same as the input. Figure 5.3.7 shows part of the output voltage from a tape head tracking along a music cassette tape.

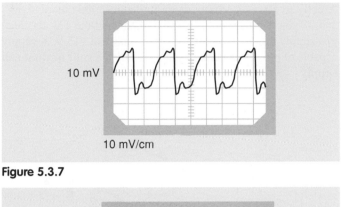

Figure 5.3.7

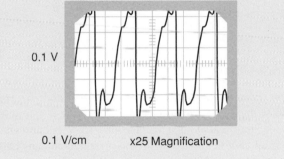

Figure 5.3.8

The maximum amplitude of the input signal is only a few millivolts. An ideal amplifier must increase the amplitude of each part of this signal so that the output is as shown in figure 5.3.8. **The shape of the input and output signals is the same but the amplitude of each part of the output is 25 times larger than the corresponding part of the input**.

Types of amplifiers

An ideal amplifier with a gain of 100 could be represented as shown in figure 5.3.9.

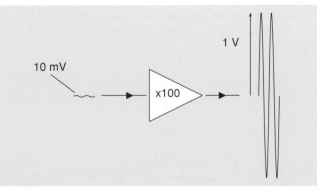

Figure 5.3.9

The input and output traces have the same shape so the frequency spectrums of the input and the output signals are identical.

Figure 5.3.10 shows another type of amplifier where the frequency spectrums of the input and output signals are the same.

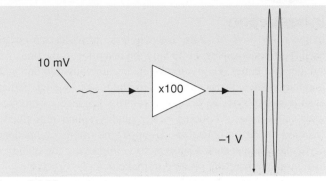

Figure 5.3.10

This amplifier produces negative output voltages when the inputs are positive and positive outputs when the inputs are negative. This is called an **inverting amplifier**. Each time the input is +10 mV the output is –1 V so the gain is stated as –100. Some engineers would say that figure 5.3.10 shows an amplifier with a gain of –100, while others would describe it as an inverting amplifier with a gain of 100. The number 100 shows the magnitude of the gain while the negative sign indicates that this is an inverting amplifier.

If the output and input voltages have the same sign, as in figure 5.3.9, the amplifier is called a **non-inverting amplifier**.

A potential problem

For a non-inverting amplifier with a gain of 100 it would seem sensible to suggest that an input of 1 V would produce an output voltage of 100 V. This would in fact only happen if the e.m.f. of the power supply to the amplifier is 100 V or greater. If a 6 V battery is used as the supply for the amplifier the maximum output is 6 V. An input of 0.05 V will give an output of 5 V and an input of 0.06 V will give an output of 6 V. But for an input above 0.06 V the calculated output would be higher than the voltage powering the amplifier. This cannot happen so the amplifier will in fact give its maximum output voltage of 6 V. When the amplifier's input and gain want to give an output voltage greater than the supply voltage we say that the amplifier is **saturated**. Figure 5.3.11 shows input and output voltage traces for an amplifier circuit where saturation occurs. The lower amplitude parts of the input waveform are still amplified as expected. When the amplifier is saturated the output stays at the saturation level. **The saturation level is assumed to be the same as the voltage of the supply powering the amplifier**.

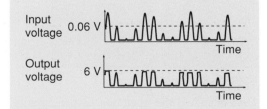

Figure 5.3.11

In the amplifier circuit of figure 5.3.11 the input and gain for certain parts of the signal combine to give saturation. Clearly the output is no longer an amplified copy of the input so saturation is undesirable in high quality audio amplifiers.

5.4 Operational amplifiers

Introduction

In the early days of radio and wireless, amplification circuits used thermionic valves. Following the invention of the transistor, smaller and better quality amplifiers became possible but as the popular music industry grew the demand for yet smaller, perhaps even portable, systems became a priority. Engineers were faced with a dilemma; producing better quality sound reproduction required more complicated circuitry and more components, particularly transistors. Consequently, improving the quality of hi-fi systems therefore seemed to make them less portable.

Figure 5.4.1 A silicon chip

The whole electronics industry changed following the invention of the microchip in 1958 by J. Kilbey at Texas Instruments. These very small microchips contain circuits equivalent to many thousands of transistors. Standard circuits, previously made from discrete transistors and joined together by wires, could be produced on a single integrated circuit (i.c.). One of the most useful groupings of standard circuits to be incorporated onto an i.c. is known as an **operational amplifier or op-amp**. This i.c. was called an op-amp because it was designed to perform mathematical operations such as multiplication, division, addition, subtraction and even differentiation and integration.

Figure 5.4.2 Op-amp

Understanding the operation of the circuits inside a modern operational amplifier i.c. is a very complex task! However, just as it is possible to drive a car without a detailed understanding of the principles of the internal combustion engine, we can use op-amp i.c.s without knowing precisely why they work. We simply need to know how to use them.

Power connections to the op-amp

The circuit symbol for an op-amp is as shown in figure 5.4.3. An op-amp is an *active* component so it must be connected directly to a power supply using the connections marked $+V_s$ and $-V_s$ on the diagram. It is usual to connect the $+V_s$ terminal to a supply of +15 V and the $-V_s$ terminal to a -15 V. A supply capable of providing these voltages is called a **split or dual rail supply**.

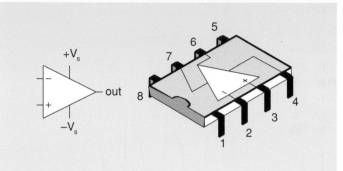

Figure 5.4.3 Op-amp symbol

If the potential difference between the $+V_s$ and $-V_s$ terminals is 30 V, the $+V_s$ terminal is +15 V above the 0 V terminal and the $-V_s$ terminal is at a voltage of -15 V because it is at a potential 15 V below 0 V. With this arrangement, any part of the input signal which has a voltage of 0 V will give an output of 0 V while other non zero parts of the output can only have values between +15 V and -15 V.

Inputs and outputs

The op-amp has one output and two input terminals. The output terminal is labelled 'out' and the inputs are labelled '+' and '−'. **The terminal labelled '+' is called the non-inverting input while the '−' terminal is known as the inverting input.**

The voltage at the output terminal, V_{out}, depends on the voltages at the inputs, V_+ and V_-, and on the other components connected to the op-amp. Altering the arrangement of the components around the op-amp makes it perform different tasks. For example the circuits of figure 5.4.4 and 5.4.5 use identical components and identical input voltages yet in figure 5.4.4 the output voltage will be 10 V whereas in figure 5.4.5 the output voltage is 2.5 V.

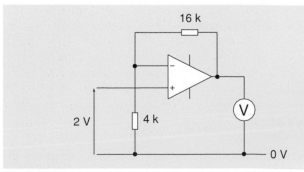

Figure 5.4.4

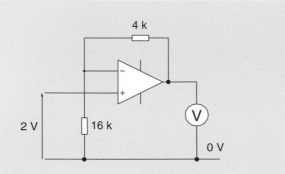

Figure 5.4.5

In figure 5.4.4 the output voltage equals the input voltage multiplied by 5 whereas in figure 5.4.5 the output is only 1.25 times the input. Both of these circuits perform the mathematical operation of multiplication. If the input voltage variation of figure 5.4.6 is applied to the circuit of figure 5.4.4 the output will be as shown. The circuit of figure 5.4.4 is an amplifier with a gain of 5.

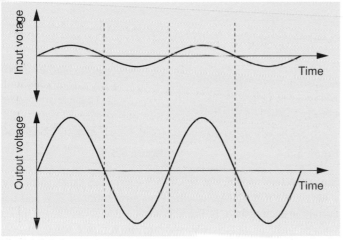

Figure 5.4.6

Principles of operation

In circuits figure 5.4.4 and 5.4.5 the *output is connected back to the inverting input via a resistor*. This causes **negative feedback**. All of the op-amp circuits that you meet in this course will have negative feedback. **The operation of op-amp circuits, with stabilising negative feedback**, can be explained using two basic rules.

Rule 1

The op-amp will keep the inverting and non-inverting inputs (V_ and V+) at the same voltage. This can also be expressed by saying that there will be no p.d. between the inverting and non-inverting pins on the op-amp chip.

In the circuit of figure 5.4.7 the output is connected, via a resistor, to the inverting input so providing the essential *negative feedback*. The non-inverting input is connected to zero so, as a consequence of rule 1, the op-amp will maintain the inverting input at 0 V.

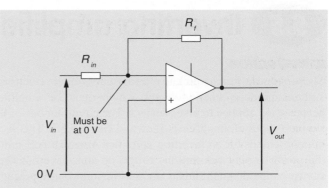

Figure 5.4.7 Example

Rule 2

The inverting and non-inverting terminals of an op-amp have an infinite input resistance. This means that no current will flow into the inverting or non-inverting terminals.

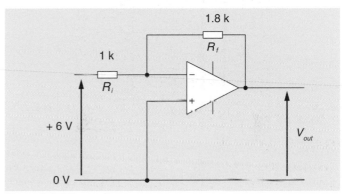

Figure 5.4.8 Example

Figure 5.4.8 is similar to figure 5.4.7 except that we now know that a 6 V battery pack is connected to the resistor joined to the inverting input. There is still negative feedback and the non-inverting input, V_+, is still at 0 V, so Rule 1 means that the inverting input is at 0 V. Since one end of the 1 kΩ resistor is at 0 V and the other side is at 6 V we can say that the potential difference across the resistor is 6 V. Therefore the current from the supply in resistor, R_i, is given by;

$$I = \frac{V}{R}$$

$$I = \frac{6-0}{1k}$$

$$I = 6 \ mA$$

Since Rule 2 tells us that no current flows into the input terminals of the op-amp we can say that the current through the 1.8 kΩ feedback resistor must also be 6 mA.

EXPERIMENT You can check this by setting up the circuit of figure 5.4.8 and by using an ammeter to confirm that the current in the feedback and input resistors is the same. Altering the input between 6 V and 0 V will change the ammeter readings but when the input current changes the current in the feedback resistor will change to the same value.

5.5 Inverting amplifiers

Introduction

Microphones, guitar pickups and many other electronic systems produce small voltages which must be amplified before being applied to other devices. The circuit for one of the simplest types of amplifier is shown in figure 5.5.1. This is an op-amp circuit for an inverting amplifier. You will notice that the power supply connections to the op-amp have not been shown on the diagram; this is not an oversight. Since practical circuits involve many operational amplifiers, showing the power supply connections to each would make circuit diagrams unnecessarily complex. By convention the power connections are not shown on the diagrams, however, we must remember that op-amps are active components and must be connected to split rail supplies for amplifiers to function properly.

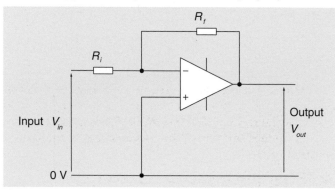

Figure 5.5.1

We can identify the circuit of figure 5.5.1 as an inverting amplifier by noticing that;
1. the non-inverting input is joined directly to the 0 V line
2. part of the output signal is fed back via resistor R_f to the inverting input
3. the input signal is also connected to the inverting input via resistor R_i.

Only when all of these conditions are satisfied is an op-amp circuit an inverting amplifier.

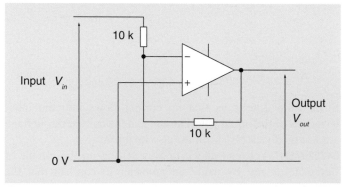

Figure 5.5.2

The circuit diagram of figure 5.5.2 looks different to figure 5.5.1, but careful checking will show that it meets the three requirements listed above. On the other hand the circuit of figure 5.5.3 looks fairly similar to figure 5.5.1 but it fails on the first and third requirements, so is not an inverting amplifier.

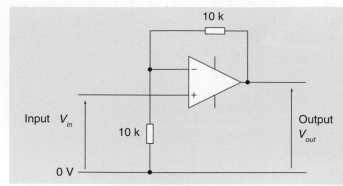

Figure 5.5.3

Revising voltage dividers

Being able to identify inverting amplifiers is the first step in understanding how they work. Revising earlier work on voltage dividers will also help us to understand operational amplifier circuits better.

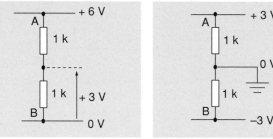

Figure 5.5.4   **Figure 5.5.5**

Since the two resistors in figure 5.5.4 have the same value, the midpoint of this voltage divider is at a potential of 3 V and the current in each resistor is 3 mA. Point A is 3 V above the midpoint voltage while point B is 3 V below the midpoint voltage.

The potential difference across resistors A and B in figure 5.5.5 is also 6 V and the current in each of the resistors is 3 mA. This assumes that all of the charges flowing through resistor A, flow through resistor B. We can express this another way by saying that no current flows in the earthed connection joining the midpoint of A and B to 0 V.

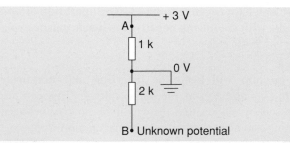

Figure 5.5.6

In the circuit of figure 5.5.6 the p.d. across the top resistor is 3 V creating a current of 3 mA in the 1 kΩ resistor. If no charge leaves the voltage divider via the earthed connection, the current in the 2 kΩ resistor is also 3 mA. A current of 3 mA in a 2 kΩ resistor means that the voltage across the resistor is 6 V. Therefore point B in figure 5.5.6 must be at a potential of –6 V.

Inverting amplifier operation

In analysing the circuits of figures 5.5.5 and 5.5.6 we had to assume that the midpoint of each voltage divider is at 0 V and that the currents in the top and bottom resistors are the same. Ensuring that these conditions apply is precisely the function of the op-amp in the circuit of figure 5.5.7.

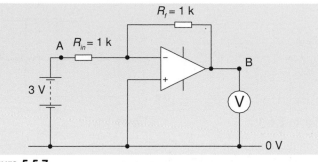

Figure 5.5.7

The first rule governing the action of an op-amp, as stated in the previous chapter, tells us that the inverting and non-inverting inputs are at the same voltage. Since the non-inverting input is at 0 V the inverting input will also be fixed at 0 V. The potential difference across the 1 kΩ input resistor, labelled R_{in}, is 3 V and the current in it is 3 mA.

The second rule governing the action of the op-amp tells us that no charge flows into the inverting or non-inverting inputs of the op-amp. Therefore the current in the input and feedback resistors must be the same so the current in the feedback resistor is also 3 mA. The voltage across the 1 kΩ feedback resistor must therefore be 3 V so point B is at –3 V. The inverting amplifier in figure 5.5.7 has a gain of 1.

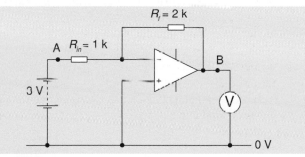

Figure 5.5.8

If the 1 kΩ feedback resistor of figure 5.5.7 is replaced by a 2 kΩ resistor as shown in figure 5.5.8 the current in the input resistor would still be 3 mA. The current in the feedback resistor is therefore also 3 mA, so the potential difference across the 2 kΩ feedback resistor is 6 V. The voltmeter would show a reading of –6 V. We say that this amplifier has a gain of –2.

Setting the gain of an inverting amplifier

In the inverting amplifier circuit of figure 5.5.9 the inverting input of the op-amp is fixed at 0 V so the current, i, in the input resistor, R_{in}, is given by;

$$i = \frac{V_{in}}{R_{in}}$$

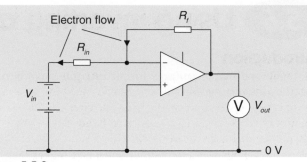

Figure 5.5.9

Since all of this current passes through the feedback resistor, generating a negative output voltage, $-V_{out}$, we can say;

$$-V_{out} = i\,R_f$$

or

$$-V_{out} = \frac{V_{in}}{R_{in}} \times R_f$$

$$V_{out} = -\frac{R_f}{R_{in}} \times V_{in}$$

This formula is called the **inverting mode gain expression**. The output voltage is the input voltage multiplied by the ratio of the external resistances. **The ratio R_f/R_{in} is called the gain and determines the size of the output signal in relation to the input signal.**

Verifying the inverting mode gain expression

EXPERIMENT By setting the input voltage in figure 5.5.10 to precisely 1 V and using resistance substitution boxes to provide the input and feedback resistances, R_{in} and R_f, the output voltages can be measured for a number of different combinations of R_{in} and R_f.

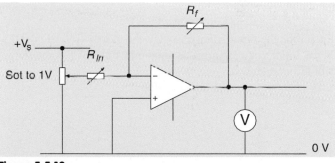

Figure 5.5.10

If V_{in} is precisely 1 V then;

$$V_{out} = -\frac{R_f}{R_{in}} \times V_{in}$$

becomes

$$V_{out} = -\frac{R_f}{R_{in}}$$

By measuring V_{out} for different combinations of R_f and R_{in} we can verify the expression for the gain of an inverting amplifier.

5.6 Using inverting amplifiers

Introduction

The input and output voltages for an inverting amplifier are linked by the relationship;

$$V_{out} = -\frac{R_f}{R_{in}} \times V_{in}$$

If R_f is greater than R_{in} the output voltage is bigger than the input. The size of the output is calculated by multiplying the input voltage by R_f/R_{in}. If R_f is smaller than R_{in} the output voltage is a fraction of the input. This type of circuit is called an attenuator.

The gain, G of an inverting amplifier is given by the equation;

$$G = \frac{V_{out}}{V_{in}} = \frac{R_f}{R_{in}}$$

The circuit of figure 5.6.1 shows how an op-amp can be made into an inverting amplifier with a gain of –10.

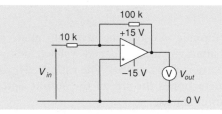

Figure 5.6.1

If the voltage signal shown in figure 5.6.2 is applied to the input of this amplifier the output will be as shown.

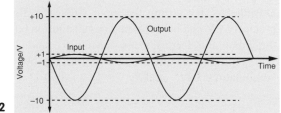

Figure 5.6.2

Each time the input voltage is +1 V the output voltage is –10 V. If the input voltage varies as shown in figure 5.6.3 the output would be as shown.

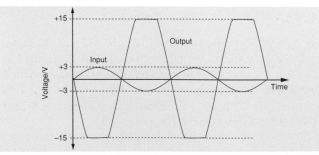

Figure 5.6.3

The amplifier saturates when the input is above 1.5 V. Since the op-amp is being powered by a split rail +/–15 V supply, the output cannot be more positive than +15 V or more negative than –15 V. As this amplifier has a gain of –10, only inputs between –1.5 V and +1.5 V will be amplified without distortion.

Generating square waves

Distortion can occur even with very small input voltages, if the gain of the amplifier is very high.

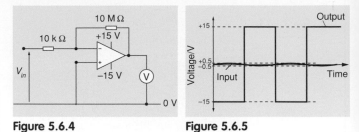

Figure 5.6.4   **Figure 5.6.5**

For the circuit of figure 5.6.4, if the input voltage is more positive than +15 mV or more negative than –15 mV the output saturates. If the input to this amplifier is a voltage varying continuously between +0.5 V and –0.5 V the output would be as shown in figure 5.6.5.

Since the input signal changes between +15 mV and –15 mV very quickly, the output changeover from –15 V to +15 V is almost instantaneous (as shown by the vertical sections on the output graph). This type of output variation is called a **square wave**.

A little further…

Producing digital timing pulses

In the circuit of figure 5.6.6 an alternating mains voltage from a step down transformer is used as the input voltage for a very high gain inverting amplifier. The input to the amplifier is a variable voltage with a frequency of 50 Hz.

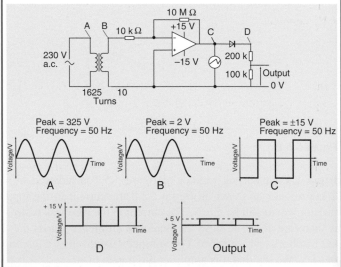

Figure 5.6.6

The frequency of the square wave output is also 50 Hz. The diode ensures that the output of the op-amp can only be +15 V or 0 V. The resistors connected to the output reduce the amplitude of the square wave output to 5 V. These single frequency square pulses with an amplitude of 5 V are widely used in electronic timing and counting circuits.

More advanced amplifier circuits

In this section we use the principles outlined so far to analyse the type of circuits that you will meet in more advanced courses. The op-amp circuits studied so far have been able to perform the mathematical operations of multiplication and division. The relative sizes of the input and output voltages are set by the input and feedback resistance values.

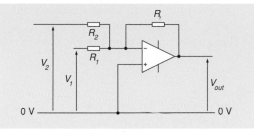

Figure 5.6.7

The circuit of figure 5.6.7 has two input voltages, V_1 and V_2, two input resistors, R_1 and R_2 and a single feedback resistor, R_f. Although this looks a little more complicated than the circuits met so far we can still use our knowledge of the principles governing the behaviour of an op-amp to determine how the output voltage depends on the inputs.

For the circuit of figure 5.6.7;

1 the non-inverting input is joined to the 0 V line
2 part of the output signal is fed back via a resistor to the inverting input
3 input voltages are connected to the resistors joined to the inverting input.

As these three requirements are met, figure 5.6.7 is a type of inverting amplifier.

Since the non-inverting input is joined to 0 V and there is no p.d. between the inverting and non-inverting inputs, the inverting input is also at 0 V. The voltage across resistor R_1 is V_1, so the current in R_1 is;

$$i_1 = \frac{V_1}{R_1}$$

In addition, we can say that the voltage across resistor R_2 is V_2 so the current in R_2 is given by;

$$i_2 = \frac{V_2}{R_2}$$

Since no current passes into the inverting input of the op-amp the current through the feedback resistor, i_f, is given by;

$$i_f = i_1 + i_2$$

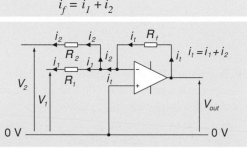

Figure 5.6.8

With the circuit diagram of figure 5.6.8 the output voltage, which will be negative, is given by;

$$-V_{out} = (i_1 + i_2)R_f$$

But since;

$$i_1 = \frac{V_1}{R_1} \ and \ i_2 = \frac{V_2}{R_2}$$

$$-V_{out} = \left(\frac{V_1}{R_1} + \frac{V_2}{R_2}\right)R_f$$

$$-V_{out} = \frac{R_f}{R_1}V_1 + \frac{R_f}{R_2}V_2$$

With resistors R_f, R_1 and R_2 having the same value this equation further simplifies to;

$$V_{out} = -(V_1 + V_2)$$

This type of circuit is called a summing amplifier.

Summing and scaling

For the amplifier circuit of figure 5.6.9 one of the input resistances has been halved to 50 kΩ. The output of this circuit is given by;

$$-V_{out} = \frac{R_f}{R_1}V_1 + \frac{R_f}{R_2}V_2$$

$$-V_{out} = \frac{100}{100}V_1 + \frac{100}{50}V_2$$

$$V_{out} = -(V_1 + 2V_2)$$

The input voltage V_2 is scaled by a factor of two before being added to V_1. Scaling and adding is also used in the circuit of figure 5.6.10.

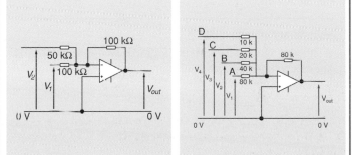

Figure 5.6.9   **Figure 5.6.10**

This circuit is called a four bit digital to analogue converter. Four bits of digital information (1 or 0), stored in binary form, are simultaneously applied to the inputs labelled D, C, B and A. An analysis similar to that shown above can show that if D, C, B and A have an input voltage of 1 V or 0 V, the output voltage is given by;

$$-V_{out} = A + 2B + 4C + 8D$$

Placing different combinations between 0000 and 1111, at the inputs creates a varying voltage at the output. Information stored in digital form on compact discs can be converted by digital to analogue converters into analogue voltages. These analogue voltages can be very precise copies of the signal that created them. Music recorded onto CD can be reproduced accurately. Not surprisingly, therefore, very fast digital to analogue converters are at the heart of modern CD players.

Worked example 1

A signal generator produces the output voltage waveform shown in figure 5.7.1.

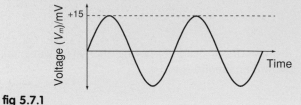

fig 5.7.1

This signal is used as the input to the amplifier shown in figure 5.7.2.

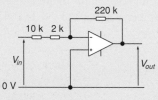

fig 5.7.2

a) In which mode is the op-amp shown in figure 5.7.2 being used? Explain your answer.
b) Calculate the gain of the amplifier.
c) Copy the waveform of figure 5.7.1. Add another line to show how the output of the amplifier alters as the input changes as shown.

Answers and comments

a) With this amplifier the non-inverting input is joined to 0 V. The input voltage is connected to the resistor joined to the inverting input and the feedback is from the output to the inverting input. These features make the circuit of figure 5.7.2 an inverting amplifier.
b) The gain of an inverting amplifier can be calculated from;

$$G = \frac{R_f}{R_{in}}$$

In this circuit the input resistance is made up from the 10 kΩ and 2 kΩ resistors in series. The total input resistance is 12 kΩ.
Therefore;

$$G = \frac{R_f}{R_{in}} = \frac{220k}{12k} = 18.3$$

c) The output voltage is an inverted and amplified copy of the input. Since the peak input voltage is shown as 15 mV the peak output voltage will be 15 × 18.3 = 275 mV. This is below the saturation limit so the output waveform is as shown in figure 5.7.3.

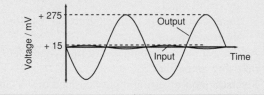

fig 5.7.3

Worked example 2

A student sets up the circuit as shown in figure 5.7.4 in an attempt to make a light meter.

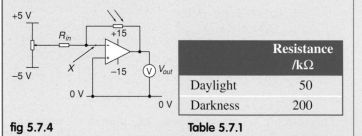

fig 5.7.4

	Resistance /kΩ
Daylight	50
Darkness	200

Table 5.7.1

The resistance of the LDR in light and darkness is as shown in table 5.7.1. She uses the potentiometer to set the input voltage at –1 V and notices that when the LDR is covered the voltmeter shows a reading of 2 V.

a) What is the voltage at the point marked X in figure 5.7.4. Explain your answer.
b) Calculate the resistance of R_{in}.
c) Calculate the reading shown on the voltmeter when the LDR is placed in daylight.

Answers and comments

a) The circuit of figure 5.7.5 is an inverting amplifier. The output is 2 V so the amplifier is not saturated. The negative feedback ensures that the inverting and non-inverting inputs of the op-amp are at the same voltage. Therefore point X is at 0 V.
b) The output and input voltages of the inverting amplifier are related by the equation;

$$V_{out} = -\frac{R_f}{R_{in}} \times V_{in}$$

Substituting values gives;

$$2 = -\frac{200k}{R_{in}} \times (-1) \quad R_{in} = \frac{200}{2} \times 1$$

$$R_{in} = 100 \text{ k}\Omega$$

c) When the LDR is in daylight its resistance is 50 kΩ. The input resistance is still 100 kΩ so we can calculate the output voltage using;

$$V_{out} = -\frac{R_f}{R_{in}} \times V_{in}$$

$$V_{out} = -\frac{50k}{100k} \times (-1) = 0.5 \text{ V}$$

Questions

1 Figure 5.7.5 shows part of the block diagram of an electronic system shown in an electronics magazine.

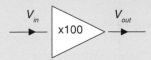

fig 5.7.5

 a) What does this symbol represent?
 b) A voltage of 100 mV is used as the V_{in} signal. Calculate the output voltage.
 c) Calculate the input voltage required to give an output voltage of 2.5 V.
 d) The article in the electronics magazine states that the amplifier saturates when the input voltage is greater than 150 mV.
   (i) Explain the meaning of the term saturates.
   (ii) Explain what this information tells us about the voltage of the power supply connected to the amplifier.

2 Figure 5.7.6 shows a number of circuits involving op-amps. Identify which of these are inverting amplifiers and calculate the gains of these inverting amplifiers.

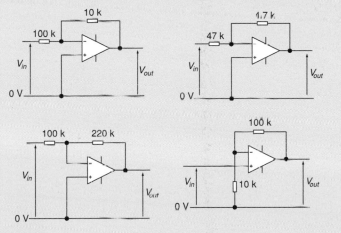

fig 5.7.6

3 Figure 5.7.7 shows a graph which was drawn in a student's notebook. The graph is labelled 'Amplifier Characteristics'.

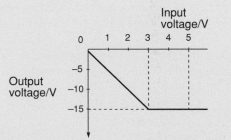

fig 5.7.7

 a) The student's notes states that this graph is for an inverting amplifier with a gain of 5. Is this statement correct? Justify your answer.
 b) Over what range of input voltages will the student's circuit amplify without distortion? Explain your answer.

4 The following waveform is applied to the input of the circuit shown below.

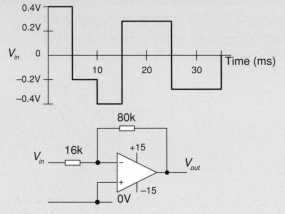

fig 5.7.8

 a) Calculate the gain of the inverting amplifier.
 b) Calculate the output voltage from the amplifier when the input is 0.4 V.
 c) Sketch a graph to show how the output of the amplifier changes when the input changes as shown in figure 5.7.8.

5 Figures 5.7.9 and 5.7.10 show inverting amplifier circuits constructed using op-amps.

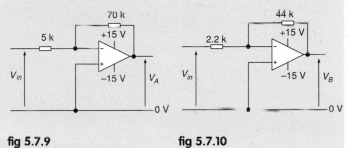

fig 5.7.9　　　　　　**fig 5.7.10**

 a) Calculate the gain of the amplifier circuit shown in figure 5.7.9.
 b) The voltage at the output of the circuit of figure 5.7.10 is 3.5 V. Calculate the size of the input voltage.
 c) The output of figure 5.7.9 is now connected to the input of figure 5.7.10.
   (i) Calculate the size of the input voltage to circuit 5.7.9 that will produce a 3.5 V output from the second circuit.
   (ii) Calculate the overall gain of the system.
   (iii) How is the overall gain of the system (G_{Tot}) related to the individual gains (G_a and G_b) of circuits 5.7.9 and 5.7.10?
   (iv) Calculate the maximum input voltage that can be amplified when the circuits are joined together.
   (v) Is it possible to use a single inverting amplifier to produce a circuit with the gain that you have calculated in part c(ii)? Justify your answer.

Past paper question

A health physicist builds an electronic stethoscope system to monitor heart beats.

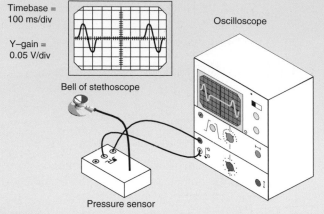

fig 5.8.1

The trace shown is obtained when the oscilloscope has a Y-gain setting of 0.05 V/division.

a) What is the peak voltage of the signal shown on the oscilloscope?

b) The following circuit is designed to amplify the signal from the electronic stethoscope to enable a doctor to hear the heart beats.

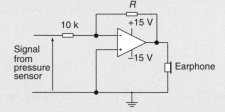

fig 5.8.2

A peak output of 0.9 V is required to operate the earphone. Calculate the value of the resistor R.

SQA 1992

Answers and comments

a) From the trace on the oscilloscope we can see that the peak of the signal is approximately 1.5 divisions above the middle line. From the Y-gain calibration we can therefore say that this corresponds to a peak voltage given by;

$$Peak\ voltage = 1.5 \times 0.05$$
$$Peak\ voltage = 0.075\ V.$$

b) The output voltage from the pressure sensor is now used as the input to the inverting amplifier. From this inverting amplifier we want to get an output voltage of 0.90 V for the earphone. To get this positive output from an inverting amplifier the input voltage must be negative. From the trace on the oscilloscope we can see that the heart beat produced a symmetrical trace so we can calculate how to produce an output of 0.9 V from an input of –0.075 V.

$$V_{out} = -\frac{R_f}{R_{in}} \times V_{in}$$

$$0.90 = -\frac{R}{10\ k} \times (-0.075)$$

$$R = \frac{0.9 \times 10}{0.075}$$

$$R = 120\ k\Omega$$

The feedback resistor has a value of 120 kΩ.

Worked example

Figure 5.8.3 shows a cell with an e.m.f., E, connected to an inverting amplifier. The amplifier is not saturated and ammeters A_1 and A_2 are used to measure the current at certain points in the circuit.

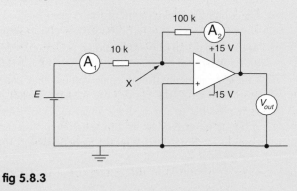

fig 5.8.3

Answers and comments

a) The amplifier is not saturated. The negative feedback ensures that the inverting and non-inverting inputs of the op-amp are at the same voltage. Therefore point X is at 0 V.

b) The current in the 10 kΩ input resistance is 7 μA so we can calculate the voltage across this resistor using;

$$V = I R$$
$$V = 7 \times 10^{-6} \times 10 \times 10^3$$
$$V = 0.07\ V$$

c) One of the rules governing the behaviour of the op-amp tells us that the input resistance of the op-amp is so high that no charge flows into the op-amp's input. This means that all of the charge flowing in the input resistor also flows through the feedback resistor. The current in the feedback resistor is therefore 7 μA.

a) What is the potential at the point marked X in the circuit?

b) Ammeter A_1 shows a reading of 7 μA. Calculate the potential difference across the 10 kΩ input resistor.

c) What is the reading on ammeter A_2? Explain your answer.

d) Calculate the potential difference across the 100 kΩ feedback resistor.

e) What is the value of the output voltage? Justify your answer.

d) The current in the 100 kΩ feedback resistor is 7 μA so we can calculate the voltage across this resistor using;

$$V = I\,R$$
$$V = 7 \times 10^{-6} \times 100 \times 10^3$$
$$V = 0.7\,V$$

e) The input voltage to this inverting amplifier is positive so the output voltage is negative. We have calculated that the voltage across the feedback resistance is 0.7 V and have stated that point X is at a voltage of 0 V therefore the output voltage of the op-amp is –0.7 V.

Questions

1 Figure 5.8.4 shows a 'two stage amplifier' circuit. This circuit uses two inverting amplifiers. The input resistance to the first amplifier has a fixed 20 kΩ resistor and a 10 kΩ variable resistor.

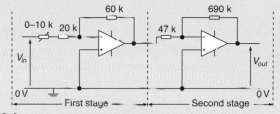

fig 5.8.4

a) Calculate the gain of the second stage amplifier.

b) Calculate the largest gain of the amplifier making up the first stage of the two stage amplifier circuit.

c) The input voltage V_{in} is +0.3 V. Show that the output of the second stage of the amplifier will vary between 8.8 V and 13.2 V as the variable resistor is altered.

2 Figure 5.8.5 shows an LED connected to the output of an inverting amplifier. This LED lights at normal brightness when the voltage across it is 1.5 V and the current in it is 10 mA.

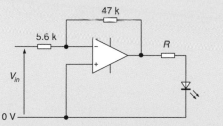

fig 5.8.5

a) Calculate the gain of the inverting amplifier.

b) The input voltage V_{in} is set at –0.9 V. Calculate the output voltage of the inverting amplifier.

c) Calculate the current in the input resistor.

d) The LED lights at normal brightness. What is the voltage across resistor R? Justify your answer.

e) Calculate the value of resistor R needed to light the LED at normal brightness when the input to the inverting amplifier is – 0.9 V.

f) Explain how the LED can light at normal brightness even though the current in the input and feedback resistors is much less than 10 mA.

3 A student investigating inverting amplifier circuits sets up the circuit shown in figure 5.8.6.

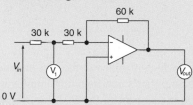

fig 5.8.6

a) Calculate the gain of the inverting amplifier shown in figure 5.8.6.

b) Using only the resistor values shown in figure 5.8.6 redraw the circuit to show how the student could make an inverting amplifier with the following gains;

(i) 3   (ii) ⅓   (iii) 4   (iv) 1.5

4 A student flicking through the pages of an electronics textbook finds the circuit shown in figure 5.8.7 in the chapter entitled 'Inverting amplifier circuits'.

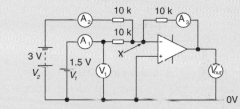

fig 5.8.7

a) What is the voltage at the point marked X in this amplifier circuit.

b) The 1.5 V and 3 V cells have negligible internal resistance.

(i) What is the potential difference across each of the 10 kΩ input resistors?

(ii) Show by calculation that the readings on ammeters A_1 and A_2 are 0.15 mA and 0.3 mA respectively.

c) What is the current in the feedback resistor? Explain your answer.

d) Calculate the potential difference across the feedback resistor.

e) What is the output voltage of this circuit?

f) Describe how the output voltage V_{out} is related to input voltages V_1 and V_2.

g) Describe and explain the effect of changing the feedback resistance to 20 kΩ.

5.9 Difference amplifiers

Introduction

In all of the amplifiers studied so far the non-inverting input of the amplifier is connected directly to the 0 V line. These amplifiers are not saturated so the negative feedback ensures that the inverting input is also at zero volts.

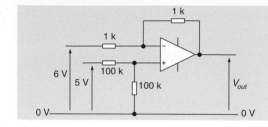

Figure 5.9.1

In the circuit of figure 5.9.1 a 5 V signal is connected to a 100 kΩ resistor which is joined to the non-inverting input of the amplifier. The amplifier's non-inverting input is itself connected by another 100 kΩ resistor to the 0 V line. These two resistors form a voltage divider as shown in figure 5.9.2, with the non-inverting input of the amplifier being connected to the midpoint.

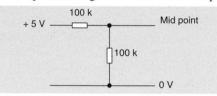

Figure 5.9.2

The voltage divider causes a p.d. of 2.5 V across each of the 100 kΩ resistors. The non-inverting input pin of the amplifier of figure 5.9.1 is therefore *not* at 0 V, as you would expect for an inverting amplifier. The non-inverting input is at a voltage of 2.5 V. Since there must be *no p.d.* between the non-inverting and inverting inputs of the amplifier, the inverting input is also at 2.5 V.

The two 1 kΩ resistors in figure 5.9.1 form a voltage divider where the midpoint is connected to the op-amp's inverting input. In this example the midpoint of the voltage divider of figure 5.9.3 is held at 2.5 V by the voltage at the non-inverting input. Therefore, there will be a p.d. of 3.5 V (6 – 2.5 V) across the 1 kΩ input resistor and consequently the current in the 1 kΩ input resistor is 3.5 mA.

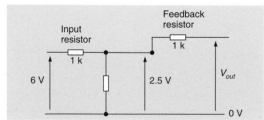

Figure 5.9.3

The current in the 1 kΩ feedback resistor will also be 3.5 mA and so the voltage across the feedback resistor is 3.5 V. The voltage V_{out} must therefore be 3.5 V below the midpoint of the voltage divider shown in figure 5.9.3. So;

$$V_{out} = 2.5 - 3.5 = -1 \text{ V}$$

For this example with simple numbers we have shown that the magnitude of the output voltage is the difference between the input voltages to the amplifier. The circuit of figure 5.9.1 is called a **difference or differential amplifier**.

Practical difference amplifiers

The analysis of the circuit of figure 5.9.1 was simple because of the resistor values chosen. A more involved analysis can be done for cases where all four of the resistors have different values but, in general, difference amplifiers commonly use the arrangement of resistors shown in figure 5.9.4.

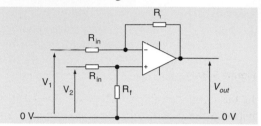

Figure 5.9.4

The resistors to which the inputs are connected are called input resistors, R_{in}, and the other resistors are called feedback resistors, R_f. In general for a difference amplifier it can be shown that;

$$V_{out} = \frac{R_f}{R_{in}} (V_2 - V_1)$$

From this equation we can see that the circuit of figure 5.9.4 amplifies the difference between the voltage signals at the inputs and rejects any part of an input signal common to both inputs. This circuit is called a difference or differential amplifier.

Using difference amplifiers

Figure 5.9.5 shows two voltage dividers arranged so that the midpoint of each is providing one input for a difference amplifier. Recalling the principles of the Wheatstone bridge will convince you that there is no potential difference between the two inputs when;

$$\frac{R_1}{R_2} = \frac{R_3}{R_4}$$

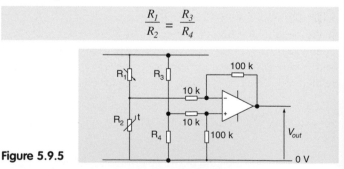

Figure 5.9.5

In figure 5.9.5, R_2 is a thermistor whose resistance decreases as its temperature rises. R_1 is a variable resistor which can be altered so that the bridge is balanced at one particular temperature. If the temperature of the thermistor changes by a small amount, the bridge becomes slightly out of balance. The difference amplifier will give a reading on the voltmeter which is ten times bigger than the voltage between the midpoints of the voltage dividers.

Difference amplifiers increase the sensitivity of bridge circuits by amplifying the small out-of-balance voltages so that changes in the surroundings can be shown easily on meters or indicators.

5.10 Op-amp drive circuits

Introduction

In the inverting and difference amplifier circuits studied so far we have shown how the output voltage is controlled by the external resistors connected to the op-amp. In the circuit of figure 5.10.1 an input voltage of –2 V will make the voltmeter connected to the output show a reading of 12 V.

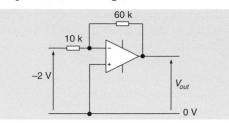

Figure 5.10.1

In the circuit of figure 5.10.2 a potential difference of 1 V between the inputs will make the voltmeter connected to the output show a reading of 12 V.

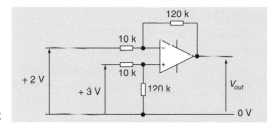

Figure 5.10.2

It would be easy to imagine that connecting a car headlamp bulb to the outputs of each of these circuits and selecting appropriate voltages for the op-amp circuits, would make the headlamp bulb light at full brightness.

However, neither of these circuits would light the car headlamp bulb, **Although they are capable of producing the required output voltages, the op-amps would not be able to supply all the current needed to light the lamp.**

Op-amps and current

If the headlamp bulb is rated 12 V, 24 W it means that when connected to a 12 V supply it will light at normal brightness and it will dissipate 24 W of power. Since electrical power is related to current and voltage by the equation;

$$P = V I$$

we can calculate the current required for normal brightness;

$$24 = 12 \times I$$

$$I = \frac{24}{12}$$

$$I = 2\ A$$

Operational amplifiers are low current devices and can only produce maximum output currents of around 20 mA for a limited period. There is therefore no possibility of the lamp lighting properly.

For op-amp circuits to drive devices needing a lot of electrical power we can use the type of circuit shown in figure 5.10.3.

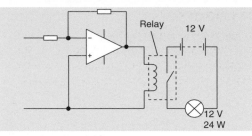

Figure 5.10.3

The small currents supplied by the output of the op-amp are sufficient to activate the relay. When the relay is activated the 2 A current for the lamp comes directly from the 12 V supply. With the circuit shown in figure 5.10.3 the bulb is either totally on or totally off. Figure 5.10.4 shows a circuit which will allow the brightness of the bulb to be varied by the output voltage from the op-amp.

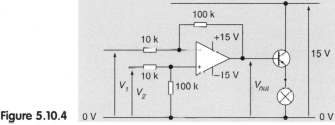

Figure 5.10.4

The transistor circuit connected to the differential amplifier is called an **npn emitter follower**. When the output voltage from the difference amplifier is positive the voltage at the emitter terminal of the transistor will be 0.7 V below the voltage at the base. Hence the voltage across the bulb will be 0.7 V less than the output voltage from the op-amp. As V_{out} changes from +0.7 V to +13 V the voltage across the bulb will increase from 0 V to +12.3 V and the bulb's brightness will change accordingly.

Measuring the output current limit

EXPERIMENT In the circuit of figure 5.10.5 the 10 kΩ variable resistor is connected across the output of a differential amplifier.

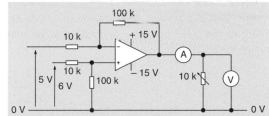

Figure 5.10.5

The 10 V output from the amplifier will cause a current of 1 mA in the 10 kΩ variable resistor. As the output resistance is lowered, more and more current is demanded from the op-amp's output. The output voltage will stay the same until the resistance is reduced below approximately 500 Ω. At this value the current from the op-amp has reached its maximum of 20 mA. Lower the resistance further and the output voltage will fall below the calculated value.

If the output voltage from an amplifier is less than expected it is possible that too much current is being demanded at the output. Engineers can use this overloading effect to detect short circuits in complex op-amp circuitry.

5.11 Consolidation questions

Worked example 1

Figure 5.11.1 shows a circuit where the op-amp is being used in differential mode

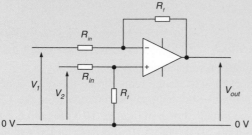

fig 5.11.1

a) State the differential mode gain expression linking the output voltage V_{out} with the input voltages, V_1 and V_2 and the resistances.

b) Calculate the output voltage for the input voltages and resistance values shown in figure 5.11.2.

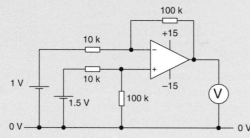

fig 5.11.2

c) Describe how the output of this circuit can be used to activate a relay to switch on a lamp connected to a 230 V a.c. mains supply.

Answers and comments

a) The differential mode gain expression is;

$$V_{out} = \frac{R_f}{R_{in}} (V_2 - V_1)$$

b) Comparing this expression with the diagram of figure 5.11.2 we can see that R_{in} and R_f are 10 kΩ and 100 kΩ respectively. V_2 is the voltage joined by the input resistor to the op-amp non-inverting input and V_1 is the voltage joined by the resistor to the inverting input of the op-amp. We can calculate the output voltage from;

$$V_{out} = \frac{R_f}{R_{in}} (V_2 - V_1)$$

$$V_{out} = \frac{100\,k}{10\,k} (1.5 - 1.0)$$

$$V_{out} = 5\ V$$

c) The output of the op-amp is connected to the coil of the relay. The current sourced by the op-amp activates the relay, powering the lamp from the 230 V supply.

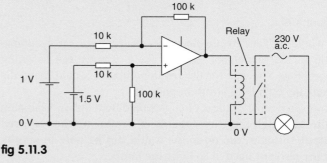

fig 5.11.3

Worked example 2

a) The Wheatstone bridge circuit shown below is used to investigate how the resistance of an LDR varies with light intensity.

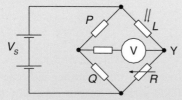

fig 5.11.4

At a certain light intensity a zero reading is obtained on the meter, when $P = 1$ kΩ, $Q = 100$ kΩ and $R = 220$ kΩ. What is the resistance of the LDR at this light intensity?

b) A pupil who is a keen photographer wishes to design a circuit which will warn her when the light conditions are such that she should be using the flash on her camera. She sets up the circuit shown in figure 5.11.5.

Answers and comments

a) When the bridge is balanced we can say that;

$$\frac{P}{Q} = \frac{L}{R}$$

$$\frac{1k}{100k} = \frac{L}{220k}$$

$$L = 2.2\ kΩ$$

b) (i) This amplifier is operating in the differential mode.

(ii) When the light intensity decreases the resistance of the LDR increases. Consequently the voltage at point Y reduces. In the expression for the differential amplifier $(V_2 - V_1)$ becomes positive. The difference between V_2 and V_1 is then amplified by 100. The positive output voltage switches on the transistor and the LED lights.

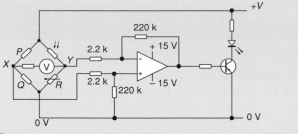

fig 5.11.5

(i) In what mode is the operational amplifier working?
(ii) Explain what happens when the light intensity on the LDR starts to decrease and how the circuit would indicate that a flash is required.
(iii) Under certain conditions when the LED is on, the output voltage of the operational amplifier is 1.2 V. What is the potential difference between the points X and Y in the diagram under these conditions?

SQA 1991

Questions

1 A student investigating difference amplifiers sets up the circuit shown in figure 5.11.6. Ammeter A_2 shows a reading of 0.15 mA.

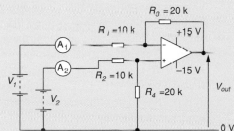

fig 5.11.6

a) Calculate the voltage across the resistor R_2.
b) Explain why the current in resistor R_4 is 0.15 mA.
c) Calculate the voltage across resistor R_4.
d) What is the voltage V_2? Justify your answer.
e) When the voltage V_1 is 2.5 V, calculate the value of the output voltage of the difference amplifier.

2 Figure 5.11.7 shows part of a circuit as shown in an electronics magazine. The amplifier has a gain of 100. A student sets an input voltage of 0.06 V and measures an output voltage of 6 V.

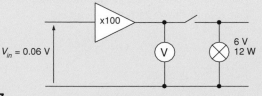

fig 5.11.7

When the student closes the switch she notices that the lamp, which will light at normal brightness when connected to a 6 V battery, lights only very dimly.
a) Explain this observation.
b) Describe how she could add a transistor to this circuit to light the lamp at normal brightness when the input to the amplifier is 0.06 V.

(iii) In this case we want to calculate $(V_2 - V_1)$;

$$V_{out} = \frac{R_f}{R_{in}}\ (V_2 - V_1)$$

$$1.2 = \frac{200k}{2.2k}\ (V_2 - V_1)$$

$$(V_2 - V_1) = 0.012\ V$$

The p.d. between points X and Y is 12 mV.

3 A student sets up the Wheatstone bridge circuit shown in figure 5.11.8.

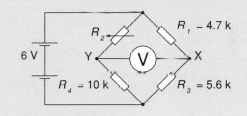

fig 5.11.8

a) Calculate the voltage across the 5.6 kΩ resistor. The variable resistor is adjusted until the bridge is balanced.
b) (i) What is the voltage across the 10 kΩ resistor when the bridge is balanced? Explain your answer.
(ii) What is the value of the variable resistor when the bridge is balanced.
The bridge is now connected to the differential amplifier shown in figure 5.11.9.

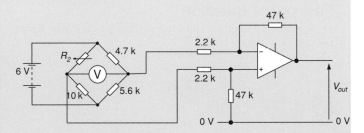

fig 5.11.9

c) When the voltmeter in the Wheatstone bridge shows a reading of 0 V what is the output voltage from the difference amplifier? Explain your answer.

Past paper question

a) In order to compare the brightness of a number of low voltage lamps, a solar cell is used to detect the light from the lamps. An op-amp, working in the inverting mode, is used to amplify the solar cell voltage.

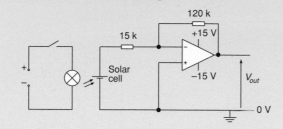

fig 5.12.1

The apparatus is set up near to a window and, with the lamp switched off, there is an output voltage V_{out} of -1.75 V

   (i) Explain why the output voltage V_{out} of the op-amp is not zero.

   (ii) Calculate the solar cell voltage.

b) With the solar cell in the same position, the circuit is now altered so that the op-amp is working in the differential mode as shown in figure 5.12.2.

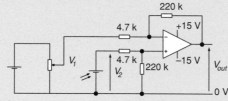

fig 5.12.2

   (i) With the lamp still unlit, the potentiometer setting is adjusted until the output voltage is zero. Explain how this circuit enables the output voltage to be set to zero volts.

   (ii) With V_1 unchanged, the lamp is switched on and the output voltage V_{out} is now 1.50 V. Calculate the voltage which the solar cell now produces.

SQA 1994

Answers and comments

a) (i) Background light from the surroundings falling on the solar cell creates an input voltage for the inverting amplifier.

   (ii) The equation governing the behaviour of the inverting amplifier is;

$$V_{out} = -\frac{R_f}{R_{in}} \times V_{in}$$

$$-1.75 = -\frac{120k}{15k} \times V_{in}$$

$$V_{in} = \frac{1.75 \times 15}{120}$$

$$V_{in} = 0.219\ V$$

b) (i) The potentiometer is used to alter the value of voltage V_1. When the value of V_1 equals the voltage produced by the solar cell the output of the difference amplifier is zero. The difference between the input voltages is now zero so the output will be zero regardless of the gain of the amplifier. The difference amplifier gives an output of 0 V when V_1 is 0.219 V.

   (ii) Stating that the value of V_1 is unchanged tells us that V_1 is 0.219 V. The output voltage from the difference amplifier is given by the equation;

$$V_{out} = \frac{R_f}{R_{in}}\ (V_2 - V_1)$$

$$1.5 = \frac{220k}{4.7k}\ (V_2 - 0.219)$$

$$V_2 = 0.251\ V$$

The solar cell is producing a voltage of 0.251 V.

Past paper question

a) A pupil connects an op-amp circuit as shown in figure 5.12.3.

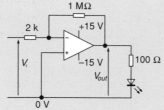

fig 5.12.3

   (i) In which mode is this op-amp working?

   (ii) Calculate the gain of the op-amp in this mode.

   (iii) The input voltage V_1 varies with time as shown in figure 5.12.4 (see over). Draw a graph showing how the output voltage V_{out} varies with time.

Answers and comments

a) (i) The op-amp is working in the inverting mode.

   (ii) The gain of the amplifier is given by;

$$G = \frac{R_f}{R_{in}} \qquad G = \frac{1000\ k}{2\ k}$$

$$G = 500$$

   (iii) The output voltage varies as shown in figure 5.12.5.

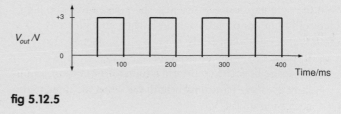

fig 5.12.5

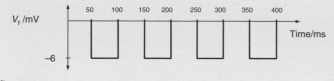

V_1 /mV

fig 5.12.4

(iv) Describe fully the behaviour of the LED in the above circuit when supplied by this output voltage.

SQA 1993

(iv) The output of the difference amplifier will alternate between 3 V for 50 ms and 0 V for 50 ms. This will make the LED flash 10 times per second.

Questions

1 A student studying the behaviour of inverting amplifiers sets up the circuit shown in figure 5.12.6.

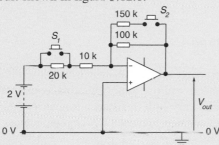

fig 5.12.6

The input voltage is 2 V.

a) State the gain expression for an inverting amplifier in terms of the input and output voltages V_{in} and V_{out} and the input and feedback resistances R_{in} and R_f.

b) Copy and complete the following table to show the output voltages when switches S_1 and S_2 are set as described in the table.

S_1	S_2	Output voltage V_{out}
open	open	
open	closed	
closed	open	
closed	closed	

Table 5.12.1

c) For which combination of the switches is the gain of the inverting amplifier greatest? Explain your answer.

2 A student sets up the circuit shown in figure 5.12.7.

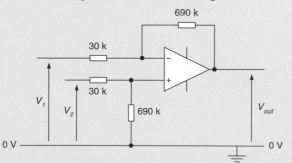

fig 5.12.7

a) In what mode is the amplifier operating?

b) State the differential mode gain expression linking the output voltage V_{out} with the input voltages V_1 and V_2 and the resistances.

The amplifier's input voltages, V_1 and V_2 are provided by function generators and are as shown in figure 5.12.8.

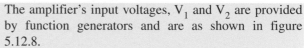

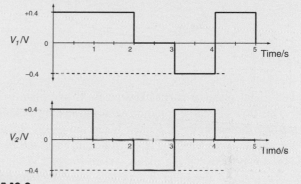

fig 5.12.8

c) Draw a graph to show how the output voltage V_{out} varies during the 5 seconds when the input voltages are as shown in the graphs for V_1 and V_2.

3 The circuit of figure 5.12.9 shows an op-amp connected as an inverting amplifier.

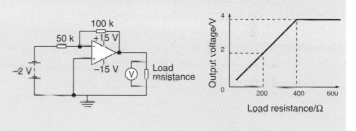

fig 5.12.9   fig 5.12.10

Different values of load resistor, R, are connected to the op-amp and the output voltages are measured. The graph of figure 5.12.10 shows how the output voltage varies with resistance.

a) Explain why the output voltage of the op-amp is less than 4 V when the load resistor has a low value.

b) Calculate the output current from the op-amp when the load resistor has a value less than 400 Ω.

c) Copy figure 5.12.9 and add the extra circuitry to show how a relay activated by a current of 5 mA could switch on a 100 W lamp, powered by a 230 V a.c. power supply, when the input to the inverting amplifier is −2 V.

6.1 Storing charge

Introduction

Earlier in this book (section 4.1) the idea of static and current electricity was introduced with an experiment using the apparatus shown in figure 6.1.1.

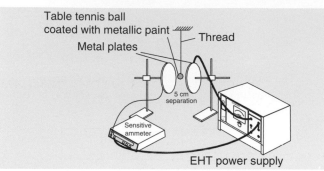

Figure 6.1.1

As the ball shuttles backwards and forwards between the charged metal plates it carries charge from one plate to the other. This flow of charge could be demonstrated by looking at the reading shown on the ammeter. When the ball is shuttling quickly the ammeter reading is greater than when the ball is moving slowly between the plates. When the ball is moving from one plate to the other it transfers some of the charge stored on the plates. This section will look in more detail at the ability of a system to charge up, store charge and discharge.

Storing charge

Two metal plates separated by air as shown in figure 6.1.2 form the simplest type of **capacitor**. When the switch is closed charges are stored on parallel metal plates on either side of an air gap. The plate connected to the negative terminal of the battery receives electrons to become negatively charged while the plate connected to the positive terminal of the battery loses electrons to become positively charged.
Increasing the charge on the plates is called charging. There is *no movement* of charge across the air gap between the plates.

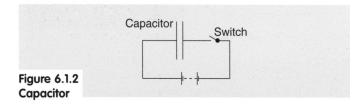

Figure 6.1.2
Capacitor

As the charge on the plates increases a *potential difference* is created across the gap. **Charging stops when this potential difference equals the e.m.f. of the supply**. In figure 6.1.3

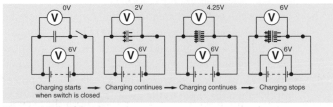

Figure 6.1.3 Charging

charging stops when the reading on the voltmeter connected across the plates equals the reading on the voltmeter connected across the power supply.
When the p.d. across the plates equals the e.m.f of the power supply the plates are said to be **fully charged**. When the switch in the circuit of figure 6.1.4 is closed you might, (wrongly) expect to be able to watch the reading shown on the voltmeter increase slowly as the capacitor charges.

Figure 6.1.4

In practice when the switch is closed the reading on the voltmeter will rise almost instantaneously to 6 V. In this circuit there is very little resistance to restrict the flow of charge, so the plates will become fully charged almost immediately.

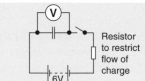

Figure 6.1.5

In figure 6.1.5 a resistor is added to restrict the rate at which the plates can charge. **The larger the value of the resistor, the longer it takes for the capacitor to become fully charged**.

How much charge can the plates hold?

EXPERIMENT Charge stops flowing onto the plates of the capacitor when the voltage across the plates equals the e.m.f of the supply. The capacitor is then fully charged. With the apparatus shown in figure 6.1.6 we can measure how much charge is actually stored by the capacitor when it is fully charged.

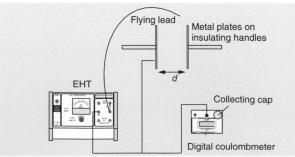

Figure 6.1.6

When the flying lead from the power supply is touched against the metal plate charges are transferred and the system is fully charged almost instantaneously. The p.d. across the capacitor plates will equal the e.m.f of the power supply. If the flying lead is then removed, the capacitor retains its charge until the digital coulombmeter is touched against this plate.

The charge stored by the system is transferred to the coulombmeter. The digital readout shows the quantity of charge stored when the capacitor is charged to a specific voltage. Repeating the experiment for the same charging voltage and with different charging voltages will produce the type of results shown in table 6.1.1.

Charging voltage (V)/Volts	Charge stored (Q)/nC					Mean charge stored (Q)/nC	Uncertainty
500	8	7	8	6	9	7.6	±0.6
1000	15	14	16	17	16	15.6	±0.6
1500	23	22	25	24	24	23.6	±0.6
2000	33	32	33	34	35	33.4	±0.6
2500	40	41	42	43	42	42	±0.6
3000	49	52	51	51	52	51	±0.6

Table 6.1.1

The mean charge stored for each charging voltage is as shown and the approximate random uncertainty in each of the measurements can be calculated using;

$$Uncertainty = \frac{Maximum\ reading - Minimum\ reading}{Number\ of\ readings}$$

Even allowing for the small uncertainty in each of the mean values, the results show a definite trend. **The quantity of charge stored increases as the charging voltage increases**. Figure 6.1.7 shows a graph of these results.

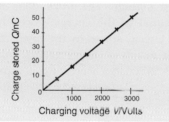

Figure 6.1.7

Since the graph of charge stored, Q, against charging voltage, V, is a straight line passing through the origin we can say that;

$$Q \propto V$$

Therefore;

$$\frac{Q}{V} = constant$$

The constant in this relationship is called the **capacitance**, C so we can say that;

$$Q = CV$$

Capacitance

The capacitance of a system is a measurement of how easily it can store charge. A more formal definition of the capacitance of a system can be derived from the relationship;

$$C = \frac{Q}{V}$$

The capacitance of a system is numerically the same as the charge stored when the p.d. across the capacitor is 1 volt.
So the capacitance of the capacitor can be defined as **the charge stored per volt**.

Capacitance is measured in units called Farads (F) but one Farad is often too large a unit for practical purposes. In the experiment outlined earlier the plates stored 42 nanocoulombs of charge when the p.d. was 2500 V. The capacitance of the plates can therefore be calculated using;

$$C = \frac{Q}{V}$$

$$C = \frac{42 \times 10^{-9}}{2500}$$

$$C = 16.8 \times 10^{-12}\ F$$

The capacitance of the plates is 16.8×10^{-12} F. This would more usually be quoted as 17 picofarads (pF).

A little further…

Practical capacitors

The capacitance of the parallel plate system shown in figure 6.1.6 was indeed very small, so large voltages had to be used for it to store measurable quantities of charge. In most applications such large voltage supplies are not used, so methods need to be found for storing appreciable quantities of charge at moderate voltages.

You can use the apparatus of figure 6.1.6 to show that the charge stored for any particular voltage can be increased by;
1 moving the plates closer together
2 using plates that have a larger area
3 separating the plates with an insulating material other than air.

By proper selection and arrangement of the materials between the conductors as shown in figure 6.1.8, practical capacitors can be manufactured with a wide range of capacitance values.

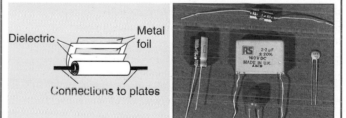

Figure 6.1.8 Parallel plate capacitor   **Figure 6.1.9 Capacitors**

Modern electronic circuits use capacitors to produce time delays. If an adjustable time delay is required a variable capacitor like that shown in figure 6.1.10 can be used.

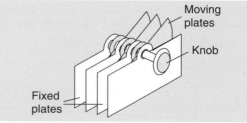

Figure 6.1.10

As the knob is turned the area of overlap between the plates changes, so the capacitance changes.

6.2 Capacitors storing energy

Introduction

When the switch in the circuit of figure 6.2.1 is moved from position B to position A the capacitor charges.

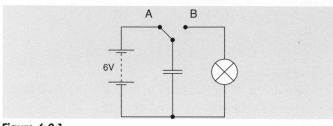

Figure 6.2.1

When the switch is changed back again to position B the lamp lights up momentarily because electrical energy is transferred from the capacitor to the lamp.

The capacitor can only transfer energy to the lamp after it has been charged up by the battery. We can therefore conclude that a **charged capacitor stores electrical energy** and that **the energy to charge the capacitor comes from the battery**.

Charging capacitors

In an uncharged capacitor like that shown in figure 6.2.2 each of the plates are electrically neutral. There are equal numbers of protons and electrons in the metal of which the plates are made.

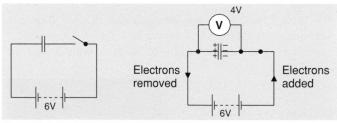

Figure 6.2.2

When the switch is closed electrons move to one plate from the negative terminal of the battery and electrons move from the other plate to the positive of the battery.

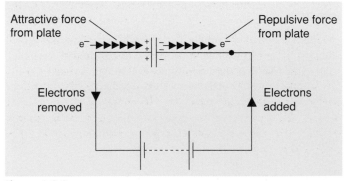

Figure 6.2.3

At one particular time the situation will be as shown in the diagram of figure 6.2.3 where the capacitor is partially charged. Any further electrons which the negative terminal of the battery wants to add to the negative plate will experience a repulsive

force from the electrons already on the negative plate. Similarly, to make the positive plate more positively charged, the positive terminal of the battery will have to exert an attractive force on the electrons remaining in the positive plate. The attractive force from the battery will have to be greater than the attractive force from the positive charges on the plate. To move electrons against these attractive and repulsive forces the *battery must use up some of its energy*. The battery is *doing work* by moving the electrons and energy is being stored in the electric field between the capacitor plates.

When the capacitor is fully charged no more electrons are transferred, so the battery is not adding energy to the capacitor. The quantity of charge held by a fully charged capacitor depends on the voltage across its plates when fully charged. Increasing this voltage will force the capacitor to store more charge and so more work must be done by the battery. Consequently more energy is stored by a capacitor when the p.d. across it increases.

Quantitative energy

As work is done by a battery when increasing the charge on an initially uncharged capacitor, the voltage across the plates of the capacitor increases.

The amount of work done when charging a capacitor to a specific voltage is given by the area under the graph of charge against voltage. The work done when charging a capacitor is equal to the energy stored by the capacitor. Therefore the energy stored by the capacitor also equals the area under the graph of charge against voltage.

The quantities of charge stored by a capacitor charged to different voltages are summarised by the graph of figure 6.2.4.

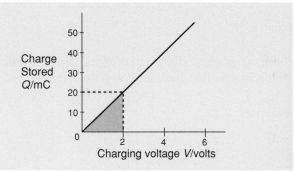

Figure 6.2.4

To calculate the quantity of energy stored by this capacitor when there is a potential difference of 2 V across its plates, we must calculate the area of the shaded section of figure 6.2.4;

$$Area\ of\ triangle = \frac{1}{2}(Base \times Height)$$

$$Energy\ stored = \frac{1}{2} \times 2 \times 20 \times 10^{-3}$$

$$Energy\ stored = 20 \times 10^{-3}\ J$$

By a similar calculation we can show that this capacitor stores 125 mJ when it has a p.d. of 5 V across its plates.

In general, when a capacitor is charged to a p.d., V, and in doing so stores charge, Q, we can calculate the energy stored;

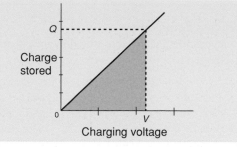

Figure 6.2.5

$$Energy\ stored = \frac{1}{2}(Base \times Height)$$

$$Energy\ stored = \frac{1}{2} \times Q \times V$$

$$Energy\ stored = \frac{1}{2}QV$$

Related expressions

In the previous section you saw that the quantity of charge, Q, stored by a capacitor of capacitance, C, when charged to a voltage, V, is given by the equation;

$$Q = CV$$

Combining this with the energy expression;

$$E = \frac{1}{2}QV$$

Gives;

$$E = \frac{1}{2}(CV) \times V$$

$$E = \frac{1}{2}CV^2$$

Similarly we can combine the charge and energy equations in another way;

$$Q = CV\ so\ V = \frac{Q}{C}$$

and;

$$E = \frac{1}{2}QV$$

$$E = \frac{1}{2}Q \times \frac{Q}{C}$$

Therefore;

$$E = \frac{1}{2}\frac{Q^2}{C}$$

So;

$$E = \frac{1}{2}QV \quad E = \frac{1}{2}CV^2 \quad E = \frac{1}{2}\frac{Q^2}{C}$$

These are all equivalent expressions. The most appropriate one to use depends on the information given in any particular example.

Q How much energy is stored by a 10,000 µF capacitor, when fully charged from a 10 V power supply?

A In this question we are given values for the capacitance, C, and the charging voltage, V. The simplest way to find the energy stored is using the equation;

$$E = \frac{1}{2}CV^2$$

$$E = \frac{1}{2} \times 10,000 \times 10^{-6} \times (10)^2$$

$$E = 0.5\ J$$

In this example the capacitor stores 0.5 Joules of energy.

Measuring the energy stored in a charged capacitor

EXPERIMENT With the apparatus shown in figure 6.2.6 the capacitor is charged when the switch is in position A.

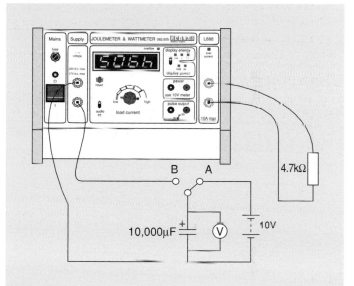

Figure 6.2.6

Moving the switch to position B discharges the capacitor and causes a current in the 4.7 kΩ resistor. The quantity of energy transferred from the capacitor to the resistor is measured by the joulemeter. The voltmeter will show when the capacitor is fully discharged and at this time the joulemeter will record the total energy which was previously stored in the capacitor.

The variation of energy stored with charging voltage can be investigated by setting the p.d. of the supply to different values before charging the capacitor. Discharging the capacitor through the joulemeter and resistor will measure the energy stored by the capacitor when the p.d. across its plates is set to a particular value.

$$Since;\ E = \frac{1}{2}CV^2$$

$$E \propto V^2$$

A graph of energy stored versus charging voltage squared will result in a straight line having a gradient of ½C.

Worked example 1

The electronic flash gun in a photographer's studio produces a very bright flash of light for a short period of time. The electronic flash circuitry uses capacitors to store energy which delivers power to the flash tube.

The 47 µF capacitor in the flash circuitry is charged to 333 V.

a) Calculate the charge stored by the capacitor.
b) Calculate the energy stored by the fully charged capacitor.
c) In one particular setting the flash lasts 2.6 ms. Calculate the average power dissipated by the flash tube.

Answers and comments

a) We can calculate the charge stored by the capacitor when it is charged to a particular voltage using the equation;

$$Q = CV$$
$$Q = 47 \times 10^{-6} \times 333$$
$$Q = 0.0157\ C$$

b) We can calculate the energy stored by the capacitor when it is charged to a particular voltage using the equation;

$$E = \frac{1}{2}CV^2$$

$$E = \frac{1}{2} \times 47 \times 10^{-6} \times (333)^2$$

$$E = 2.61\ J$$

c) To calculate the average power knowing the time taken to dissipate a certain quantity of energy we can use;

$$Average\ power = \frac{Energy\ stored}{Time} = \frac{2.61}{2.6 \times 10^{-3}}$$

$$Average\ power = 1004\ W$$

Worked example 2

The capacitor shown in figure 6.3.1 is connected to a power supply which gives a constant current of 50 µA.

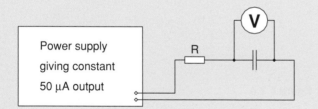

fig 6.3.1

a) Calculate the quantity of charge transferred to the capacitor from the power supply in the first 20 seconds after the power supply is switched on.
b) There is a p.d. of 6 V across the capacitor 150 seconds after the power supply is switched on. Calculate the value of the capacitor used in the circuit of figure 6.3.2.
c) Calculate the energy stored by the capacitor when it is charged to a p.d. of 9 V.

Answers and comments

a) Since the current in the circuit is constant we can use the equation $Q = It$ to calculate the quantity of charge transferred to the capacitor.

$$Q = It$$
$$Q = 50 \times 10^{-6} \times 20$$
$$Q = 1.0\ mC$$

b) The quantity of charge stored by the capacitor after 150 seconds is;

$$Q = It$$
$$Q = 50 \times 10^{-6} \times 150$$
$$Q = 7.5\ mC$$

When the capacitor stores 7.5 mC of charge the p.d. across its plates is 6 V so its capacitance can be found using;

$$C = \frac{Q}{V} = \frac{7.5 \times 10^{-3}}{6}$$

$$C = 1.25 \times 10^{-3}\ F$$

c) We can calculate the energy stored by the capacitor when it is charged to a particular voltage using the equation;

$$E = \frac{1}{2}CV^2$$

$$E = \frac{1}{2} \times 1.25 \times 10^{-3} \times (9)^2$$

$$E = 50.6\ mJ$$

Questions

1

a) A capacitor stores 0.04 C of charge when the p.d. across its plates is 8 V. Calculate the capacitance of the capacitor.

b) A 1000 µF capacitor is connected to a 6 V battery. What is the p.d. across the fully charged capacitor? Explain your answer.

c) An initially uncharged 470 µF capacitor is charged using a constant current of 8.15 mA. Calculate the p.d. across the capacitor after 75 seconds.

2 A fully charged photographic flash gun discharges 0.4 C of charge in a flash lasting 0.2 seconds.

a) Calculate the average current during the discharge.

b) The p.d. across the capacitor in the fully recharged flash gun is 100 V. Calculate the capacitance of the capacitor in the flash gun.

c) Calculate the energy stored by the fully charged capacitor.

d) Calculate the average power dissipated in each flash.

e) The photographer purchases a new type of flash lamp that connects to the original capacitor in his flash gun. This new lamp provides the same brightness of flash but requires only 10 J of energy. Calculate the voltage to which the capacitor in the flash gun must now be charged to provide the same level of flash.

3 During an investigation of capacitors a student charges a 47 µF capacitor so that the p.d. across its plates is 8 V.

a) Calculate the charge stored by the capacitor.

b) Calculate the energy stored when the p.d. across the plates of the capacitor is 8 V.

c) In the second part of the experiment he partially discharges the capacitor so that it stores only 164.5 µC of charge. Show that the p.d. across the plates of the capacitor is now 3.5 V.

d) What fraction of the initial energy stored by the capacitor remains after the second part of the student's experiment?

4 A capacitor is a device for storing charge.

a) Draw a diagram to show the essential features of the construction of a capacitor.

A pupil investigating the charging of a capacitor sets up the circuit shown in figure 6.3.2.

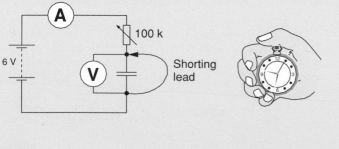

fig 6.3.2

When the shorting lead is connected across the capacitor the reading on the voltmeter is 0 V. The variable resistor is now adjusted until the reading on the ammeter is 55 µA.

b) Calculate the resistance of the variable resistor when the current in it is 55 µA.

The shorting lead is removed and the clock is started. During the experiment the variable resistor is adjusted to ensure that the current remains at 55 µA as the capacitor charges. The student records the time taken for the capacitor to charge to a number of different voltages.

Time/s	0	16	30	47	61
Voltage/V	0	1.5	3.0	4.5	6.0

Table 6.3.1

c) Calculate the charges stored by the capacitor for each of the voltages across the capacitor shown in table 6.3.1.

d) Plot a graph of charge stored against charging voltage.

(i) What does this graph tell you about the relationship between the quantity of charge stored by the capacitor, Q, and the p.d. across the plates of the capacitor, V. Give reasons for your answer.

(ii) From your graph show that the capacitor has a capacitance of approximately 570 µF.

(ii) Add a line to your graph to represent the results that the student would obtain when the experiment is repeated with a 220 µF capacitor.

5 Capacitor X in figure 6.3.3 has a capacitance of 100 µF and is charged to a voltage of 10 V. Capacitor Y has a capacitance of 50 µF and stores 0.4 mC of charge.

fig 6.3.3

a) Calculate the p.d. across capacitor Y.

b) Calculate the energy stored by X and Y.

c) What should be the p.d. across X in order to store the same energy as you have just calculated for Y?

6.4 Charging capacitors

Introduction

Closing the switch in the circuit of figure 6.4.1 causes the capacitor to charge up.

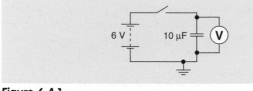

Figure 6.4.1

The reading on the voltmeter rises almost immediately to 6 V showing that when fully charged the p.d. across the plates of the capacitor equals the e.m.f. of the supply. The capacitor charges very quickly because there is no resistance in the circuit to limit the flow of charge. The final charge stored by the capacitor can be calculated using;

$$Q = CV$$
$$Q = 10 \times 10^{-6} \times 6$$
$$Q = 60 \times 10^{-6} \ C$$

Voltage and current variation during charging

In the circuit of figure 6.4.2 a resistor has been added to the circuit so that the build up of charge can be monitored.

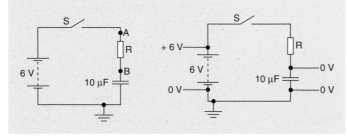

Figure 6.4.2

Initially just before the switch is closed we can assume that the plate of the capacitor connected to the negative side of the battery is at 0 V. Since the capacitor is initially uncharged there is no p.d. across its plates so the other plate is also at 0 V. The instant the switch is closed, side B of the resistor connected to the uncharged capacitor is at 0 V and side A of the resistor is at 6 V. This potential difference across the resistor causes a current which can be calculated using Ohm's Law;

$$I = \frac{V}{R} = \frac{6}{100}$$
$$I = 0.06 \ mA$$

The current in the resistor causes charges to flow onto the capacitor, so the p.d. across the capacitor increases.

If at one stage during the charging process there is a potential difference of 2 V across the capacitor we can calculate the quantity of charge stored using;

$$Q = CV$$
$$Q = 10 \times 10^{-6} \times 2$$
$$Q = 20 \times 10^{-6} \ C$$

Because the bottom plate of the capacitor is connected to the negative terminal of the battery, we can still assume that it is at 0 V. Therefore when the p.d. across the capacitor is 2 V, the top plate must be at a voltage of +2 V.

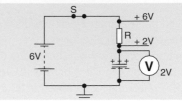

Figure 6.4.3

Now the lower end of the resistor is at 2 V and the top end is at 6 V so the p.d. across the resistor is only 4 V. This causes a current, I, given by;

$$I = \frac{V}{R} = \frac{4}{100}$$
$$I = 0.04 \ mA$$

The new, lower charging current means that charge is flowing onto the capacitor at a *slower rate*.

As the capacitor charges, the difference between the voltage across the capacitor, V_C, and the e.m.f. of the battery reduces. Consequently the charging current decreases and so the p.d. across the resistor is smaller.

Charging can be described by the diagram of figure 6.4.4.

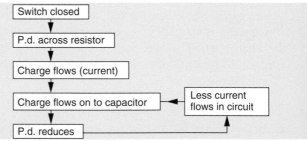

Figure 6.4.4

The gradual reduction in the rate at which the capacitor charges can be investigated using the apparatus of figure 6.4.5.

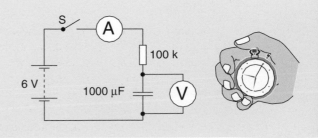

Figure 6.4.5

The reading shown on the voltmeter can be recorded at 20 second intervals after the switch is closed. Table 6.4.1 shows results typical of this experiment and the voltage-time graph is shown in figure 6.4.6.

Time/s	0	20	40	60	80	100	120	140	160	180
Voltage/V	0	1.1	2	2.6	3.2	3.7	4.1	4.4	4.7	4.95

Table 6.4.1

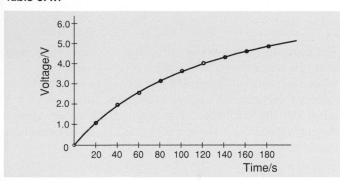

Figure 6.4.6

Since the e.m.f. of the battery, E, remains constant during the experiment we can calculate the voltage across the resistor, V_R, each time the voltage across the capacitor, V_C, is measured using;

> *p.d. across resistor = Supply voltage – p.d. across capacitor*
> $$V_R = E - V_C$$

Knowing the voltage across the 100 kΩ resistor at a particular instant allows us to calculate the current in the circuit at that instant. Table 6.4.2 shows the calculated currents for the charging of the capacitor in figure 6.4.5. Figure 6.4.7 shows the corresponding graph.

Time/s	0	20	40	60	80	100	120	140	160	180
Current/μA	59.5	48	39	33	27	22	18	15	12	10

Table 6.4.2

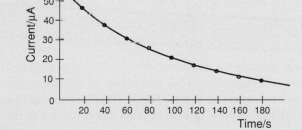

Figure 6.4.7

Changing the circuit resistance or capacitance

The rate at which the capacitor charges is set by the size of the current in the circuit. Using a *larger resistance* in the charging circuit causes *lower charging currents* so the capacitor takes *longer* to charge to a certain voltage.

Increasing the capacitance, C, means that more charge will be stored on the capacitor when it is charged to a p.d. of V volts ($Q=CV$). This means that charging with a particular resistor will take longer than for a similar circuit with a lower capacitance.

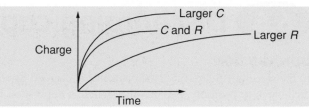

Figure 6.4.8

 In the circuit of figure 6.4.9 the 220 μF capacitor is initially uncharged.
What is the maximum current in the resistor?

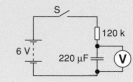

Figure 6.4.9

 The maximum current in the resistor will occur the moment the switch is closed because the potential difference across the resistor will be at its highest value (6 V).

$$I = \frac{V}{R} = \frac{6}{120k}$$

$$I = 0.05 \ mA$$

 If after about 25 seconds the current is 0.02 mA, how much charge is now stored on the capacitor?

 If the current in the resistor is 0.02 mA we can calculate the p.d. across the resistor using;

$$V = IR$$
$$V = 0.02 \times 10^{-3} \times 120 \times 10^3$$
$$V = 2.4 \ V$$

If the voltage across the resistor is 2.4 V, the p.d. across the capacitor can be found from;

> *p.d. across capacitor = Supply voltage – p.d. across resistor*
> $$V_C = 6 - 2.4$$
> $$V_C = 3.6 \ V$$

Knowing the p.d. across the capacitor the charge stored can be found from;

$$Q = CV$$
$$Q = 220 \times 10^{-6} \times 3.6$$
$$Q = 792 \times 10^{-6} \ C$$

When the current in the circuit of figure 6.4.9 is 0.02 mA the charge stored on the capacitor is 792 μC.

Warning

The information in this question gives a value for the current, I, at a certain time, t, and asks you to calculate the charge stored, Q. Being given values for I and t and asked to calculate Q there is a tendency to want to use $Q = It$. However, $Q = It$ can only be used in situations where there is a constant current, I, for a time, t. Clearly during capacitor charging and discharging the *currents change* so $Q = It$ cannot be used.

6.5 Discharging capacitors

Introduction

In the circuit of figure 6.5.1 the capacitor is charged by setting the switch to position A. When the switch is at position B the capacitor is *isolated* and the quantity of charge stored remains constant. When the switch is moved to position C the capacitor will discharge by passing charge through the lamp.

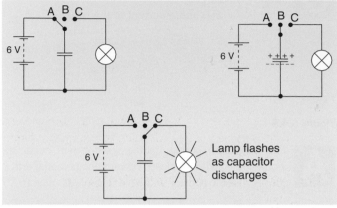

Figure 6.5.1

The lamp in figure 6.5.1 lights briefly when the switch is moved to position C because the energy stored in the capacitor is transformed into light and heat by the filament in the lamp. When a capacitor with a larger value is used in the circuit of figure 6.5.1, the flash of the bulb will last longer. More energy is stored by the larger value capacitor. With the circuit of figure 6.5.2 the 10 μF capacitor is charged by moving the switch to position A.

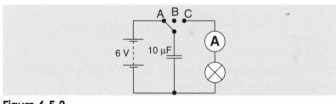

Figure 6.5.2

When the switch is moved to position C the discharging capacitor causes the lamp to flash briefly and the ammeter indicates how the current in the circuit changes during the discharge of the capacitor. **Initially, just after the switch is changed over from A to C the current is large but reduces as the capacitor discharges.**

Voltage and current changes during discharge

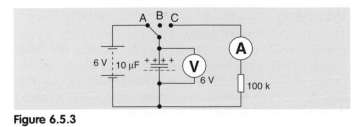

Figure 6.5.3

When the 10 μF capacitor shown in figure 6.5.3 is fully charged using a 6 V battery it stores a quantity of charge which can be calculated from;

$$Q = CV$$
$$Q = 10 \times 10^{-6} \times 6$$
$$Q = 60 \times 10^{-6} \ C$$

The capacitor stores 60 μC of charge because the voltage across its plates rises to 6 V when it is fully charged.

When the switch in figure 6.5.3 is moved to position C, the capacitor immediately applies a potential difference of 6 V across the 100 kΩ resistor. This potential difference causes a current in the resistor. The size of the current can be found from Ohm's Law;

$$I = \frac{V}{R} = \frac{6}{100k}$$

$$I = 0.06 \ mA$$

The initial discharge current in the circuit of figure 6.5.3 is 0.06 mA or 60 μA. This means that, *for an instant,* charge flows off the plates of the capacitor at a rate of 60 μC per second. As a consequence of this current;

1 the charge remaining on the capacitor reduces
2 the p.d. across the plates of the capacitor decreases
3 the current in the 100 kΩ resistor decreases.

If after a few seconds the current has reduced to 40 μA, we can calculate the p.d. across the capacitor by first calculating the p.d. across the resistor;

$$V = IR = 40 \times 10^{-6} \times 100 \times 10^{3}$$
$$V = 4 \ V$$

At the instant when the reading shown on the ammeter of figure 6.5.3 is 40 μA there is a p.d. of 4 V across the capacitor and the charge remaining on the capacitor can be calculated using;

$$Q = CV = 10 \times 10^{-6} \times 4$$
$$Q = 40 \times 10^{-6} \ C$$

Charge leaving the capacitor during discharge reduces the potential difference across the plates of the capacitor so there is less current in the circuit. Smaller currents mean that the rate of flow of charge from the capacitor is less so as discharge progresses the rate at which the capacitor discharges slows down. The argument for the current and voltage changes during discharge of a capacitor can be summarised by steps 1 to 4 in figure 6.5.4.

By adding step 5 as shown in figure 6.5.5 the linked nature of the voltage and current changes can be summarised. Smaller currents cause the capacitor to discharge slower and produce lower voltages across the capacitor which in turn reduces the size of the currents even further. Although discharge, once started, continues, the rate of discharge reduces.

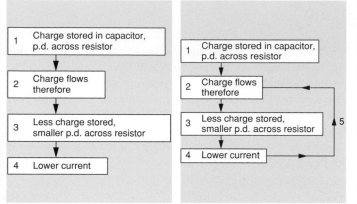

Figure 6.5.4   Figure 6.5.5

EXPERIMENT The gradual slowing down in the rate of discharge can be seen experimentally using the apparatus shown in figure 6.5.6.

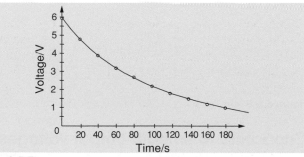

Figure 6.5.6

The capacitor is charged by moving the switch to position A and the reading on the voltmeter is noted. Once the switch is moved to position C further voltmeter readings can be taken at 20 second intervals for about 3 minutes. Table 6.5.1 shows typical results for this experiment.

Time/s	0	20	40	60	80	100	120	140	160	180
Voltage/V	5.95	4.8	3.9	3.3	2.7	2.2	1.8	1.5	1.2	1.0

Table 6.5.1

Figure 6.5.7 shows the voltage-time graph for the discharge of this capacitor.

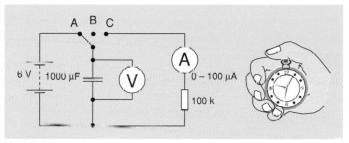

Figure 6.5.7

The experiment could be repeated and the value of the *current* in the circuit recorded at 20 second intervals for a period of 3 minutes. However, since we know the voltages across the 100 kΩ resistor at regular intervals we can calculate, using Ohm's Law, the current in the resistor at these times. The calculated values are shown in table 6.5.2.

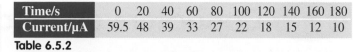

Time/s	0	20	40	60	80	100	120	140	160	180
Current/µA	59.5	48	39	33	27	22	18	15	12	10

Table 6.5.2

Figure 6.5.8 shows the current-time graph for the discharge of this capacitor.

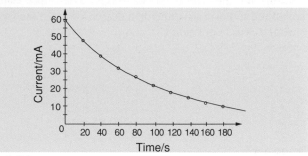

Figure 6.5.8

Because the voltage across the resistor and the current in the circuit are directly proportional, it is not surprising that figures 6.5.7 and 6.5.8 have the same shape.

Changing the circuit resistance or capacitance

EXPERIMENT In the circuit of figure 6.5.9 the resistances can be changed. This allows us to investigate how the rate of discharge depends on the resistance.

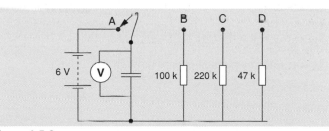

Figure 6.5.9

Since the same p.d. will cause a larger current in a circuit with lower resistance we should not be surprised at the shapes of the graphs of figure 6.5.10. Capacitors, charged to a particular voltage, will discharge faster through lower resistances.

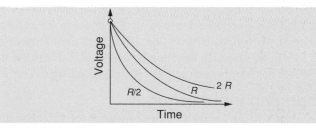

Figure 6.5.10

Capacitors, charged to a particular voltage, will take *longer* to discharge through *larger* resistances.

Using capacitors with a larger capacitance will also take longer to fully discharge. This is because for any particular voltage more charge will be stored so charge will flow for a longer time. By selecting appropriate values of capacitance, *C*, and resistance, *R*, **accurate *CR*-time delay** circuits can be constructed.

6.6 Capacitors and a.c.

A simple charge/discharge circuit

In the capacitor charge and discharge circuit of figure 6.6.1 the reading on the voltmeter increases when the switch is connected to position A and then decreases when the switch is changed over to position C.

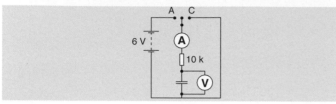

Figure 6.6.1

When the capacitor is uncharged and the switch is set to position A the capacitor charges. Initially, just when charging starts, the current in the resistor is at its maximum value of 0.6 mA. This decreases during charging as shown in figure 6.6.2.

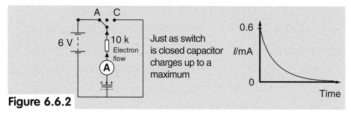

Figure 6.6.2

When the charging current has reduced to zero the capacitor is fully charged. Moving the switch to position C now discharges the capacitor by passing charge through the resistor.

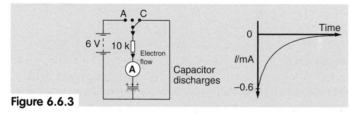

Figure 6.6.3

Once again the current value is largest at the instant the switch is closed and the initial size of this current is again 0.6 mA. However, when the capacitor is discharging the p.d. across the resistor is in the opposite direction, so the current in the ammeter is also in the opposite direction. The maximum current of 0.6 mA in the opposite direction is shown as –0.6 mA on the graph. As discharge continues the current gets closer to zero.

Changing input voltages

The input voltage for the circuit of figure 6.6.4 is a changing voltage which has alternate 6 V and 0 V sections each lasting for 1 second.

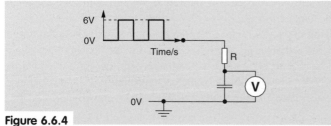

Figure 6.6.4

For the first second the output from the supply is 0 V so the capacitor is uncharged and the voltmeter will show a reading of 0 V. After 1 second the input voltage rises instantly to 6 V. This puts a p.d. across the circuit, and the capacitor starts to charge. As it charges, the reading shown on the voltmeter increases. If suitable values of capacitance and resistance are used the voltage across the capacitor can reach 6 V in 1 second as shown in figure 6.6.5.

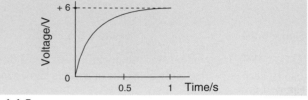

Figure 6.6.5

At the end of the first second, just as the voltage across the capacitor reaches +6 V, the voltage from the supply returns to 0 V.

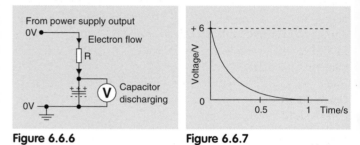

Figure 6.6.6   **Figure 6.6.7**

The charged capacitor now starts to discharge. The reading on the voltmeter will reduce as shown in figure 6.6.7.
The input and output voltage variations for this circuit can be summarised as shown in figure 6.6.8.

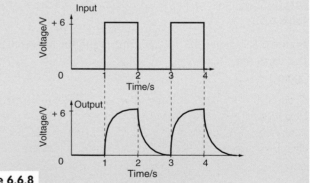

Figure 6.6.8

Capacitors to smooth output voltages

In the circuit of figure 6.6.4 the input voltage involved, suddenly changes between 6 V and 0 V. The output voltage for the circuit changed more gradually. The output voltage is smoother than the input.
The input to the circuit of figure 6.6.9 is an alternating voltage varying continuously between peaks of +330 V and –330 V. The four diodes form a circuit called a diode bridge. The output from this diode bridge varies between 0 V and +330 V reaching its peak value every 0.01 seconds.

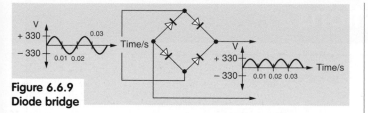

**Figure 6.6.9
Diode bridge**

If a resistor is connected across the output of this circuit the voltage across it rises and falls between 0 V and +330 V at regular intervals. By adding a capacitor across the output resistor as shown in figure 6.6.10 the output voltage can be smoothed.

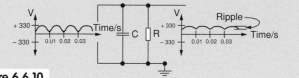

Figure 6.6.10

When the input voltage rises to 330 V the capacitor charges up and the p.d. across its plates becomes 330 V. When the input falls towards 0 V the capacitor discharges but only has time to partly discharge before the rising input voltage returns the p.d. across the capacitor to +330 V. **Consequently the voltage across the resistor only changes by a small amount known as the *ripple*.** The smaller the size of this ripple voltage, the smoother the output voltage.

The diode bridge circuit with smoothing resistor and capacitor can be used to convert an a.c. supply into one providing an almost steady d.c. voltage.

Capacitors, current and frequency

In the circuit of figure 6.6.10 the charged capacitor only discharges partially before the input voltage recharges it. In the circuit of figure 6.6.11 a continuously alternating voltage supply is connected across a capacitor.

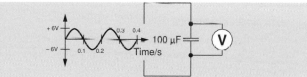

Figure 6.6.11

There is no resistor in this circuit so the voltage across the capacitor follows instantly the voltages set by the power supply. After 0.05 seconds the voltage across the capacitor is 6 V. The charge stored on the 100 μF capacitor can be found from;

$$Q = CV$$
$$Q = 100 \times 10^{-6} \times 6$$
$$Q = 600 \times 10^{-6} \text{ C}$$

After a further 0.05 s the voltage across the capacitor is 0 V so the capacitor has given up its entire 600×10^{-6} C. If an ammeter is included in the circuit it will register a current because of this movement of charge.

The current at any instant during the discharge will vary, but the *average* current during the 0.05 seconds taken to discharge can be found from;

$$Average\ current = \frac{Total\ charge}{Time}$$

Therefore;

$$Average\ current = \frac{600 \times 10^{-6}}{0.05}$$

$$Average\ current = 12\ mA$$

Figure 6.6.12 shows an alternating voltage with the same peak value as in figure 6.6.11 but the frequency has doubled.

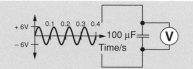

Figure 6.6.12

The fully charged capacitor will again store 600×10^{-6} C of charge but in this case it fully discharges in 0.025 seconds. The average current during the discharge can again be found using;

$$Average\ current = \frac{600 \times 10^{-6}}{0.025}$$

$$Average\ current = 24\ mA$$

Doubling the frequency has doubled the current in the circuit because the same quantity of charge has left the capacitor in half the time.

From this analysis we can say that the current, I, in the circuit of figure 6.6.13 is directly proportional to the frequency, f, at which the p.d. across the capacitor varies ($I \propto f$).

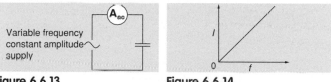

Figure 6.6.13   **Figure 6.6.14**

Setting up the apparatus of figure 6.6.13 and recording the currents over a range of frequencies produces results such as those shown in figure 6.6.14. For a fair test the amplitude of the output voltage must be kept the same for each frequency.

From the graph we can see that low frequencies produce little current. This is similar to a d.c. circuit with a high resistance. However when the input voltage varies at higher frequencies there are larger currents in the circuit. This is similar to a d.c. circuit with a low resistance.

In capacitor circuits we say that capacitors block low frequency or d.c. signals whilst allowing high frequencies to pass through. In the circuit of figure 6.6.15 the input has both a high frequency and a low frequency component. The high frequency part will be able to pass through the capacitor to earth and only the low frequency part 'appears' at the output. The capacitor has filtered out the higher of the two frequencies.

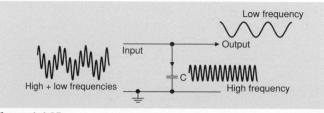

Figure 6.6.15

Past paper question

The circuit of figure 6.7.1 is set up to investigate the charging of a capacitor.

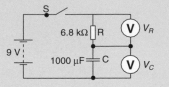

fig 6.7.1

At the start of the experiment the capacitor is uncharged.

a) The graph of figure 6.7.2 shows how the p.d., V_c, across the capacitor varies from the instant the switch S is closed.

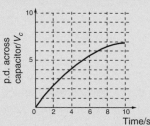

fig 6.7.2

Sketch a graph showing how the p.d., V_R, across the resistor varies during the first 10 seconds of charging.

b) When the capacitor is fully charged, it is removed from the circuit and connected across a 10 Ω resistor. What is the total energy dissipated in the resistor?

c) In another experiment, the fully charged capacitor is connected across a 20 Ω resistor instead of the 10 Ω resistor. How does the energy dissipated in this resistor compare with that calculated in part (b)? You must justify your answer.　　**SQA 1992**

Answers and comments

a) The total voltage across the circuit is a constant 9 V. Therefore we can say that;

$$V_R + V_c = 9$$
$$V_R = 9 - V_c$$

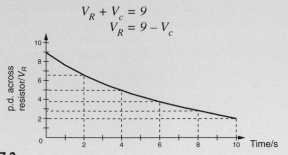

fig 6.7.3

Therefore the shape of the graph is as shown in figure 6.7.3

b) When the capacitor is connected to the resistor all of its energy is dissipated in the resistor. We can calculate the energy stored by the capacitor using;

$$E = \frac{1}{2}CV^2 = \frac{1}{2} \times 1000 \times 10^{-6} \times (9)^2$$

$$E = 40.5 \, mJ$$

c) The quantity of energy stored by the capacitor does not depend on the resistance through which the capacitor is discharged. Using both the 10 Ω and 20 Ω resistors the quantity of energy dissipated will be the same. However the fully charged capacitor will discharge more quickly when connected across the 10 Ω resistor. So, even though the energy dissipated in both resistors is the same, the power dissipated will be greater in the 10 Ω resistor.

Worked example 2

The circuits of figure 6.7.4 and figure 6.7.5 have identical signal generators both supplying the same constant amplitude output voltage at a frequency of 500 Hz.

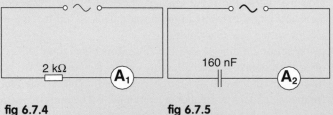

fig 6.7.4　　　　　　**fig 6.7.5**

Ammeters A_1 and A_2 both show an r.m.s. current reading of 2 mA.

a) Calculate the r.m.s. voltage output of the supply.

b) What is the 'resistance' of the capacitor? Explain your answer.

c) Sketch graphs to show how ammeter readings A_1 and A_2 vary as the supply frequency is increased to 5 kHz.

Answers and comments

a) The ammeters are both recording the r.m.s. currents in the circuits. For the circuit of figure 6.7.4 we know that the output p.d. from the supply is causing a 2 mA current in the 2 kΩ resistor. We can therefore calculate the output p.d. of the supply using;

$$V = I \times R = 2 \, mA \times 2 \, k\Omega = 4 \, V$$

The r.m.s. output voltage of the supply is 4 V.

b) In figure 6.7.5 a supply providing the same output voltage causes the same 2 mA current in the circuit. Therefore we can say that the 'resistance' of the capacitor at this frequency must also be 2 kΩ.

c) The current in the resistor is independent of the frequency of the output voltage from the supply. The current in the capacitor circuit is directly proportional to the frequency so the graphs are as shown in figure 6.7.6.

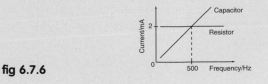

fig 6.7.6

Questions

1 In the circuit of figure 6.7.7 an uncharged 220 µF capacitor, a 2 kΩ resistor and a switch are connected in series and joined to a 6 V power supply which has negligible internal resistance.

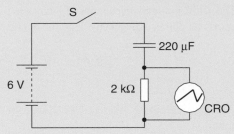

fig 6.7.7

An oscilloscope is connected across the resistor.
a) Calculate the initial current in the resistor after the switch is closed.
b) Sketch the trace that will be shown on the suitably adjusted oscilloscope as the capacitor charges.
c) How much charge will be stored by the fully charged capacitor?
d) Switch S is now opened. The 220 µF capacitor is replaced with a 100 µF capacitor. Sketch the graph shown on the oscilloscope when the switch is closed again. Indicate any similarities and differences compared to your answer to (b).

2 A student sets up the circuit shown in figure 6.7.8 to study the discharge of different capacitors.

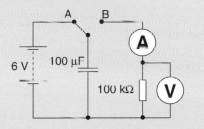

fig 6.7.8

The student changes the switch to position B and records the voltmeter reading at 10 second intervals. The results are shown in table 6.7.1.

Time/s	10	20	30	40
Voltage/V	2.2	0.81	0.3	0.11

Table 6.7.1

a) What is the p.d. across the capacitor when the switch is initially changed over from A to B?
b) Draw a graph to show how the p.d. across the capacitor changes with time over the first 40 seconds of the discharge.
c) Calculate the charge remaining on the capacitor 15 seconds after the discharge begins.
d) How much charge leaves the capacitor between the 5th and 35th second after discharge begins?
e) The 100 µF capacitor is replaced with an uncharged 220 µF capacitor and the experiment is repeated. Add a line to your graph in (b) to show how the p.d. across the capacitor varies with time during the first 40 seconds of discharge.

3 As part of a practical test a student sets up the circuit shown in figure 6.7.9. The capacitor is initially uncharged.

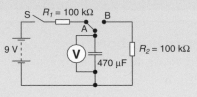

fig 6.7.9

The student is told to set the changeover switch to A, close switch S and record the voltmeter reading at 20 second intervals for 2 minutes. After 2 minutes the student has to change the switch to position B and record the voltage readings for a further 2 minutes. The results obtained by the student are as shown in table 6.7.2.

Time/s	0	20	40	60	80	100	120	140	160	180	200	220	240
Voltage/V	0	3.1	5.2	6.5	7.4	7.9	8.3	5.4	3.5	2.3	1.5	1.0	0.55

Table 6.7.2

a) During the first 2 minutes is the capacitor charging or discharging? Justify your answer.
b) Draw a graph to show how the voltage varies during the 4 minutes of the student's investigation.
c) The student is now asked to repeat the investigation with R_1 changed to 50 kΩ. Sketch on your graph to show how the voltage would now vary with time.
d) R_1 is reset to 100 kΩ and R_2 is changed to 50 kΩ. Sketch on your graph to show how the voltage would now vary with time if the experiment is repeated.

4 In the circuit shown in figure 6.7.10 the initially uncharged capacitor is connected in series with an ammeter, resistor, switch and power supply.

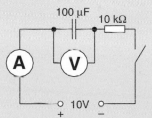

fig 6.7.10

a) Calculate the current in the circuit just after the switch is closed.
b) At one point during the charging process, the capacitor stores 0.4 mC of charge.
   i) Calculate the p.d. across the capacitor.
   ii) What is the voltage across the resistor when the capacitor stores 0.4 mC of charge? Explain your answer.
   iii) Show that the current in the resistor is 0.6 mA when the capacitor stores 0.4 mC of charge.
c) At a later time in the charging process the current in the resistor is 0.25 mA. Calculate the charge stored by the capacitor.

Past paper question

Figure 6.8.1 shows the flash-lamp circuit for a camera.

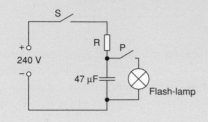

fig 6.8.1

Initially the capacitor is uncharged and switches S and P are open. Switch S is now closed.

a) Calculate the energy stored in the fully charged capacitor.

b) When the camera shutter is operated, switch S is opened and switch P is closed. The capacitor now fully discharges through the flash-lamp in a time of 1.6 ms. What is the average power developed in the flash?

c) The following information is marked on the capacitor; 47 μF 300 V. This means that the maximum voltage that should be applied across the capacitor is 300 V. Could this capacitor be connected safely to a 240 V a.c. supply? Justify your answer. **SQA 1991**

Answers and comments

a) The fully charged capacitor will have a p.d. of 240 V across its plates. We can calculate the energy stored using;

$$E = \frac{1}{2} CV^2 = \frac{1}{2} \times 47 \times 10^{-6} \times (240)^2$$

$$E = 1.35 \ J$$

b) We can calculate the power developed using the equation;

$$Power = \frac{Energy \ stored}{Time}$$

$$Power = \frac{1.35}{1.6 \times 10^{-3}} = 844 \ W$$

c) For safety reasons the p.d. across the capacitor must not exceed 300 V. The r.m.s. output of the a.c. supply is 240 V so we must calculate the peak voltage of this 240 V supply.

$$V_{peak} = \sqrt{2} \times V_{r.m.s.} = \sqrt{2} \times 240$$

$$V_{peak} = 339 \ V$$

Since the peak voltage of the supply exceeds the rating of the capacitor we would conclude that it is not safe to connect this capacitor to the supply.

Past paper question

A pupil sets up the apparatus shown to measure how long a ball takes to travel between two strips of metal foil on a track. The ball breaks each foil in turn. A computer with an appropriate interface is used to monitor and display the voltage across the capacitor.

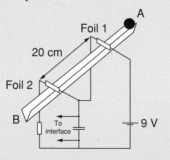

fig 6.8.2

a) What is the voltage across the capacitor before foil 1 is broken?

b) What happens in the circuit after foil 1 is broken?

c) When foil 2 is broken the voltage across the capacitor is 2 V. Draw a graph of voltage against time to show the computer display as the ball rolls from A to B. Numerical values are required on the voltage axis.

d) Indicate on the time axis of your graph, the region which corresponds to the ball travelling between the foils. **SQA 1992**

Answers and comments

a) Before foil 1 is broken, it completes a circuit joining the capacitor directly to the 9 V power supply. Therefore the p.d. across the capacitor is 9 V before the foil 1 is broken.

b) After foil 1 is broken a p.d. of 9 V from the capacitor is applied across the resistor. The p.d. across the capacitor will decrease as the capacitor discharges through the resistor.

c,d) As the ball rolls from A to foil 1 the voltage across the capacitor is constant at 9 V. In the time that the ball is rolling from foil 1 to foil 2 the p.d. across the capacitor reduces from 9 V to 2 V. After the ball passes through foil 2 the p.d. across the capacitor stays constant at 2 V.

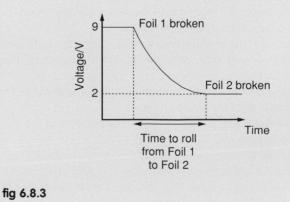

fig 6.8.3

Questions

1 An initially uncharged capacitor is charged from a power supply that provides a constant charging current of 100 μA. After 100 seconds the potential difference across the capacitor is 5 V.
   a) Calculate the charge that will be stored by the capacitor after 100 seconds.
   b) Calculate the energy that will be stored by the capacitor after 100 seconds.
   c) Calculate the value of the capacitor used in this experiment.
   d) (i) The capacitor is now replaced with a 1000 μF capacitor. How does the charge now stored after 100 seconds compare with that in the first capacitor? Explain your answer.
      (ii) How does the energy now stored after 100 seconds compare with the energy stored in the first capacitor? Explain your answer.
   e) What is the p.d. across the 1000 μF capacitor after 100 seconds? Explain your answer.

2 A student sets up the circuit shown in figure 6.8.4 to study the discharge of different capacitors.

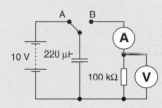

fig 6.8.4

The student changes the switch from position A to position B and records the voltmeter reading at 10 second intervals. The results are shown in table 6.8.1.

Time/s	0	10	20	30	40	50	60
Voltage/V	10	6.3	4	2.6	1.6	1.0	0.7

Table 6.8.1

   a) What is the p.d. across the capacitor when the switch is initially changed over from A to B?
   b) Draw a graph to show how the charge stored by the capacitor changes with time over the first 60 seconds of the discharge. Show numerical scales on both axes.
   c) The 100 kΩ resistor is replaced with a 50 kΩ resistor and the experiment repeated. Add a line to your graph to show how the charge stored by the capacitor would vary with time as the capacitor discharges.
   d) The 100 kΩ resistor is replaced in the circuit. A 100 μF capacitor is substituted for the 220 μF capacitor. The experiment is repeated. Add a line to your graph to show how the charge stored by this capacitor would vary with time as the capacitor discharges.

3 The circuit of figure 6.8.5 shows a capacitor connected in series with a power supply and a lamp. The voltage across the output terminals of the power supply has a constant amplitude of 6 V.

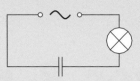

fig 6.8.5

When the frequency of the a.c. supply is 500 Hz the bulb is dimly lit.
   a) State what happens to the brightness of the lamp when the frequency of the supply is increased. Explain your answer.
   b) When the capacitor is replaced with a 1 kΩ resistor the bulb is also dimly lit. State what happens to the brightness of the lamp when the frequency of the supply is increased. Explain your answer.

4 In the circuit of figure 6.8.6 a 500 Ω resistor is connected in series with an uncharged 220 μF capacitor and a switch.

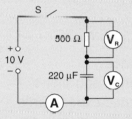

fig 6.8.6

Voltmeters are connected across the resistor and capacitor.
   a) Calculate the current in the circuit immediately after the switch is closed.
   b) At the instant when the current in the circuit is 14 mA what is the reading on the voltmeter connected across the resistor?
   c) How much charge is stored by the capacitor when the current in the circuit is 9 mA?
   d) Calculate the energy stored by the capacitor when the capacitor is fully charged.
   e) How much energy is still to be transferred from the power supply to the capacitor when the current in the circuit is 4.5 mA?

Waves & radiation

PART 3

$$a = \frac{dv}{dt}$$

$v = \text{lift} \, s$ $\text{drag } s$

$a = \frac{dv}{dt}$ weight $\frac{d}{dt}\left(\frac{ds}{dt}\right)$

thrust

$= \frac{d^2s}{dt^2}$

B

A

7.1 Waves – revisited

Introduction

Figure 7.1.1 shows children holding the ends of a rope.

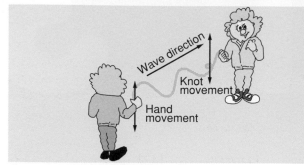

Figure 7.1.1 Waves on a rope

If one of the children moves her end of the rope up and down quickly, some of the kinetic energy she creates is transferred to the other end of the rope. **Energy is transferred even though no mass moves from one end of the rope to the other.** Indeed when the motion stops each point on the rope is still in the same place as it was before it moved. Scientists say that, in this case, the energy is transferred from one place to the other by a *wave* motion. The motion in figure 7.1.1 represents a **transverse wave**. In this type of wave each part of the material through which the wave travels moves **at right angles** to the direction in which the wave travels. If a knot is tied in the rope it would be seen to move up and down as the wave travels from one girl to the other. The disturbance moves along even though the individual parts only move up and down.

Figure 7.1.2 A Mexican wave

Sometimes, at large sporting events the spectators entertain themselves by starting a Mexican wave. For this type of wave the participants stand up and sit down again in sequence. From a distance it appears that the disturbance is moving along the crowd. Again, this is typical of a transverse wave where the disturbance moves *along* even though the individuals only move *up and down*.

Terminology

To explain waves properly we need to introduce a number of new terms;

The **wave frequency** is the number of complete waves passing any point each second. Frequency is measured in Hertz, Hz, and once a wave leaves the source its frequency cannot be altered.

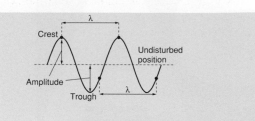

Figure 7.1.3 Wave

The words **crest** and **trough** are used to describe the top and bottom respectively of a transverse wave as shown in figure 7.1.3.

The **amplitude** of the wave is the distance between the peak of the crest and the undisturbed position. If the wave is symmetrical this distance will be the same as the distance between the bottom of the trough and the undisturbed position.

The **energy** associated with a wave of a certain frequency depends on its amplitude. If the girl in figure 7.1.1 creates waves with a bigger amplitude then more energy is transferred to the material through which the wave passes.

The **wavelength** is the distance from a point on a wave to the identical point on the next wave. This can be stated as the distance from a crest on a wave to the next crest on the wave.

The material through which a wave travels is called the **medium**. For waves arriving on earth from the sun the medium through which they pass is the vacuum in space!

Longitudinal waves

The waves described so far are transverse waves and they transmit energy at right angles to the direction in which particles of the medium move. This is *not* the case with sound waves.

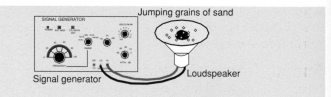

Figure 7.1.4 Loudspeaker and sand demonstration

When some granules of dry salt, sugar or sand are placed in the upturned cone of a loudspeaker which is emitting sound, the granules jump up and down. If the frequency increases the granules jump more often and if the amplitude of the sound is increased they jump higher. We can explain sound waves by assuming that **sound energy is transmitted by molecules colliding with neighbouring molecules.**

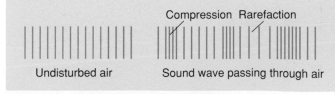

Figure 7.1.5 Compressions and rarefactions

The sound energy is transmitted by the movement of the air molecules to and fro *along the direction in which the energy is moving.* **This type of wave is called a longitudinal wave.** Sound, ultrasound and seismic (shock) waves are the most common examples of longitudinal waves.

Sound in a vacuum

When an electric bell is placed inside a gas jar and the air is extracted using a vacuum pump, the noise from the bell decreases and even becomes too faint to hear.

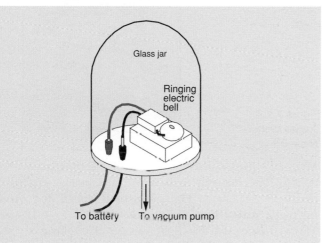

Figure 7.1.6 Bell in a jar

At first it might seem quite strange to be able to see the hammer hitting the bell and not hear any sound. From this observation we can conclude that the **longitudinal** sound waves need a **medium** to travel through whereas the transverse light waves do not.

Thunder and lightning are produced simultaneously but we always see the lightning a few seconds before hearing the thunder. Similarly when a sonic boom is created by an aeroplane, it appears to come from a position behind the aircraft. **These observations show us that light waves travel faster than sound waves.**

Speed, frequency and wavelength

Figure 7.1.7 shows a loudspeaker producing sound waves of frequency, f. The microphone is detecting the sound waves and turning them into an electrical signal which is used to produce a trace on the screen of an oscilloscope. If the signal generator is emitting waves at a frequency of 100 Hz it is creating 100 waves per second. The time interval between peaks is therefore one hundredth of a second.

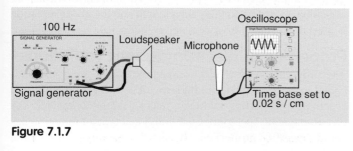

Figure 7.1.7

Generally, if a signal source is producing continuous waves of frequency, f, the time to produce one wave is given by $1/f$. This is called the **periodic time** or **period** and is given the symbol T.

$$Periodic\ time = \frac{1}{Frequency}$$

$$T = \frac{1}{f}$$

When the sound waves leave a source, one complete wave will pass any point in one period. This enables us to determine the average speed at which the wave is travelling through a medium.

$$Average\ speed = \frac{Distance\ travelled}{Time\ taken}$$

Since the front of the wave will travel a distance of one wavelength, which is given the symbol λ, in one period we can say;

$$Average\ speed = \frac{\lambda}{T}$$

$$But\ since;\ T = \frac{1}{f}$$

$$Average\ speed = \frac{\lambda}{1/f}$$

$$Average\ speed = f\lambda$$

This equation is written as,

$$v = f\lambda$$

Sound waves travel at different speeds in different materials. The speed also depends on the temperature of the material. Generally sound travels through air at room temperature at approximately $330\ m\,s^{-1}$. Its speed through helium gas at the same temperature is about five times faster.

Direct measurement of the speed of sound

EXPERIMENT Using the apparatus in figure 7.1.8 the hammer strikes the metal block producing a sound and starting the timer.

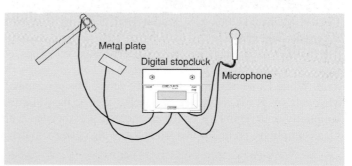

Figure 7.1.8 Direct measurement of the speed of sound

When the sound waves are detected by the microphone the timer stops. By varying the distance between the source of the sound and the detector it is possible to measure the time taken to travel different distances. The speed of sound in air can be calculated from the gradient of a distance-time graph.

Introduction

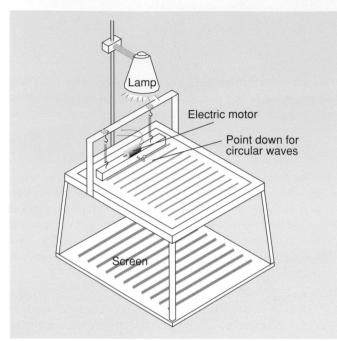

Figure 7.2.1 Ripple tank

Circular or straight waves can be produced by dipping a pointed rod or a straight edge into the water of a ripple tank. The motion of the waves along the tank can be seen by looking at the bright patches or dark shadows on the screen below the transparent base of the tank.

The properties which can be demonstrated using water waves are characteristic of other types of transverse waves. The wavelength of water waves in a ripple tank is such that we can observe changes easily.

James Clerk Maxwell

In 1831 Michael Faraday discovered that a changing or moving *magnetic* field could set up a changing *electric* field. In 1864 the Scottish mathematician James Clerk Maxwell proposed that a *changing electric field could set up a changing magnetic field*. While Faraday's work was easily verified by experiment, Maxwell's was not. However Maxwell's theory, based on the simplicity and symmetry of the mathematics associated with his assumptions, predicted that when a moving charge oscillates

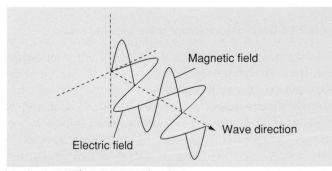

Figure 7.2.2 Electromagnetic wave

it also radiates an *electromagnetic wave*. This is made up from magnetic and electric fields of the same frequency at right angles to each other and to the direction of travel. The conventional way of showing such a wave is illustrated in figure 7.2.2.

**Figure 7.2.3
James Clerk
Maxwell**

Maxwell's theory links the speed of electromagnetic waves through any material with other fundamental electric and magnetic constants. **The calculated speed for electro-magnetic waves in a vacuum is close to 3×10^8 m s^{-1}.** Maxwell proposed that **light** was an electromagnetic wave. Following his work, the search for other electromagnetic waves was taken up and in 1887 waves having the same speed as light but with wavelengths of a few metres were discovered. The experimental work on the discovery of these radio waves was done by Henrich Hertz.

Members of the electromagnetic family

During the 100 years following Maxwell's work, other types of radiation were detected which can be classed as electromagnetic. To be classed as electromagnetic, waves must;

1. be transverse
2. travel at a speed of 3×10^8 m s^{-1} in a vacuum
3. be unaffected by external magnetic or electric fields.

Although all members of the electromagnetic spectrum travel at the same speed (3×10^8 m s^{-1}) in a vacuum, their frequencies and wavelengths differ. It is useful to remember that the speed, frequency and wavelength of any type of wave are connected by the equation $v = f\lambda$. The speed of light through a vacuum is given the symbol c, so for all electromagnetic waves travelling through a vacuum we can say;

$$c = f\lambda$$

Generally the different types of electromagnetic waves are classified according to their wavelength and the above equation is used to calculate the associated frequencies.

Properties of electromagnetic radiations

The different types of electromagnetic waves are sometimes called electromagnetic radiations. As well as having the common properties which make them part of the electromagnetic spectrum, each individual type of radiation has some specific properties. It should be understood that electromagnetic waves form a **spectrum** where the divisions between one type of radiation and its neighbour is not sharp or

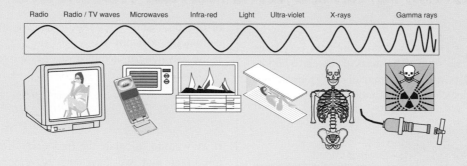

Figure 7.2.4 Electromagnetic spectrum

sudden. The following descriptions are of the typical properties associated with each type of radiation.

γ-rays (Gamma rays)

These are produced following rearrangements within the nuclei of certain radioactive materials or following nuclear reactions. The γ-rays produced have very high energies and can penetrate many centimetres of human tissue. For this reason cobalt-60 which is a γ emitter, is used to destroy deep-lying cancerous tumours. The gamma rays can also penetrate metal sheets up to 4 cm thick. This allows them to be used to monitor the thickness of sheet metals produced in continuous casting processes. The thicker the metal sheet the lower the intensity of the radiation passing through.

X-rays

Since their discovery by Röntgen in 1895, X-rays have been used extensively in medicine. They are produced when very fast moving electrons are rapidly decelerated by striking targets of tungsten or molybdenum. The use of X-rays to show breaks in bones demonstrates that their penetration of matter depends upon the density of the material through which they are passing. Penetration is lowest for dense materials such as bone.

X-rays also affect photographic film. This is a convenient way of showing the location of breaks in bones or of locating imperfections in welded metal joints and

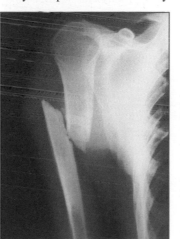

Figure 7.2.5 X-ray picture of a broken bone

castings. A source of X-rays is placed inside the metal joint to be tested and photographic film wrapped around the outside. Cracks or weak spots will allow more of the radiation to pass through the joint. When the photographic film is developed the position of a crack can be established. The short wavelength of X-rays makes them a useful tool for probing the structures of certain crystals or organic molecules.

Ultraviolet

Ultraviolet radiation is sometimes called ultraviolet light because its range of wavelengths starts from the violet end of the visible spectrum. Ultraviolet radiation produces a change in the colour of skin and can also cause damage to skin cells. Ultraviolet radiation can promote other chemical reactions and can make certain materials fluoresce.

Visible radiation

The retina of the eye can detect the narrow band of electromagnetic radiation with wavelengths between about 400 and 700 nanometres.

Colour	Red	Yellow	Green	Blue
Typical wavelengths/nm	640	590	510	480

Table 7.2.1

As with all parts of the electromagnetic spectrum, visible light has a range of wavelengths. When white light is dispersed we can see the constituent colours of this range of wavelengths. A beam of white light passing through a prism will be *dispersed* and if a suitable source is used it may be possible to detect some UV radiation near the blue end of the spectrum of colours and perhaps a little infrared just beyond the red end.

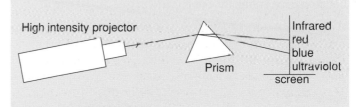

Figure 7.2.6 Producing a spectrum

Infrared radiation

This is not visible to the human eye. It has a range of wavelengths longer than visible red light and is best known for its ability to heat human skin.

Microwaves and radio waves

The modern perception of microwaves centres around the ability of one particular wavelength in the microwave region to heat molecules of water. These microwaves can be used to re-heat or cook food. Other wavelengths in the microwave region and wavelengths in the radio part of the spectrum are used to transmit mobile telephone signals or similar communications data.

Introduction

Some of the quantities we study in physics are very very large while others are very very small. The mass of the earth and its distance from other stars are astronomically large quantities while the mass of an electron and the diameter of a proton are both extremely small.

Figure 7.3.1 A large galaxy and an atom

The radiations in the electromagnetic spectrum have different wavelengths and frequencies. The wavelengths range from a few kilometres down to fractions of a nanometre. The kilometre and the nanometre are both measurements of distance but before we can compare their relative sizes we need to know how they are defined.

Showing large and small numbers

The basic unit of distance is the metre. In modern science the metre is defined as the distance that electromagnetic waves travel, in a vacuum, in 1/299,792,458th of a second. Dividing the metre into 100 equal parts creates 100 centimetres and for many measurements metres and centimetres are satisfactory units. The height of adults, the length of fabric samples and the size of the rooms in a flat or house can be described in metres and centimetres.

The centimetre is 1/100th part of a metre. In mathematical notation one centimetre can be written as 1×10^{-2} m.

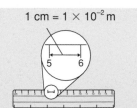

Figure 7.3.2
1 cm = 1×10^{-2} m

One millimetre is a 1/1000th part of a metre and in mathematical notation this is written as 1×10^{-3} m.

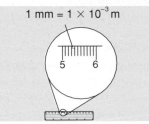

Figure 7.3.3
1 mm = 1×10^{-3} m

When dealing with shorter lengths there are other commonly used multipliers. One millionth of a metre is called a micrometre or micron (μm) where 1 μm = 1×10^{-6} m. A distance

of one thousand times *smaller* than a micrometre is called a nanometre, so 1 nm = 1×10^{-9} m.

If the wavelength of microwaves is quoted as 100 μm we can say that;

$$\lambda = 100 \ \mu m$$
$$\lambda = 100 \times 10^{-6} m$$
$$\lambda = 1 \times 10^{-4} m.$$

If we want to calculate the frequency of these microwaves it is important that we have the wavelength in metres before substituting numbers into the equation.

$$c = f\lambda \text{ and } c = 300,000,000 \ m \, s^{-1}$$

$$f = \frac{c}{\lambda} = \frac{300,000,000}{1 \times 10^{-4}}$$

$$f = 3,000,000,000,000 \ Hz$$

This number, while correct, is cumbersome and can be expressed more simply as,

$$f = 3 \times 10^{12} \ Hz$$

We can also use prefixes to simplify the way that large numbers are expressed.

One kilometre is a distance of 1000 m and 1 kilohertz is a frequency of 1000 Hz.

One megahertz (1 MHz) = 1,000,000 Hz or 1×10^6 Hz

One thousand megahertz is called one gigahertz (GHz) and is written mathematically as 1×10^9 Hz.

Summary

We have already seen that distances and frequencies are often written more simply using prefixes. Other quantities can also be expressed in a similar manner. The common prefixes are listed in table 7.3.1 along with an example of their usage.

Prefix	Name	Multiplier	Example
p	pico	1×10^{-12}	Capacitance = 2.2pF = 2.2×10^{-12}F
n	nano	1×10^{-9}	Wavelength of blue light = 400 nm = 4×10^{-7} m
μ	micro	1×10^{-6}	Current of 50 μA = 5×10^{-5} A
m	milli	1×10^{-3}	Charge of 19 mC = 1.9×10^{-2} C
k	kilo	1×10^{3}	Energy of 4.2 kJ = 4200 J
M	mega	1×10^{6}	Power of 650 MW = 6.5×10^{8} W
G	giga	1×10^{9}	Frequency of 4.5 GHz = 4.5×10^9 Hz

Table 7.3.1

When substituting numbers into equations it is important that the quantities are expressed in standard units. If this is done the units of the resulting calculation will also be in their most basic form. In calculations involving $c = f\lambda$ expressing the frequency in Hz, and the wavelength in m will give units of $m \, s^{-1}$ for the calculated speed. Using kHz or cm would not give the speed in $m \, s^{-1}$.

7.4 Consolidation questions

1 Straight waves travel across the surface of a ripple tank.
   a) The waves travel 33 cm in 1.5 s. What is their speed?
   b) The distance between successive wave crests is 4 cm. What is their frequency?

2
   a) A sound source with a frequency of 1 kHz emits waves with a wavelength of 20 cm. What time do these waves take to travel 330 m?
   b) Light travels at 3×10^8 m s^{-1}. Calculate the time the light waves take to travel 330 m.
   c) Explain why, during a storm, lightning is seen before thunder is heard.

3 Figure 7.4.1 shows the displacement of one point in a ripple tank at different times as a transverse wave passes travelling at 5 cm s^{-1}.

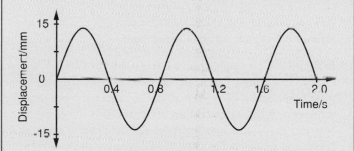

fig 7.4.1

   a) What is the amplitude of the wave?
   b) What is the frequency of the wave?
   c) What is the wavelength of the wave?
   d) What is the average speed of the point in the ripple tank as it moves between crests and troughs of these waves?

4 A student making notes on transverse and longitudinal waves writes down the following statements;
I   both transverse and longitudinal waves transfer energy
II   in the medium through which a longitudinal wave travels, the particles move at right angles to the direction in which the wave moves
III all electromagnetic waves are transverse but sound and ultrasound are longitudinal waves.
   Which of these statements is/are true?
   A III only
   B I and II
   C I and III
   D II and III
   E I, II and III.

5 A pupil studies sound waves by connecting a microphone to an oscilloscope. The oscilloscope traces produced by two different audio frequencies are drawn below. The oscilloscope settings are the same for each trace.

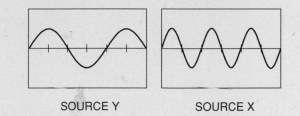

SOURCE Y   SOURCE X

fig 7.4.2

   He makes the following statements about the two sources.
I   both sources have the same amplitude
II   source X emits more energy per second than source Y
III source X has a higher frequency.
   Which of these statements is/are true?
   A I only
   B I and II
   C I and III
   D II and III
   E I, II and III.

6 An electromagnetic wave has a frequency of 1 GHz. In which part of the electromagnetic spectrum does this radiation occur?
   A visible (400 – 700 nm)
   B radio waves (1 m – 1 km)
   C microwaves (1 mm – 10 cm)
   D radar (10 cm – 1 m)
   E ultraviolet (700 nm – 1 mm)

7 Which row of the following table shows the correct wavelengths for red, green and blue light?

	Red	Green	Blue
A	500 nm	600 nm	700 nm
B	700 nm	590 nm	510 nm
C	640 nm	510 nm	480 nm
D	510 nm	700 nm	590 nm
E	640 nm	590 nm	510 nm

Table 7.4.1

7.5 Reflection

Introduction

Reflection is probably the simplest and most frequently used property of waves. Many modern fishing boats are fitted with echolocation equipment which emits a beam of sonar and 'listens' for the reflected signal. Radar stations scan the skies with electromagnetic waves and analyse reflected signals.

Figure 7.5.1 Radar listening station

Echoes

An echo is a simple reflection. A longitudinal sound wave emitted from a source travels through the air before colliding with a hard surface which makes the sound wave return along its original direction. When the distance between the source and the reflector is large the time delay between hearing the emitted and reflected sounds is noticeable.

Sonar systems measure the time delay between the emitted and reflected signals. Because the speed with which the ultrasound travels through water is known, the depth of the reflecting shoal of fish below the fishing boat can be calculated.

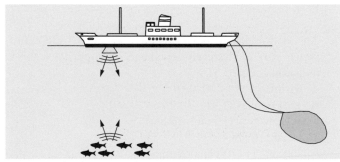

Figure 7.5.2 Sonar detection of fish

 The time between the emitted and reflected pulses in figure 7.5.2 is 1.2 seconds. Sound waves travel through the water at 1500 m s^{-1}. At what depth, d, is the reflecting shoal below the boat?

A

$$Speed = \frac{Total\ distance\ travelled}{Time\ taken}$$

$$1500 = \frac{2d}{1.2}$$

$$d = \frac{1500 \times 1.2}{2}$$

$$d = 900\ m$$

Echolocation

In the previous example, it is extremely unlikely that all of the fish in the shoal are at exactly the same depth! Different depths would give rise to the possibility of the sonar pulses being reflected from different parts of the shoal. It is therefore unlikely that the reflected pulse is exactly the same shape as the emitted pulse. The shape of the reflected pulse will therefore carry information about the shape and size of the shoal of fish.

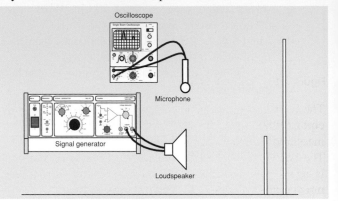

Figure 7.5.3 Double reflector experiment

In figure 7.5.3 a sound source directs pulses of sound at a pair of reflecting plates. Part of the beam is reflected from the closer plate while some of the remainder is reflected from the other reflector. If the sound takes 0.01 seconds more to travel between the source and the more distant plate, a microphone connected to an oscilloscope would display the two reflected signals as shown in figure 7.5.4.

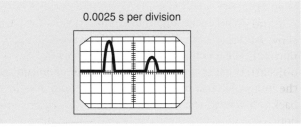

Figure 7.5.4 Oscilloscope screen

For this simple situation it is possible to reconstruct the arrangement of the reflectors by examining the trace, even if the reflectors could not be seen. The shape of more complex objects can be determined by using computers to perform the

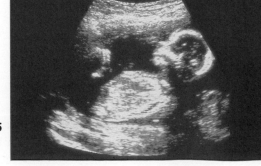

Figure 7.5.5 Ultrasound scan

reconstruction. In the screening of expectant mothers, ultrasound scans play an important part in the early diagnosis of foetal abnormalities.

The Laws of Reflection

Figure 7.5.6 shows the paths of incoming (or **incident**) and **reflected** rays of light falling on a plane mirror. The **normal** on the diagram is a construction line drawn at right angles to the reflecting surface at the point where the incident ray hits the reflector.

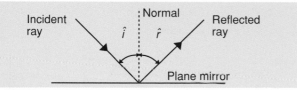

Figure 7.5.6 First Law of Reflection

The First Law of Reflection states that **the angle of incidence equals the angle of reflection**. Both of these angles are measured relative to the normal.

The Second Law of Reflection takes a little more thought but is again common sense. This states that the **incident ray, normal and reflected ray are in the same plane**.

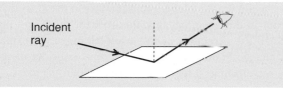

Figure 7.5.7 Second Law of Reflection

We can use these laws to determine the image position for an object placed in front of a plane mirror as in figure 7.5.8. Rays from the object are radiated in all directions but if we consider just two of these we can use the First Law of Reflection to show how the rays will diverge after the reflection from the mirror. **Since the rays emerge from the same point on the object they must appear to be emerging from that point on the image.** We can find this point by tracing the reflected rays back behind the mirror to the point where they join. You should notice that **the distance of the image behind the mirror is the same as the distance that the object is in front of the mirror.**

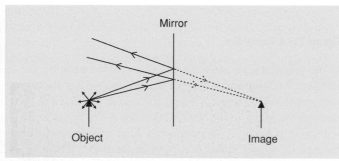

Figure 7.5.8

Curved reflectors

The Laws of Reflection also apply when the reflector is curved. Figure 7.5.9 shows how a curved reflector can be used to collect radio waves or microwaves emitted by a satellite.

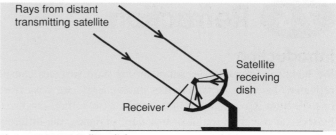

Figure 7.5.9 Satellite dish

The curvature of the reflector is such that all the collected radiation is focused onto the receiver.

Since the transmitting satellite is a long way from the receiving dish the radiation can be regarded as arriving in parallel rays. The reflector is curved in such a way that parallel rays arriving are reflected onto the receiver placed at the focus. This technique is useful for collecting wave energy from weak or distant sources. The signal from the transmitting satellite collected at the focus is amplified because all of the wave energy collected over the whole area of the dish has been concentrated at a point.

Searchlights

Not only can curved reflectors be used to collect wave energy, they can also be used to transmit energy in a specific direction. Many searchlights place the lamp at the focus of a **parabolic mirror** to produce a parallel beam of light.

Using the reflector in this way illustrates the **principle of reversibility of light**. A spherical reflector concentrates wave energy at the focus. Similarly, if a source emitting wave energy is placed at the focus the emerging beam will itself be parallel.

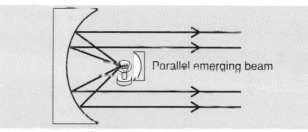

Figure 7.5.10 Parallel beam searchlight

In a whispering gallery one person whispering at the focus of a large reflector cannot be heard by passers-by but can be heard clearly by a friend whose ear is placed at the focus of the receiving reflector even though it is some considerable distance away.

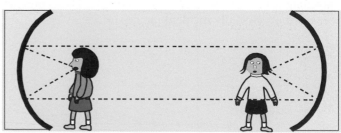

Figure 7.5.11 Whispering gallery

Once again the principle of reversibility can be demonstrated if the receiver is used as a transmitter and in this way a two way whispered conversation can take place.

7.6 Refraction

Introduction

We have seen how wave energy hitting hard surfaces can be reflected. In some situations the wave energy can pass from one medium into another. For example, visible light produced by the sun travels through the vacuum of space, the different layers of the atmosphere and eventually through glass window panes before entering a greenhouse. We now want to investigate the effect that different materials have on waves travelling through them.

Changing direction

The frequency of a wave depends *only* on the frequency of the source. This fact tells us that **the frequency of a wave does not alter when the waves are passing through different materials**. While the frequency of light passing from air into glass does not change, the speed with which the light travels through glass is less than its speed in air. From the equation;

$$v = f\lambda$$

we can see that the wavelength must also be shorter in the glass.

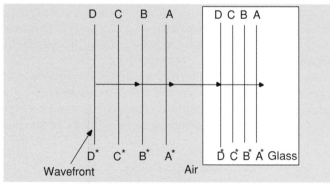

Figure 7.6.1 Wavefronts entering glass

We can represent a parallel beam of light produced by an extended source and passing into glass as shown in figure 7.6.1. The lines A-A*, B-B* etc. represent the crests of consecutive waves produced at the same time. These are called **wavefronts**. The distance between these wavefronts is less in the glass than in the air because the waves are travelling more slowly in the glass. **So in equal time intervals the wavefronts travel further in air than in the glass**.

In this particular example the rays are normal to the surface and any line drawn to represent the direction in which the waves are travelling would be parallel to the normal. These rays are described as **normally incident** on the surface.

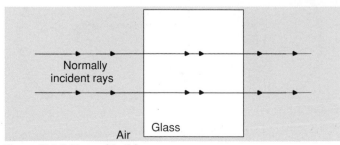

Figure 7.6.2 Normal incidence

The situation where the wavefronts are not normally incident is shown in figure 7.6.3. **In this case the angle between the direction in which the waves are travelling and the normal is called the angle of incidence, $\hat{i}$**

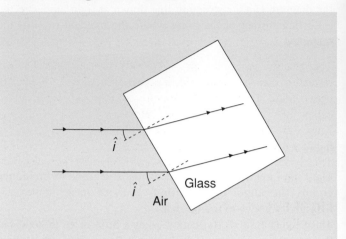

Figure 7.6.3 Angular incidence

Once again the wavelength must be less in the glass than in the air but in this case, because the incident ray is not along the normal, **the direction of the wave changes**. To explain why this change in direction occurs we must consider the progress of a wavefront as it travels from the air into the glass.

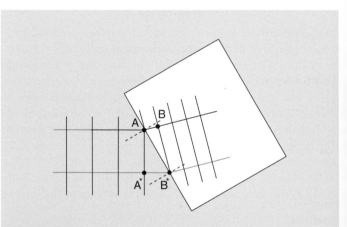

Figure 7.6.4 Refraction

In the few moments after the waves are in the position shown in figure 7.6.4 the end A* of the wavefront A-A* will continue to travel through the air whereas the end A will have to travel through the glass. In the time it takes for A* to travel to B*, A will go to position B. But since the wave travels more slowly through the glass, A will not travel as far as A*. The result of this is that the wave energy now travels in a slightly different direction.

This change in direction caused by the change in wave speed is called **refraction**.

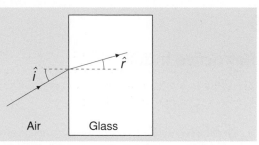

Figure 7.6.5 Angles of incidence and refraction

The changes in direction which occur as a result of this refraction can be summarised as in figure 7.6.5. The angle between the normal and the direction of the refracted waves is called the angle of refraction $\hat{r}$. In this situation where the waves are travelling into a denser material the angle of incidence will always be greater than the angle of refraction. This allows us to say that **waves entering a denser material bend towards the normal**.

Light rays leaving glass

Since light paths are reversible it would seem sensible to use the arguments from the previous section to show that waves travelling from a denser material into a less dense one are refracted away from the normal.

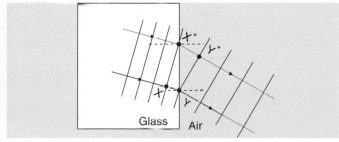

Figure 7.6.6 Wavefronts leaving glass

We can confirm that this is indeed what happens by considering figure 7.6.6. As the waves leave the denser material the end of the wave X will travel to Y in the same time that X* travels to Y*. Since the waves travel faster in the air than in the glass the distance X*-Y* is greater than the distance X-Y. As a consequence of this **the waves are refracted away from the normal on travelling from the denser to the less dense medium**.

It can also be shown that when a ray of light is incident on a parallel sided glass block, the emerging ray will be parallel to the direction of the incident ray. The amount by which the emerging ray is displaced from its original direction depends on the thickness of the block.

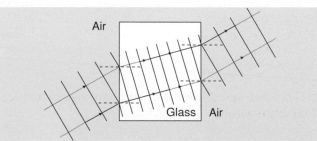

Figure 7.6.7 Parallel beams entering and emerging from glass

Lenses

Telescopes, microscopes, spectacles and many other optical instruments use specially shaped pieces of glass to refract rays of light in chosen directions.

The passage of a ray through a prism is as shown in figure 7.6.8.

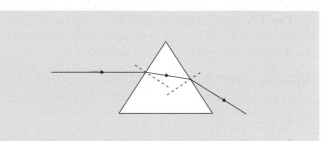

Figure 7.6.8 Passage of light through a prism

On entering the denser glass the ray of light bends towards the normal. On leaving the prism the ray bends away from the normal. This second refraction deviates the ray further from its original direction. Using two prisms and a parallel sided glass block we can create a very simple converging lens.

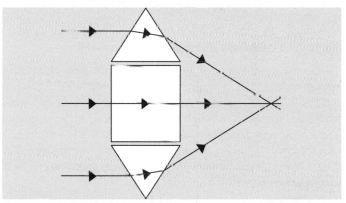

Figure 7.6.9 Simple converging lens

The refractions of the rays on entering and leaving the prisms converge the outer rays while the centre ray, being incident along the normal, travels straight through. The three rays shown can be converged to a focus.

By similar reasoning you should be able to show that the arrangement of the prism and blocks shown in figure 7.6.10 can cause the incoming parallel beam of light to diverge.

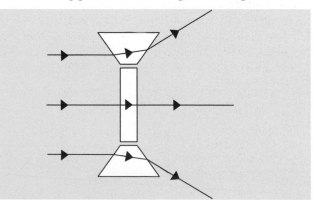

Figure 7.6.10 Simple diverging lens

7.7 Refraction equations

Introduction

When waves pass from one material into another the speed of the wave changes causing a change in wavelength. It can also alter the direction in which the waves are travelling. In this section you will see how it is possible to calculate the effect that different materials will have on waves.

Waves changing direction

In the previous section you considered the consequences of a wavefront A-A* entering an optically denser material.

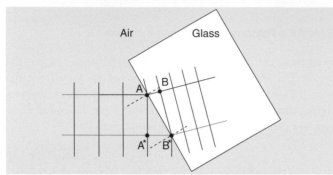

Figure 7.7.1 Wavefront entering a denser material

In the few moments after the wave enters the new material, A moves to B while A* moves to B*. The distance A* B* is greater than the distance AB because this part of the wave continues travelling through air and its speed in air is greater than in the material. The wave directions as well as the angles of incidence, $\hat{i}$, and refraction, $\hat{r}$, are shown on figure 7.7.2. The waves travel through the air with a speed of v_{air} and through the glass with a speed, v_{glass}.

A little further...

The time for point A to travel to B, t_{AB}, is the same as the time for point A* to travel to B*.

$$t_{AB} = t_{A*B*}$$

We can rearrange the relationship;

$$Speed = \frac{Distance\ travelled}{Time\ taken}$$

to give;

$$Time\ taken = \frac{Distance\ travelled}{Speed}$$

Therefore since;

$$t_{AB} = t_{A*B*}$$

we can say

$$t_{AB} = \frac{Distance\ AB}{v_{glass}} \qquad t_{A*B*} = \frac{Distance\ A*B*}{v_{air}}$$

$$\frac{Distance\ AB}{v_{glass}} = \frac{Distance\ A*B*}{v_{air}}$$

$$\frac{Distance\ A*B*}{Distance\ AB} = \frac{v_{air}}{v_{glass}}$$

Refractive index

The ratio;

$$\frac{v_{air}}{v_{glass}}$$

is called the **refractive index** of the glass and is given the symbol n.

A little further...

Not only can the refractive index be equated to the ratio of the speeds, it can also be found from;

$$n = \frac{v_{air}}{v_{glass}} = \frac{Distance\ A*B*}{Distance\ AB}$$

Figure 7.7.2 is similar to figure 7.7.1 except that other angles, geometrically equal to the angles of incidence and refraction have been identified.

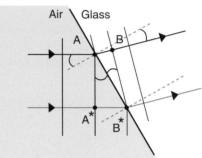

Figure 7.7.2

From this diagram we can state that;

$$Distance\ A*B* = AB* \sin \hat{i}$$
$$Distance\ AB = AB* \sin \hat{r}$$

$$n = \frac{Distance\ A*B*}{Distance\ AB} = \frac{AB* \sin \hat{i}}{AB* \sin \hat{r}}$$

$$n = \frac{\sin \hat{i}}{\sin \hat{r}}$$

This relationship linking $\sin \hat{i}$ and $\sin \hat{r}$ is called **Snell's Law**.

$$n = \frac{\sin \hat{i}}{\sin \hat{r}}$$

Refractive index and wavelength

By expressing the refractive index of glass as the ratio of the velocities;

$$n = \frac{v_{air}}{v_{glass}}$$

and recalling that;

$$v = f \times \lambda$$

we can say;

$$n = \frac{v_{air}}{v_{glass}} = \frac{f\lambda_{air}}{f\lambda_{glass}}$$

But since the frequency of the waves is the same in both materials we can also conclude that;

$$n = \frac{\lambda_{air}}{\lambda_{glass}}$$

Since rays slow down on moving from air into glass we should be able to understand from the above equations that the refractive index of glass is greater than 1.

The actual refractive index depends on the composition of the glass. Crystal glass has a higher refractive index than the glasses used for making windows. However the refractive indices for most types of glass are in the region 1.40 – 1.60.

Refractive index and different materials

So far we have concentrated on the refraction that occurs when a ray of light passes from air into glass. We have seen how the concept of refractive index can quantitatively link speeds and wavelengths before and after refraction.

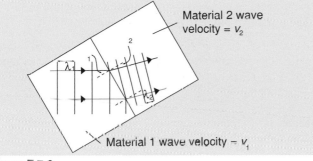

Figure 7.7.3

For more general situations where the waves are moving from material 1 into material 2 we can state the refractive index as;

$$\underset{1 \to 2}{n} = \frac{\sin \theta_1}{\sin \theta_2} = \frac{v_1}{v_2} = \frac{\lambda_1}{\lambda_2}$$

For very accurate work we need to consider **absolute refractive indices**. For these values the incident ray moves from a vacuum into the material. Since the speed of light is slightly greater in a vacuum than it is in the surrounding air, the absolute refractive index for a material is slightly different to the refractive index for rays moving from air into the material.

However, for practical purposes the difference between absolute refractive indices and values for air into materials are so small as to be negligible. (The absolute refractive index of a vacuum is by definition equal to 1, and that for air at normal atmospheric pressure and a temperature of 20°C is 1.0003.)

We must also take account of the fact that **the refractive index varies slightly as the frequency of the light changes**. Typical glasses will have a refractive index value about 1% higher for blue light than for red. Light of a single frequency is called **monochromatic**. Approximate refractive indices for some common materials for monochromatic light are shown in table 7.7.1.

Material	Refractive index
Air	1.00
Glass	1.51
Water	1.33
Diamond	2.42

Table 7.7.1

Q A ray of monochromatic light is incident from air onto the surface of a glass block as shown in figure 7.7.4. If the refractive index of the glass block is 1.5 what is the angle of refraction in the block?

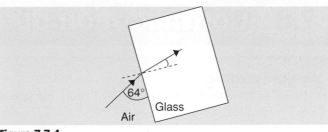

Figure 7.7.4

A We must firstly notice that the angle marked on the diagram is NOT the angle of incidence. The angle of incidence is 26°.

$$n = \frac{\sin \hat{i}}{\sin \hat{r}} \qquad 1.5 = \frac{\sin 26°}{\sin \hat{r}}$$

$$\sin \hat{r} = \frac{\sin 26°}{1.5}$$

$$\hat{r} = 17°$$

The angle of refraction in the glass is 17°.

Refraction and Snell's Law

EXPERIMENT In this experiment you will produce narrow rays of white light using a ray box and you will shine these rays onto the centre of the straight side of a semicircular glass block. Each ray will then refract into the block. Since the incident ray strikes the centre of a semicircle the refracted ray travels along a radius, before hitting the circular side along a normal and emerging into the air without further change in direction.

You should place the semicircular block on a piece of white paper and trace around the block. Find the centre of the straight side and draw a series of lines to represent incident rays as shown in figure 7.7.5.

Point the narrow rays from the ray box towards the centre of the straight side along the lines that you have drawn and plot the paths of the rays as they emerge from the circular side of the block.

For each of the rays measure the angles of incidence and refraction and calculate the ratio;

$$n = \frac{\sin \hat{i}}{\sin \hat{r}}$$

Using your refractive index values calculate the mean and the uncertainty.

Approximate to random uncertainty in n = $\dfrac{\text{Range of readings}}{\text{Number of attempts}}$

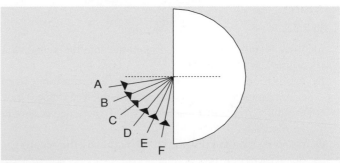

Figure 7.7.5

7.8 Consolidation questions

Worked example 1

A beam of monochromatic light passing from air into a liquid is refracted as shown in figure 7.8.1.

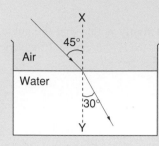

fig 7.8.1

a) What is the line marked XY called?
b) What is the angle of incidence?
c) What is the angle of refraction?
d) Calculate the refractive index of the liquid.
e) What is the speed of the light in air before entering the liquid?
f) What is the speed of the light in the liquid?

Answers and comments

The line marked as XY is at right angles to the surface at the point where the ray enters the water.

a) The line marked XY is called the normal to the surface and all angles are measured to this line.
b) The angle of incidence is the angle between the incident ray and the normal, so the angle of incidence, as marked on the diagram is 45°.
c) The angle in the material between the ray and the normal is the angle of refraction, so the angle of refraction is 30°.
d) The refractive index is calculated from;

$$n_{air \to material} = \frac{sin\ \hat{i}}{sin\ \hat{r}} = \frac{sin\ 45°}{sin\ 30°}$$

$$n_{air \to material} = 1.41$$

e) Light is an electromagnetic radiation and travels at $3 \times 10^8\ m\,s^{-1}$.

f)

$$n_{air \to material} = \frac{v_{air}}{v_{glass}}$$

$$1.41 = \frac{3 \times 10^8}{v_{liquid}}$$

$$v_{liquid} = \frac{3 \times 10^8}{1.41}$$

$$v_{liquid} = 2.13 \times 10^8\ m\,s^{-1}$$

Worked example 2

Figure 7.8.2 shows a communications satellite relaying data over a large part of Scotland.

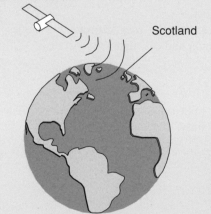

fig 7.8.2

a) Why is only a weak signal from the satellite detected at a point on the ground?
b) Dishes designed to receive information from communications satellites have a distinctive shape. Draw the shape of the dish and mark where the receiver should be placed.
c) Draw rays on the diagram to explain why the receiver is placed where it is.

Answers and comments

a) There is only a certain amount of energy that the satellite can radiate and this has to cover a large area. The signal at any point on the ground is only a very small part of the energy radiated by the satellite. For this reason the signal is said to be weak.
b) The dishes used to collect the signals from satellite transmitters are called concave reflectors. The receiver is placed at the focus of the reflector as this is where rays from distant transmitters are focused

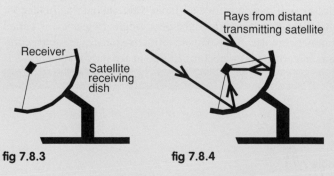

fig 7.8.3   **fig 7.8.4**

c) The rays from distant transmitters are parallel and are focused onto the receiver placed at the focal point of the curved reflector.

Questions

1 Figure 7.8.5 shows a partially completed ray diagram for a ray of light starting at A and going to B having been reflected from a plane mirror.

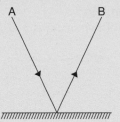

fig 7.8.5

a) Copy this diagram and mark on the normal.
b) What is the angle of incidence?
c) Mark on your diagram the angle of reflection.
d) If a ray of light started at B and hit the mirror at point P shown in figure 7.8.5 would it pass through A? Explain your answer.

2 In a shoe shop an angled mirror as shown in figure 7.8.6 is placed on the floor.

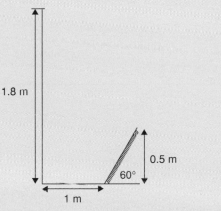

fig 7.8.6

A man 1.8 m tall stands 1 m away from the mirror. Will he be able to see the reflection of his feet in the mirror? (You may want to draw a scale diagram to answer this question.)

3 Figure 7.8.7 shows a ray of light with a frequency of 5×10^{14} Hz incident at 30° onto the surface of a block of material with a refractive index of 1.5.

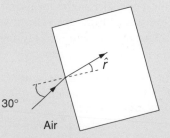

fig 7.8.7

a) Describe, in words, what happens to the direction of the ray after it enters the block.
b) What is the frequency of the light in the material?

c) Calculate the angle of refraction.
d) What is the wavelength of the light in the block?

4 In some very accurate experimental work on refraction a researcher uses a beam of red light and a beam of blue light. The glass block she uses has a refractive index of 1.58 for the red light used and 1.59 for the blue light.

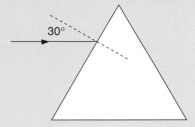

fig 7.8.8

The angle of incidence used for each of the rays is 30°. Calculate:

a) The angle of refraction for the red light.
b) The angle of refraction for the blue light.
c) Both beams are put into the block along this same path. What is the angular separation between the beams after refraction?

5

a) What is the speed of light in a material with a refractive index of 2.42?
b) The frequency of this light is 5.5×10^{14} Hz in air. What is its frequency in the material?
c) What is the wavelength of this light in the material?

6 In an optics exhibit at a science fair there is a block made from two different types of glass P and Q. A ray of monochromatic light passes through this block as shown (from X to Y).

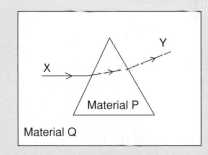

fig 7.8.9

Which of the following is true?
A The wavelength of the light is greater in material Q than in P.
B The frequency of the light is greater in Q.
C The ray bends towards the normal as it passes from material P to material Q.
D The refractive index for Q must be 1.5.
E The absolute refractive index for both materials is the same.

Worked example 1

A beam of monochromatic light of frequency 4.85×10^{14} Hz passes from air into liquid paraffin.

In liquid paraffin the light has a speed of 2.10×10^8 m s^{-1}.

a) Calculate the refractive index of liquid paraffin.

b) What is the frequency of the light in liquid paraffin?

Answers and comments

a) Since the speed of light in air is 3×10^8 m s^{-1},

$$n = \frac{v_{air}}{v_{glass}} = \frac{3 \times 10^8}{2.10 \times 10^8}$$

$$n = 1.43$$

b) The frequency of the light is set by the source and does not change during refraction so the frequency in paraffin is still 4.85×10^{14} Hz.

Worked example 2

The diagram below shows the refraction of a ray of red light as it enters and leaves a plastic prism.

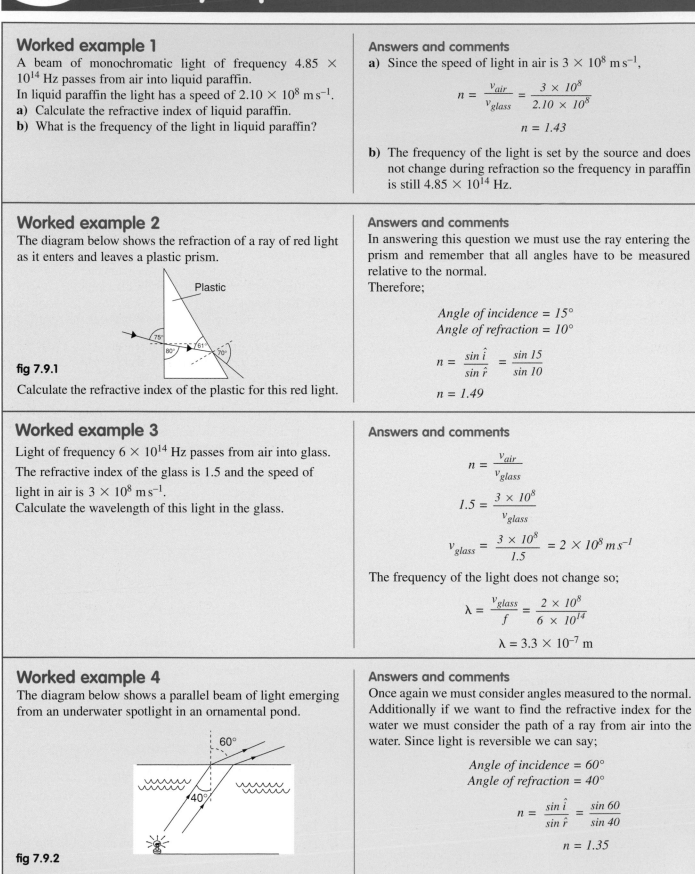

fig 7.9.1

Calculate the refractive index of the plastic for this red light.

Answers and comments

In answering this question we must use the ray entering the prism and remember that all angles have to be measured relative to the normal. Therefore;

$$Angle\ of\ incidence = 15°$$
$$Angle\ of\ refraction = 10°$$

$$n = \frac{\sin \hat{i}}{\sin \hat{r}} = \frac{\sin 15}{\sin 10}$$

$$n = 1.49$$

Worked example 3

Light of frequency 6×10^{14} Hz passes from air into glass.

The refractive index of the glass is 1.5 and the speed of light in air is 3×10^8 m s^{-1}.

Calculate the wavelength of this light in the glass.

Answers and comments

$$n = \frac{v_{air}}{v_{glass}}$$

$$1.5 = \frac{3 \times 10^8}{v_{glass}}$$

$$v_{glass} = \frac{3 \times 10^8}{1.5} = 2 \times 10^8\ m\,s^{-1}$$

The frequency of the light does not change so;

$$\lambda = \frac{v_{glass}}{f} = \frac{2 \times 10^8}{6 \times 10^{14}}$$

$$\lambda = 3.3 \times 10^{-7}\ m$$

Worked example 4

The diagram below shows a parallel beam of light emerging from an underwater spotlight in an ornamental pond.

fig 7.9.2

Calculate the refractive index of the water in the pond.

Answers and comments

Once again we must consider angles measured to the normal. Additionally if we want to find the refractive index for the water we must consider the path of a ray from air into the water. Since light is reversible we can say;

$$Angle\ of\ incidence = 60°$$
$$Angle\ of\ refraction = 40°$$

$$n = \frac{\sin \hat{i}}{\sin \hat{r}} = \frac{\sin 60}{\sin 40}$$

$$n = 1.35$$

Questions

1 A student making notes on refraction writes down the following statements;

   I   The frequency of a wave is unaltered when it enters a different medium.

   II The refractive index depends on the frequency of the incident light.

   III Waves travel slower in optically denser materials.

   Which of these statements is/are true?

   A III only

   B I and II only

   C I and III only

   D II and III only

   E I, II and III

2 A glass block has a refractive index of 1.58 for yellow light with a wavelength of 590 nm.

   a) Light enters the block at an angle of 35°. What is the angle of refraction?

   b) What is the frequency of this light in air if it travels with a speed of $3 \times 10^8 \, m\,s^{-1}$?

   c) What is the frequency of the light in the glass block after refraction?

   d) What is the wavelength of the light in the glass block?

3 A pupil investigating Snell's Law passes a ray of monochromatic light into a glass block at different angles of incidence.

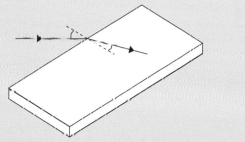

fig 7.9.3

She marks the path of the ray and records the angles of incidence and refraction as shown in the results table.

Angle of incidence	Angle of refraction
10°	6.5°
20°	13°
30°	19°
40°	25°
50°	30°
60°	35°

Table 7.9.1

   a) Calculate the sine of the angles of incidence $\hat{i}$ and refraction $\hat{r}$ and plot a graph of $\sin \hat{i}$ versus $\sin \hat{r}$

   b) Use the graph to determine the angle of refraction for an incident angle of 44°.

   c) Use the graph to determine the refractive index for the glass block.

4 A ray of monochromatic light is directed towards the centre of the straight side of a semicircular glass block which has a refractive index of 1.51.

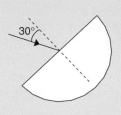

fig 7.9.4

   a) What is the angle of refraction when the angle of incidence in air is 30°?

   b) At what angle does this ray hit the semicircular side as it leaves the block?

   c) Copy figure 7.9.4 and show the path of the ray as it enters and exits the glass block.

5 The refractive index of glycerol for one particular frequency of light is 1.47.

   a) A ray of light having this frequency is incident at an angle of 60° as it enters glycerol from air. What is the corresponding angle of refraction?

   b) The angle of incidence is altered so that the angle of refraction is 30°. What is the new angle of incidence?

   c) Light travels at $3 \times 10^8 \, m\,s^{-1}$ in air. What is its speed in glycerol?

6 A ray of monochromatic light passes from glass into air at the angles shown.

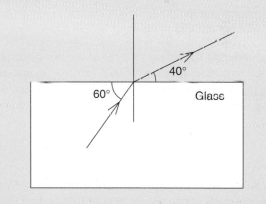

fig 7.9.5

The refractive index for the glass is;

A 0.65

B 0.74

C 1.35

D 1.53

E 1.50

7.10 Total internal reflection

Introduction

Refraction is the process where the direction in which a wave is moving changes because the waves have entered a different material and are travelling at a different speed. In the previous sections we concentrated primarily on the passage of light from air into glass.

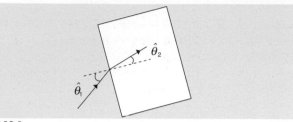

Figure 7.10.1

When defining the refractive index of a material we must consider the passage of a ray of light from air into that material. For materials optically denser than air the angle of incidence is greater than the angle of refraction. This makes the refractive index values always greater than 1.

Reversibility of light

In earlier sections you saw that if light could travel from point A to point B, the same source placed at B would retrace its path back to point A.

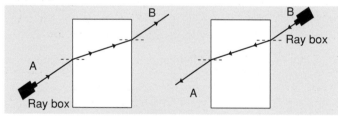

Figure 7.10.2 Reversibility of light

When considering refraction we need to use this principle with care. The refractive index of glass involves a ray of light *entering* a glass block. From figure 7.10.1 we can see that the refractive index of the glass is given by;

$$n_{glass} = \frac{sin\ \theta_1}{sin\ \theta_2}$$

Figure 7.10.3 shows a ray created within the block and retracing the path of the ray in figure 7.10.1. The reversible nature of light means that the angles in this diagram are the same as in figure 7.10.1.

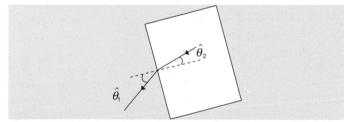

Figure 7.10.3 Ray emerging from a glass block

While the angles in figures 7.10.1 and 7.10.3 are identical, the ray in figure 7.10.1 is *entering* the glass while in figure 7.10.3 the ray is *emerging* from the glass. **To calculate the refractive index for the glass we must remember that the incident angle must be the one in air while the refracted angle is in the material.**

The similarities between the diagrams of figure 7.10.1 and 7.10.3 allow us to determine a relationship between the refractive index values for entering and emerging rays.

$$n_{air \rightarrow glass} = \frac{sin\ \theta_1}{sin\ \theta_2} \qquad n_{glass \rightarrow air} = \frac{sin\ \theta_2}{sin\ \theta_1}$$

Therefore we can say;

$$n_{air \rightarrow glass} = \frac{1}{n_{glass \rightarrow air}}$$

This equation allows us to calculate the refractive index of glass even if we are presented with information about the passage of the ray from the glass into air. The same argument can be used for finding the refractive index of any material.

$$n_{air \rightarrow material} = \frac{1}{n_{material \rightarrow air}}$$

Q A coloured special effects bulb placed at the bottom of a swimming pool emits rays in all directions. The path of one of these rays as it emerges from the water is as shown. Calculate the refractive index of the water.

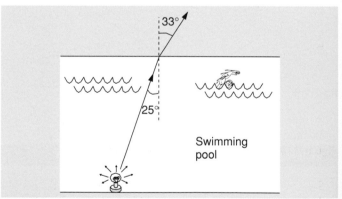

Figure 7.10.4

A From the diagram we note that the ray is travelling from water into the air. To find the refractive index of the water we must consider the path of a ray *entering* water. But since light is reversible we can say;

$$n_{water} = \frac{sin\ 33°}{sin\ 25°}$$

$$n_{water} = 1.29$$

Q A second ray is incident on the surface at an angle of 35° to the water. Calculate the angle with which this ray emerges.

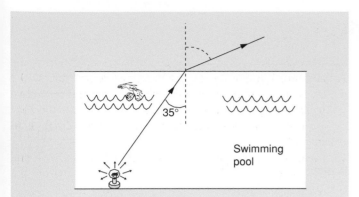

Figure 7.10.5

 To answer this part of the question we must again remember that the ray is emerging from the water into the air. We can again use the reversibility principle;

$$n = \frac{sin \; \hat{i}_{air}}{sin \; \hat{r}_{water}}$$

$$1.29 = \frac{sin \; \hat{i}}{sin \; 35°}$$

$$sin \; \hat{i} = 1.29 \times sin \; 35°$$

$$\hat{i} = 47.7°$$

Total internal reflection

The previous section has shown that we need to be careful calculating the refractive index where a ray of light is emerging from a material.

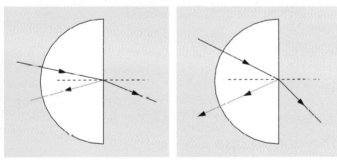

Figure 7.10.6   **Figure 7.10.7**

Figure 7.10.6 shows a ray of light incident along a radius towards the centre of the straight side of a semicircular glass block. There is no change in direction when the ray enters the block but as it emerges, the ray bends away from the normal. By comparing figures 7.10.6 and 7.10.7 we can see that when the angle of incidence increases the angle of refraction also increases.

When the angle of incidence increases sufficiently the ray will emerge at 90° to the normal as shown in figure 7.10.8. The angle of incidence needed to make this happen is called the **critical angle, θ_c**. If the angle of incidence exceeds the critical angle **total internal reflection** occurs. For angles less than the critical angle *some* of the ray is internally reflected and *some* emerges from the block. When the angle of incidence exceeds the critical angle none of the light emerges as it is *totally* internally reflected (figure 7.10.9).

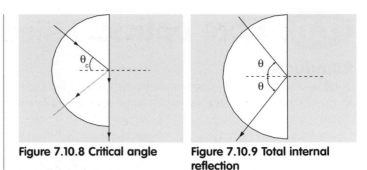

Figure 7.10.8 Critical angle   **Figure 7.10.9 Total internal reflection**

Quantitative total internal reflection

Total internal reflection occurs when a ray is travelling from a dense to a less dense material so again we need to be careful when using the refractive index values for the material. Since light is reversible, we can use the refractive index values quoted for the material if we consider the ray entering the block from air and retracing the path shown in figure 7.10.8.

$$n_{material} = \frac{sin \; 90°}{sin \; \theta_c}$$

But sin 90° = 1, Therefore;

$$n_{material} = \frac{1}{sin \; \theta_c}$$

$$sin \; \theta_c = \frac{1}{n_{material}}$$

Using this final equation we can quickly find the refractive index for the block shown in figure 7.10.8 by simply increasing the angle of incidence until the critical angle is reached.

Above the critical angle

A block of glass with a refractive index of 1.5 will have a critical angle of 41.8°. When the angle of incidence for a ray of light exceeds 41.8°, the ray will be totally internally reflected.

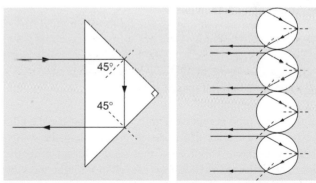

Figure 7.10.10 45° prism   **Figure 7.10.11 Reflective beads**

The angles in the glass prism are such that total internal reflection returns the ray parallel to its original direction. Reflective strips on cycle jackets and reflective road signs are coated with small plastic beads whose refractive index is such that light is reflected back along the incoming path.

Modern fibre optic communications use lasers to send high frequency pulses of light down thin plastic fibres. Total internal reflection prevents the light intensity from decreasing significantly even when the signal is transmitted along a cable a few kilometres long.

7.11 Fibre optics

Introduction

Total internal reflection occurs at an interface where light tries to travel from an optically denser material into one of lower density.

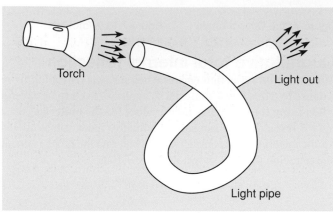

Figure 7.11.1 Light pipe

Figure 7.11.1 shows a piece of cylindrical glass bent into a particular shape. The light source puts light into one end of this light pipe and the light is detected at the other end. This demonstration may seem surprising as the light appears not to be travelling in straight lines as expected. We can explain this, however, using the ideas of total internal reflection.

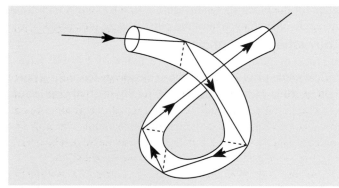

Figure 7.11.2 Total internal reflection

Since the glass pipe is relatively thin, the rays hitting the edge of the pipe have angles of incidence greater than the critical angle for the glass. This means that total internal reflection occurs. The reflected rays always travel in straight lines but frequent reflections with the walls changes the direction of the rays of light so the light appears to 'bend' its way around the corner in this piece of glass.

Practical optical fibres

The optical fibres used today are much thinner than the light pipe of figure 7.11.1. Modern optical fibres are as thin as a human hair, having a diameter of about 5 μm (5×10^{-6} m). This fibre is very flexible and has a narrow core made of very pure glass surrounded by cladding which is also made of very pure glass.

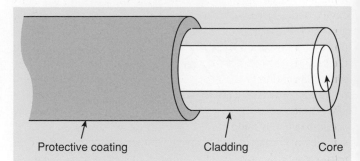

Figure 7.11.3 Construction of fibre optic cables

As in all technologies there are a variety of different ways for combining the core and cladding but each works because the cladding glass has a lower refractive index than the core. Therefore light travelling in the core is contained within the core by total internal reflection at the core cladding interface. The types of cable currently being used are really bunches of fibre optic cables.

Figure 7.11.4

A steel inner strand provides the physical strength. Each of the optical fibres, wrapped loosely in a protective polythene tube is then wrapped tightly around this steel wire. A layer of aluminium is added to provide a water barrier and the whole thing is wrapped in polythene to provide further physical protection.

The endoscope

Doctors can use bundles of thin fibres to look inside patients' bodies. **An endoscope contains two distinct bundles of optical fibres**.

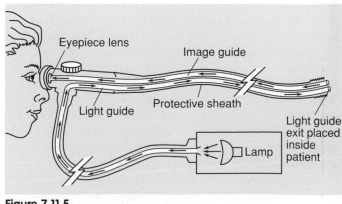

Figure 7.11.5

One of the bundles of fibres is used to take light *into* the area to be examined. The other bundle of fibres brings back the *reflected image* of the bright and dark spots within the body cavity being inspected. This bundle of fibres must be carefully arranged so that each fibre has exactly the same relative position at either end. In this way what is being viewed inside the body is recreated in the eye of the doctor. If the fibres were arranged randomly the image viewed by the doctor would not be identical to the inside of the body.

The bundles of optical fibres in the endoscope can be accompanied by biopsy probes or retrieval nets so that small samples of tissue can be taken or so that kidney stones can be retrieved. In the treatment of ulcers, pulses of laser light are guided by the endoscope onto the ulcer to burn and seal it neatly. These techniques are generally less intrusive than conventional operations where the affected part of the body is opened before surgery. Consequently **keyhole surgery** requires shorter convalescence periods and patients are less prone to infections during the recovery period. The endoscope is one of the major tools of the keyhole surgeon.

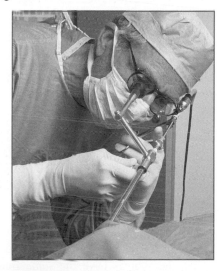

**Figure 7.11.6
An endoscope being
used in an operation**

Fibre optic communications

While only a few of us may have the opportunity to see endoscopes in use, it will be virtually impossible for anyone to avoid optical fibres used in communications, even though we may not be able to see them in operation.

Figure 7.11.7 Installation of optical cables

Those living in many cities and towns throughout Scotland will already have access to cable TV or have telephones connected to a fibre optic cable. Long distance phone calls starting in places not yet joined to the "information superhighway" will probably be at least partially routed along an optical fibre. School and other computer networks covering large distances are likely to be based around a fibre optic spine.

A little further...

Communicating through an optical fibre

Pulses of laser light are used to transmit information along very thin optical fibres. In common with all transmission systems, passing a signal through an optical fibre results in loss of signal strength (**attenuation**) and the signal emerges weaker. To overcome this loss of signal intensity, amplifiers are placed at regular intervals along the fibre optic system. As well as restoring the amplitude of the signal these also recreate its shape and in this way very accurate copies of the original signal can be transmitted over large distances.

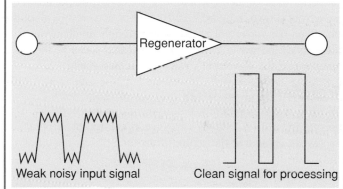

Figure 7.11.8

With optical fibres many different signals can be carried within the same glass fibre. This is made possible by a technique called **multiplexing**. In a telephone system using fibre optics short bursts of each conversation are sent in succession down the fibre many times per second. These are then decoded and sent to the appropriate destinations.

There are many different ways of cramming as much information as possible into the stream of pulses passing down a fibre optic cable and no doubt other ways of modulating the stream of pulses will be developed in the future as the volume of digital communications traffic increases.

Advantages of digital fibre optic systems

Digital transmissions through optical fibres have significant advantages over the more common metal wire communications systems. Optical fibres have a higher capacity. They can carry a wider range of frequencies so sound transmissions are of a better quality. The range of frequencies from which any signal is composed is called the **bandwidth**. TV pictures contain a lot of detail and require a large bandwidth for transmission. Television pictures travelling through optical fibres can be allocated enough bandwidth so that 25 full colour pictures per second can be transmitted. This is the basis for cable TV systems.

Fibre optic systems are also cheaper to install than metal wire systems and have a larger distance between amplifiers.

7.12 Interference

Introduction

We have already seen that energy can be transferred from a source to other points by wave motion. A transverse wave made by dipping a spherical object into the centre of a tank of water spreads energy outwards from the centre as shown in figure 7.12.1.

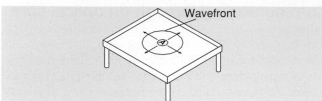

Figure 7.12.1 Ripple tank

Figure 7.12.1 shows one wavefront a certain time after it has left the source. The wavefront shows parts of the wave which are at the same stage of their oscillation. For example all points on the wavefront shown in figure 7.12.1 might be at a crest or all of the points on the wavefront might be at a trough simultaneously. If all parts of the wavefront travel at the same speed as the wave spreads out, the wavefront retains its circular shape.

Transferring energy

When a wave passes any point, energy is transferred to that point. Figure 7.12.2 shows a cross section through a ripple tank as a wave created at the centre of the tank travels outwards.

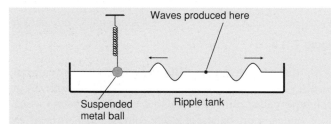

Figure 7.12.2 Wave transferring energy

The partly submerged ball attached to the end of a stretched spring will rise above its rest position and fall below that position before returning to the rest position as the wave passes. The passing of the wave transfers energy to the ball and spring.

Figure 7.12.3 shows two sources, producing straight waves in the same ripple tank. What will happen when the ball, on the end of the spring, is placed at point P?

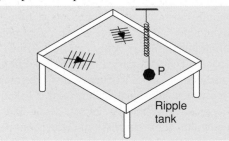

**Figure 7.12.3
Two sources in
a ripple tank**

Clearly the ball will get energy from *both* sources as their waves pass through point P. The motion of the floating ball at P will depend on a number of factors including the frequency of the waves from the sources and the amplitude of the waves.

Principle of superposition

When two waves are travelling in the same region, the resultant displacement at a point, at any given time, is the **vector sum** of the individual displacements at that point. This is known as the **principle of superposition**. If the waves from sources S_1 and S_2 are as shown in figure 7.12.4 then the resultant displacement as these waves pass P will be as shown.

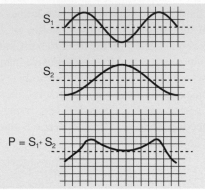

Figure 7.12.4 Superposition

This figure shows what happens when the waves from the two sources *interfere* at point P.

Sources in phase

Interference occurs at all points whenever two waves meet. Figure 7.12.4 represents a very general case where sources S_1 and S_2 have different frequencies. Consequently the crests and troughs are passing through P at different times. The resultant displacement caused by the energy from these two sources arriving at P would be difficult to observe and complicated to describe mathematically.

Much easier to understand **interference patterns** are produced when the sources have the *same* frequency. In the simplest possible case shown in figure 7.12.5 the sources of the same frequency are producing crests and troughs simultaneously. The interference pattern produced at points which are an equal distance from both sources is shown.

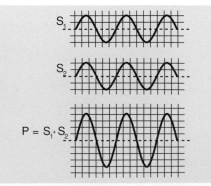

Figure 7.12.5 In phase interference

Sources such as S_1 and S_2 which continually produce crests and troughs at the same time are said to be completely **in phase**.

Constructive and destructive interference

Figure 7.12.5 shows the resultant of waves arriving in phase at a point. These waves were produced in phase with each other and have travelled exactly the same distance, so they will still be in phase at the point where they interfere. This type of interference, where waves arrive in phase at a point, is known as **constructive interference**. Positions where constructive interference occur are called **maxima**.

Sources of the same frequency but having one source producing a crest at the same time as the other source produces a trough are completely out of phase or in **anti-phase**. Such sources can also be described as having a phase difference of 180°. Sources which produce waves with a constant phase difference are described as **coherent sources**. Coherent sources may have a phase difference of 0°, 180° or any other value. The key point is that the phase difference is *constant* throughout the wave motion.

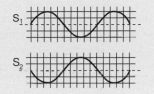

Figure 7.12.6 Coherent sources

Figure 7.12.6 shows waves produced by two sources of the same frequency. The phase difference between waves from the two sources is 180°. These sources are coherent since the phase difference is constant throughout the motion. At points, equidistant from both sources, the waves will arrive in antiphase and cause cancellation. The combined effect of the waves arriving in antiphase at a point is shown in figure 7.12.7.

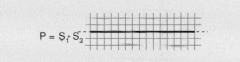

Figure 7.12.7 Destructive interference

The energy from each source arrives in antiphase and causes a cancellation. This is called **destructive inference** and any position where it occurs is called **a minimum**.

Producing coherent sources

A laser light source produces radiation with a single frequency. When laser light is incident on a double slit the two slits act as coherent sources.

The interference pattern produced by passing laser light through a double slit is seen on a screen as a series of bright and dark patches.

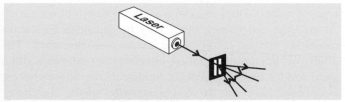

Figure 7.12.8 Laser and two slits

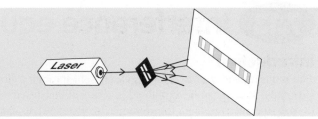

Figure 7.12.9 Interference pattern

The bright patches can be explained in terms of **constructive interference** while the dark patches are a consequence of **destructive interference.**

Other interference patterns

A ripple tank with two spherical objects dipping into the water can also produce an interference pattern.

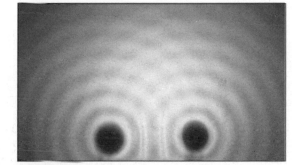

Figure 7.12.10 Ripple tank interference pattern

The bright areas in figure 7.12.10 are regions of constructive interference while the dark areas indicate places of cancellation or destructive interference.

We can explain this pattern by considering figure 7.12.11 where sources S_1 and S_2 are producing circular waves. Between each of the crests indicated there are troughs. In specific areas, crests from one source *reinforce* crests from the other while in other areas crests from one source are *cancelled* by troughs from the other.

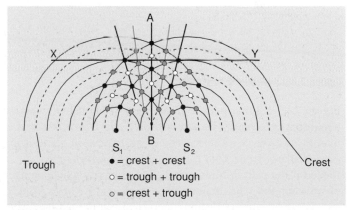

Figure 7.12.11

Along the line XY there will be a sequence of areas where constructive and destructive interference occurs.

The similarities between the sequence along XY and the pattern of bright and dark patches for the laser and double slit would suggest that both can be explained by our model of wave interference.

7.13 Interference equations

Introduction

The pattern produced by passing laser light through a double slit onto a screen can be explained in terms of constructive and destructive interference. Interference in water waves can be explained by the same model. We can use water waves to show the effect that increasing the frequency of the sources has on the arrangement of maxima and minima in an interference pattern. The frequency of both sources must be increased to the same value so that the sources remain coherent.

If the waves are travelling at a constant speed, increasing the frequency reduces the wavelength. This will result in the crests being closer together.

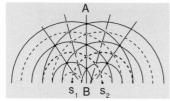

Figure 7.13.1

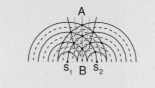

Figure 7.13.2

At the higher frequency the areas of constructive and destructive interference are closer together.

Positions of constructive interference

It is easy to understand why constructive interference occurs at the positions marked along the lines AB. At these points the waves, which are produced in phase, have travelled equal distances from each of the sources. From earlier examples it is possible to see that constructive interference also occurs at certain positions which are different distances from each source.

At any maxima in an interference pattern, the waves must arrive in phase.

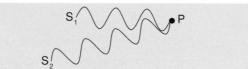

Figure 7.13.3 Maxima formation

Figure 7.13.3 shows point P in an interference pattern where a maximum is formed. The waves from S_2 have travelled further than those from S_1. The waves can still arrive in phase if the *extra* distance from S_2 to P is equal to *one wavelength*.

Indeed if the path difference is 1, 2, 3, 4 or any whole number of wavelengths a maximum will be formed at P. Crests from S_1 and S_2 are arriving at P simultaneously even though they were produced by the sources at different times. Troughs from S_1 and S_2 will also arrive together at P and will again interfere constructively.

Maxima are produced in interference patterns at all points where;

$$S_2P - S_1P = n\lambda$$

where *n* is any integer.

Positions of destructive interference

Minima are formed in interference patterns where the waves arrive in antiphase.

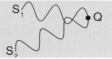

Figure 7.13.4 Minima formation

Figure 7.13.4 shows point Q in an interference pattern where a minimum is formed. The waves from S_2 have travelled further than those from S_1. The waves arrive in antiphase if the *extra* distance from S_2 to Q is equal to *half a wavelength*. If the path difference is ½, 1½, 2½, 3½ or any similar multiple of the wavelengths a minimum will be formed at Q because **crests from one source will be arriving at the same time as troughs from the other source**.

Minima are produced in all interference patterns at all points where;

$$S_2Q - S_1Q = (n + \tfrac{1}{2})\lambda$$

where *n* is any integer.

The first minimum on either side of the central maximum will be caused by a path difference of ½λ. This corresponds to a value of $n = 0$. Subsequent minima are formed when $n = 1, 2, 3$ etc. The distance between the centres of any two adjacent maxima or minima correspond to an extra path difference of one wavelength.

Q Microwaves of wavelength 2.8 cm from a single transmitter travel towards two gaps in an aluminium plate. The third maximum on one side of the straight-through position occurs at a point 13 cm from the closest slit. How far is this maximum from the other slit?

A For the third maxima we can say;

$$S_2P - S_1P = n\lambda$$
$$S_2P - 13 = 3 \times 2.8$$
$$S_2P = 21.4 \ cm$$

The maximum is 21.4 cm from the second slit.

Light: is it waves or particles?

The first demonstration of optical interference which was recognised as such, was performed by Thomas Young in 1801 using coherent sources of white light. The interference explanation of Young's experiment supported Huygen's theory that light was a wave rather than a series of corpuscles (particles) as had been suggested by Newton. Reflection, refraction, diffraction and interference are characteristic behaviours of all types of waves so they must be explained by any correct model. At the time, interference could only be explained by a wave model so Newton's corpuscular theory was discarded. **The ability of radiation to produce interference patterns is still a good test for any type of wave motion.** However more recent discoveries have led to **electromagnetic waves** also being considered as streams of particles called **photons**. These types of radiation are said to exhibit **wave particle duality**.

7.14 Interference experiments

Interference patterns with sound

EXPERIMENT In this experiment you will investigate the interference pattern produced when a signal generator is connected to two loudspeakers. A microphone is attached to an oscilloscope to identify adjacent positions of constructive interference.

You could investigate the relationship between the spacing, x, of these maxima and;
1   the distance between the loudspeakers, d
2   the distance, D, between the loudspeakers and the microphone
3   the wavelength of the sound, λ

You should be able to confirm the relationships in the equation;

$$x = \frac{\lambda D}{d}$$

You might notice that at positions of minima, there is not complete cancellation. The energy reaching such minima from each source is not quite the same. The loudspeakers might be producing signals of different amplitude. However, even if the loudspeakers do emit the same intensity of sound, at most minima the sound from one source will have travelled further and would be expected to have a smaller amplitude.

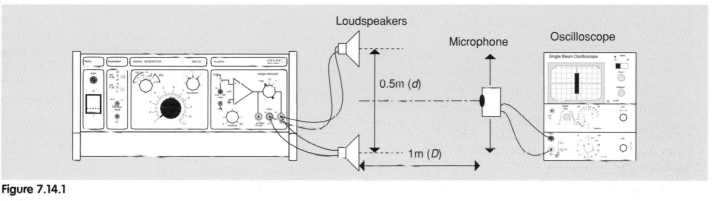

Loudspeakers
Microphone
Oscilloscope
0.5m (d)
1m (D)

Figure 7.14.1

Interference patterns for microwaves

EXPERIMENT In this experiment you will investigate the interference patterns produced when electromagnetic waves from a microwave transmitter are reflected from a metal plate. Interference is caused between the transmitted and reflected signals.

When the receiver is stationary, positions of maximum and minimum signal will be indicated by the meter when the reflecting plate is moved. If the reflecting plate is moved a distance x directly away from the probe a path difference of $2x$ is introduced between the transmitted and reflected signals.

Path difference = $2x$

You could move the reflecting plate carefully away from the

receiver and mark the positions of the plate when peaks of intensity are registered by the receiver. You can then calculate an average value for x, the separation between these positions of peak intensity.

Adjacent maxima are formed where the extra path difference between each maxima is one wavelength. Therefore for maxima;

Path difference = λ

Therefore;

$2x = \lambda$

This experiment allows us to calculate the wavelength of the microwaves. If the frequency emitted by the transmitter is known then the speed of electromagnetic waves in air can be calculated.

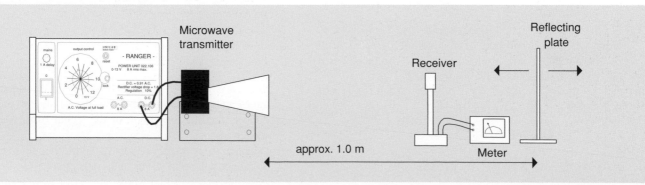

Microwave
transmitter
Reflecting
plate
Receiver
approx. 1.0 m
Meter

Figure 7.14.2

7.15 Consolidation questions

Worked example 1

Figure 7.15.1 shows two identical loudspeakers connected to a signal generator. The emitted sound has a frequency of 500 Hz.

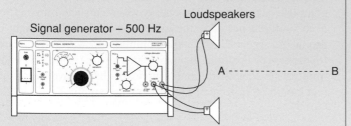

fig 7.15.1

a) A student states that at all points along the line labelled AB, waves from the two sources will interfere constructively. Explain what is meant by *constructive interference*.

b) One of the loudspeakers is moved forward by a few centimetres. Explain why the waves arriving at any point along the line AB are now unlikely to be in phase.

Answers and comments

a) All points along the line labelled AB are the same distance from both sources. Therefore waves from both sources arrive in phase causing an increase in the intensity of the sound heard.

b) Sound from one loudspeaker will now take longer than sound from the other, to reach a point along AB.

Worked example 2

Two loudspeakers are connected as shown to a signal generator which produces a signal with a frequency of 2000 Hz.

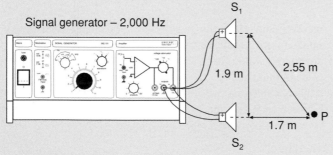

fig 7.15.2

a) Sound waves travel through air at 340 m s^{-1}. What is the wavelength of the sound produced by the loudspeakers?

b) How many complete wavelengths of this sound are there in
   (i) the path S$_2$P
   (ii) the path S$_1$P?

c) A microphone attached to an oscilloscope is placed at P. How will the amplitude of the sound detected at P compare with the amplitude of the sound produced when one of the loudspeakers is disconnected?

d) The frequency of the sound is doubled. What effect will this have on the type of interference found at P?

Answers and comments

a) From the wave speed equation $v = f\lambda$ we can say

$$\lambda = \frac{v}{f} = \frac{340}{2000}$$

$$\lambda = 0.17 \ m$$

b) The distance S$_2$P is marked as 1.7 m on the diagram. Since the wavelength of the sound is 0.17 m, S$_2$P is equal to 10 complete wavelengths. The distance S$_1$P is marked as 2.55 m, therefore;

$$n \times 0.17 = 2.55$$

$$n = \frac{2.55}{1.7} = 15$$

The distance S$_1$P equals 15 wavelengths.

c) The path difference can be calculated from;

$$S_1P - S_2P = 2.55 - 1.7$$
$$Path \ difference = 0.85 \ m$$

If this path difference equals a whole number of wavelengths then P is a point of constructive interference. We must now try to find n;

$$0.85 = n\lambda$$

$$n = \frac{0.85}{0.17} = 5$$

Since n is a whole number, P is a point of constructive interference. The amplitude of the signal is therefore greater than the amplitude of the signal from one loudspeaker alone.

d) Doubling the frequency will halve the wavelength so the path difference will be 10 wavelengths rather than 5. However, the path difference is still a whole number of wavelengths so P will still be a point of constructive interference.

Questions

1 A student uses a single microwave transmitter and the arrangement of metal plates as shown in figure 7.15.3 to investigate interference. The gaps between the plates form two slits.

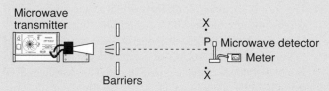

fig 7.15.3

a) She states that at point P there is a maximum of signal in the interference pattern. Explain the meaning of the term maximum of signal.

b) She states that the gaps in the barrier act as coherent sources. Explain the meaning of the term coherent.

c) Explain, in terms of waves, why a maximum of signal is produced at P.

d) When the microwave detector is moved to either of the points marked X, minima are detected. Explain, in terms of waves, why minima are produced at the points marked X.

2 A microwave transmitter is emitting waves towards two small gaps X and Y. At point P there is a minimum of intensity.

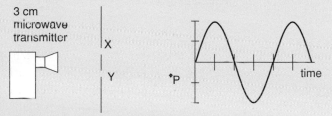

fig 7.15.4

The graph of figure 7.15.4 shows the microwave energy arriving at P from X during a certain interval of time. Which one of the following graphs shows the energy from Y arriving at P during the same interval of time.

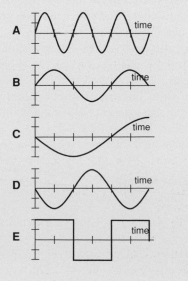

fig 7.15.5

3 A student sets up the apparatus as shown in figure 7.15.6 in a room which has been designed to minimise the reflection of sound.

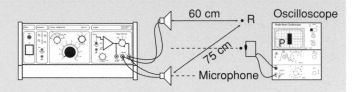

fig 7.15.6

A small sensitive microphone indicates a maximum in sound intensity when it is placed at P.

a) As the microphone is moved towards R it indicates that the sound intensity decreases and then reaches a maximum again at R. What is the wavelength of the sound waves?

b) The signal generator dial indicates that the sound waves have a frequency of 2.2 kHz. Calculate the speed of sound in air.

4 In figure 7.15.7, S_1 and S_2 represent coherent sources of monochromatic light. There are bright and dark patches in the resulting interference pattern caused by S_1 and S_2. The point marked X is at the centre of the second bright area above the central maximum.

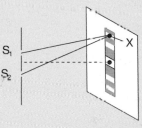

fig 7.15.7

a) The wavelength of the light source is 700 nm. What is the path difference $S_2X - S_1X$?

b) The 700 nm source is replaced by another monochromatic source emitting light with a wavelength of 400 nm. Will the second maximum in this new interference pattern be closer to or further away from the central maximum? Justify your answer.

5 A pupil sets up the apparatus shown below. A minimum of intensity is recorded when the microphone is at P.

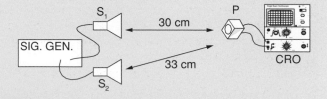

fig 7.15.8

Which of A to E could be the wavelength of the sound produced by the signal generator?

A   4 cm   B   6 cm   C   11 cm
D   30 cm   E   36 cm

Worked example 1

Figure 7.16.1 shows a ray of monochromatic light incident along the normal to side AC of a glass block.

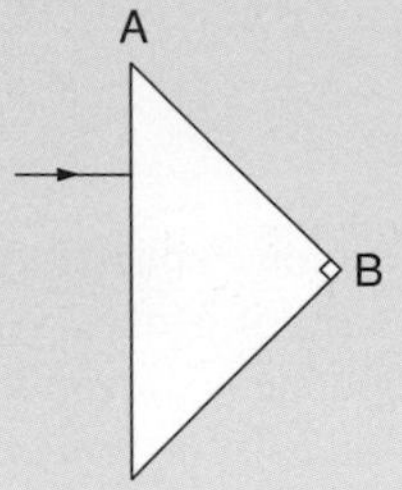

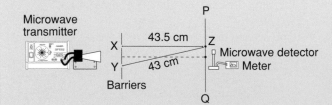

fig 7.16.1

a) What happens to the direction, speed, frequency and wavelength of the light as it enters the glass block?
b) What is the angle of incidence on the side AB?
c) The refractive index of the glass is 1.48. Calculate the critical angle for this light in the glass.
d) Copy figure 7.16.1 and complete the path of the ray until after it leaves the block.

Answers and comments

a) This ray is incident along the normal to the surface so it passes into the block without changing direction.
Since the block is optically denser than the air the light slows down. The frequency of the light does not change and as a consequence of the decrease in speed the wavelength of the light in the glass block is shorter than in the air.
b) Since the ray passes into the block without changing direction it strikes the side AB at an angle of 45°. The angle of incidence at side AB is also 45°.
c) The critical angle is calculated from;

$$n_{glass} = \frac{1}{\sin \theta_c}$$

$$\sin \theta_c = \frac{1}{1.48}$$

$$\theta_c = 42.5°$$

d) The ray is incident at AB at an angle greater than the critical angle so it is totally internally reflected. The angle of incidence on BC is also 45° so the ray is again totally internally reflected before leaving the block parallel to the incoming ray.

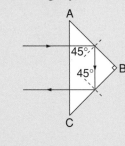

fig 7.16.2

Worked example 2

Microwaves pass through two slits X and Y in a metal plate.

Microwave transmitter

X ····· 43.5 cm ····· Z
Y ····· 43 cm ····· Microwave detector / Meter

Barriers

P ····· Q

fig 7.16.3

A microwave detector moved along the line labelled PQ indicates that the microwave signal strength is higher at some points than at others.
a) Explain, in terms of waves, why peak readings occur at a number of different points as the detector moves from P to Q.

Answers and comments

a) Peak readings occur in the interference pattern at points of constructive interference where waves from both gaps arrive in phase.
The waves will arrive in phase at any point, R, where the path difference (XR – YR) is a whole number of wavelengths.
b) At Z the waves arriving from both sources must be in antiphase so they cancel producing a minimum. Z is the first minimum because YZ – XZ = ½ λ.
c) Since Z is the first minimum we can say that;

$$YZ - XZ = ½ λ.$$

From the diagram we can see that the distances YZ = 45 cm and XZ = 43.5 cm.

$$45 - 43.5 = \frac{1}{2} λ$$

$$λ = 3 \text{ cm}$$

b) The first minimum of microwave signal on the side of the central maximum occurs at point Z as shown in figure 7.16.3. Explain, in terms of waves, why the *first* minimum is produced at this point.

c) Use the information from the diagram to calculate the wavelength of the microwaves.

d) What is the frequency of the microwaves from this transmitter?

d) Microwaves are a member of the electromagnetic spectrum and consequently travel at 3×10^8 m s^{-1}; the speed of light. Knowing their speed and wavelength we can calculate their frequency from;

$$v = f\lambda$$

$$f = \frac{v}{\lambda} = \frac{3 \times 10^8}{0.03}$$

$$f = 1 \times 10^{10} \text{ Hz}$$

Questions

1 A ray of light travels from air into glass. The angle of incidence at the glass surface is 30°.

   a) Calculate the angle of refraction. The refractive index of the glass relative to the air is 1.51.

   b) A ray of light travels from air into water. The angle of incidence in air is 45° and the angle of refraction in the water is 32°. Calculate the refractive index of the water relative to air.

   c) Does light travel faster in the glass or the water? Justify your answer with calculations.

2 The refractive index of diamond relative to air, for light from a sodium lamp is 2.42.

   a) What is the critical angle for this light in diamond?

   b) This light travels through air at 3×10^8 m s^{-1}. What is its speed in diamond?

   c) What is the wavelength of this sodium light in diamond if the wavelength in air is 589 nm?

3 Figure 7.16.4 shows the passage of a ray of monochromatic light from air into glass.

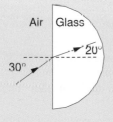

fig 7.16.4

   a) What is the refractive index of this type of glass?

   b) What is the critical angle for this light in the glass?

   c) A ray of light, starting in air, is now directed towards the glass block so that the angle of incidence is equal to the value calculated for the critical angle in the glass.

   d) Copy figure 7.16.4 and show the subsequent path of the ray in the glass.

4 Light has a speed in air of 3×10^8 m s^{-1} and in water of 2.25×10^8 m s^{-1}.

   a) What is the refractive index of the water relative to air?

   b) What is the critical angle for the water?

5 The partially completed ray diagram of figure 7.16.5 shows three rays emerging from a lamp placed 1 m below the surface of a pond full of water. The refractive index for water is 1.31.

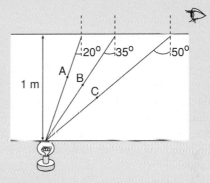

fig 7.16.5

   a) At what angle will the ray A emerge into the air?

   b) At what angle will the ray B emerge into the air?

   c) What will happen to the ray C after it strikes the surface between the water and the air?

   d) Sketch figure 7.16.5 and complete the ray diagram to explain why the observer imagines that the lamp is less than 1 m below the surface of the water.

6 In an optics exhibit at a science fair, a ray of monochromatic light travels from X to Y through a block made from two different types of glass, P and Q as shown in figure 7.16.6.

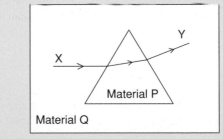

fig 7.16.6

Which of the following is true?

A The wavelength of the light is greater in Q than in P.

B The frequency of the light is greater in Q.

C The ray bends towards the normal as it passes from P to Q.

D Total internal reflection might occur at the surface where the ray travels from P to Q.

E The absolute refractive index for both materials is the same.

7 A ray of light passes from glass into air as shown.

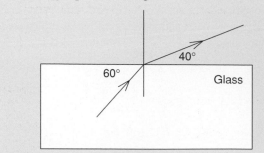

fig 7.16.7

The refractive index for this glass is;
A 0.65   B 0.74   C 1.35
D 1.53   E 1.50

8 Fibre optical cables are being used in an increasingly wide range of applications.
  a) What physical principle keeps the coded light signals within the fibre?
  b) Signals transmitted through the fibre are sometimes described as digitally encoded. What is meant by digital encoding?
  c) Doctors often use a fibre optical instrument called an endoscope to look inside patients' stomachs. Describe how this operates and outline the steps that must be taken to ensure that the image seen is an exact copy of what is being observed.

9 When a microphone is placed at point P, an equal distance from sound sources X and Y, the oscilloscope shows a zero reading. If either X or Y is covered by a solid object the reading is non zero.

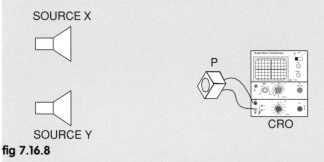

fig 7.16.8

The zero of intensity can be explained by saying;
A the sounds from X and Y have different frequencies
B the sounds from X and Y have different amplitudes
C the sources are emitting the same frequency of sound but the sounds from X and Y are in antiphase
D the phase difference between the sources is 90°
E the sources emit the same frequency in phase.

10 A student investigating the interference of sound waves sets up two loudspeakers in a room where the walls are designed to minimise sound reflections.

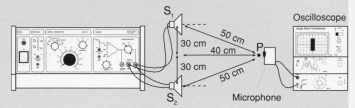

fig 7.16.9

Using a small microphone attached to an oscilloscope he detects a position of maximum intensity in the sound pattern when the microphone is at the point P, shown in figure 7.16.9.
  a) How far is the microphone from the loudspeaker labelled S_1?
  b) The sound waves from the loudspeakers have a wavelength of 10 cm. How many complete waves will fit in the distance between the loudspeaker and the microphone?
  c) Explain in terms of waves why this is a position of maximum intensity.
  d) The student moves the microphone to the position shown in figure 7.16.10.

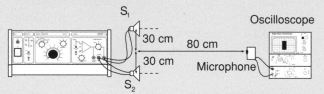

fig 7.16.10

  (i) Will the microphone indicate that this is a position of maximum or minimum intensity in the interference pattern? Justify your answers with calculations.
  (ii) Each loudspeaker is now connected to a different signal generator. The generators have the same frequency. Explain why there is no interference pattern.

11 Two sources are said to be coherent when the waves from them;
  A have the same amplitude
  B have the same frequency
  C have different frequencies but are in phase as they leave their transmitters
  D have the same frequency and have a constant phase difference throughout their motion
  E interfere constructively at all points.

12 In an experiment on interference a student sets up the apparatus shown in figure 7.16.11.

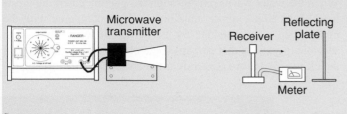

fig 7.16.11

The transmitter emits microwaves which travel through the air towards the metal reflecting plate with a speed of 3×10^8 ms^{-1}.

a) Explain, in terms of waves, how interference can occur with this arrangement of apparatus.

b) The microwave probe is moved further from the transmitter towards the reflecting plate and a number of maxima of signal are indicated on the meter. Between the maxima there are positions where the signal is a minimum.

   (i) The probe is carefully positioned where the meter indicates a maximum reading. Explain in terms of waves how this maximum is caused.

   (ii) The probe is placed at a point of maximum signal. The reflecting plate is then removed. Will the reading indicated on the meter increase or decrease? Justify your answer.

c) The plate is now replaced. State what happens when the probe is kept in one position and the plate is slowly moved further away from the transmitter.

13 A student designs an experiment to demonstrate interference using a single loudspeaker attached to a signal generator.

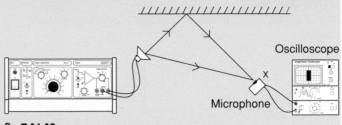

fig 7.16.12

The loudspeaker sends some waves towards a wall and some towards a microphone, placed at X. The microphone is attached to an oscilloscope.

a) Explain how interference is produced even though there is only one loudspeaker.

b) What will happen to the amplitude of the trace shown on the oscilloscope as the frequency of the sound is increased? Explain your answer.

7.17 Diffraction

Introduction

When a bright light shines through a doorway a shadow is formed.

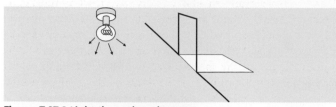

Figure 7.17.1 Light through a doorway

The light passes through the gap but is stopped by the wall surrounding the doorway.
A person standing in the doorway will block out a certain portion of the light trying to travel through the gap.

Figure 7.17.2 Shadow of a man standing in doorway

The shadow will have sharp, clearly defined edges and the same shape as the object itself. In earlier physics courses you will have met the idea of shadows as evidence that light travels in straight lines. If rays of light can get through a gap they continue along their original path. The shadow formed on a screen will therefore show some detail of the outline of the object obstructing the light.

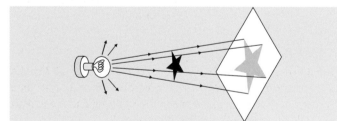

Figure 7.17.3 Ray diagram for shadow formation

Ripple tanks

In our work on interference, the ripple tank was a useful aid to studying the properties of transverse waves. When straight waves are sent towards a wide gap in a barrier they pass through as shown in figure 7.17.4. The width of the wavefront emerging is similar to the size of the gap and the regions where the wave energy has been blocked are fairly obvious.

Figure 7.17.4 Ripple tank with wide gap

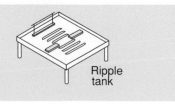

When the water waves pass through the gap the emerging crests have sharp edges whose amplitude is high compared with the undisturbed water shielded by the barrier. It would then seem sensible to suggest that perhaps some of the water at the edge of the crest would move outwards as the crest continues through the gap.

Figure 7.17.5 Mild diffraction at gap in ripple tank

The photograph in figure 7.17.5 shows that there are some effects at the edge of wavefronts after they have passed through the gap. The curving of the edges of the wavefronts becomes more pronounced as the gap gets smaller. Figure 7.17.6 shows the situation where water waves of wavelength 1 cm are incident on a 1 cm gap. The edges of the emerging waves are now sufficiently curved to make these waves circular.

Figure 7.17.6 Circular ripple tank diffraction

A considerable portion of the wave energy has now entered the geometrical shadow.

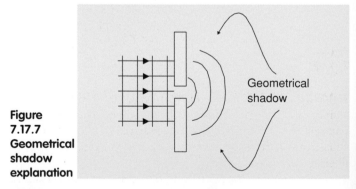

Figure 7.17.7 Geometrical shadow explanation

This wave property, where wave energy enters the geometrical shadow, is called **diffraction**.
By comparing the photographs in figures 7.17.5 and 7.17.6 you will notice that diffraction is most noticeable when the gap size is approximately the same as the wavelength of the incident waves. This helps us to explain why the shadows formed in figures 7.17.1 and 7.17.2 have sharp rather than blurred edges. In these examples the wavelength of the light is infinitely small compared to the size of the gap so effects at the edge of the shadow are not noticeable.

Diffraction of light

In earlier work we produced coherent sources of monochromatic light by allowing laser light to fall on two closely spaced slits. These were used to demonstrate the pattern of bright and dark patches that resulted from the interference of light.

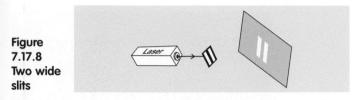

Figure 7.17.8 Two wide slits

If the slits are very much larger than the wavelength of the laser the screen will show two bright lines where the light from the source passes unaffected through the gaps.

When the slits are of approximately the same size as the wavelength of the laser light, interference occurs because of the diffraction taking place at the slits.

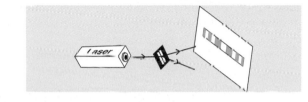

Figure 7.17.9

Laser light passing through each slit spreads into the geometrical shadow of the slit. Each slit acts as a source of coherent light. The crests and troughs in the light waves from each slit meet, causing constructive and destructive interference. As a consequence of this interference the pattern seen on the screen has more than two bright lines.

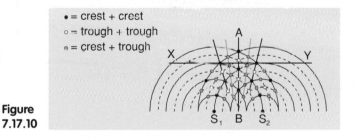

● = crest + crest
○ = trough + trough
◉ = crest + trough

Figure 7.17.10

Investigating interference patterns

It may seem strange to return to an investigation of interference patterns at this point but having seen that diffraction at each slit causes the interference patterns, we can investigate the effect of altering certain parameters.

1: If the distance between the slits decreases the distance between maxima and minima in the interference pattern increases.

2: Allowing light to pass through more than two slits produces patterns which are brighter, but if the slit separation is the same as for the 2 slits, the separation of the maxima and minima will be unaltered.

3: If coherent light of a smaller frequency (longer λ) is used the maxima and minima will be further apart.

It is possible to confirm these observations using gratings and a laser. Alternatively waves produced using spherical dippers in a ripple tank will confirm the same trends. Placing the dippers closer together gives maxima more widely spaced. Using more dippers reduces spacing while higher frequencies of vibration also results in patterns where the maxima and minima are closer together.

The grating equation

A grating is an arrangement of equally spaced scores ruled onto a piece of glass. These are ruled with great precision so that they are equally spaced. It is not uncommon for a grating to have as many as 1500 lines scored on each 1 mm of its width. Light incident on the grating will be diffracted by the clear spaces between the rulings. The rulings on the grating have to be very close together so that the spacing between the lines is approximately equal to the wavelength of light.

1500 lines per mm is the same as 1,500,000 lines per metre. Therefore the distance between these lines is

$$\frac{1}{1,500,000}\ m = 667\ nm$$

With the spacing between slits so minute in comparison with the distance between the grating and the screen it is very difficult to draw diagrams to properly explain the grating equation. If we consider figure 7.17.11 we know that a constructive interference maxima will be produced at the central position. This is called the zeroth order maxima. The first maxima on either side of the central one are called the **first order** maxima. These are formed at a particular angle θ from the straight-through position where the waves from one slit have travelled one wavelength further than the waves from the neighbouring slit.

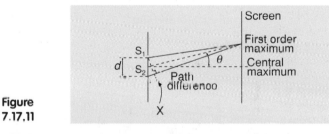

Figure 7.17.11

The path difference for the first order maximum is S_2X. For the first maximum this path difference must be equal to one wavelength. If the distance between the grating and the screen is large the angle $\widehat{S_2S_1X}$ is almost equal to θ.

If d is the distance between the slits, then;

$$S_2X = d\ \sin\theta$$

Therefore for this first maximum

$$d\ \sin\theta = \lambda$$

For further maxima the path differences are 2λ, 3λ ... $n\lambda$. So in the general case the grating forms maxima at angles given by the equation;

$$d\ \sin\theta = n\lambda$$

Where n is the order of the maximum and d the separation of the slits.

7.18 The wavelength of laser light

Introduction

A grating with two slits can produce interference patterns on a screen. However, a grating with just two narrow slits ruled very close together will allow only a small quantity of the laser light through so the resulting interference patterns are often faint.

Although two slit gratings are available, practical gratings have many more slits. If the spacing of the slits on these gratings is the same as that for the two slit grating the maxima and minima in the interference pattern will be formed at the same angular separations. However, **gratings with multiple slits will produce much brighter patterns because more of the incident light is reaching the screen**.

Once again we need to give a little thought to how light from a large number of slits can form a maximum at one point.

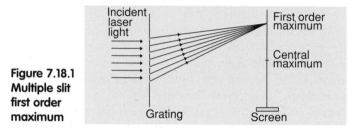

Figure 7.18.1 Multiple slit first order maximum

Each of the slits acts as a coherent source and diffraction of light passing through the slits causes interference patterns. Figure 7.18.1 shows the position of the first order maximum relative to the slits on the grating. As shown in figure 7.18.2 light which arrives at a point of maximum brightness on the screen has come from each of the slits.

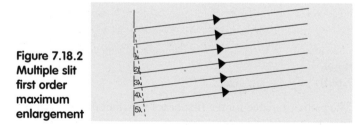

Figure 7.18.2 Multiple slit first order maximum enlargement

Constructive interference occurs on the screen because the path difference between light from neighbouring slits differs by one wavelength.

For the second order maximum the path difference between light from neighbouring slits differs by 2λ;

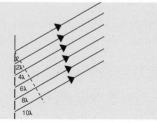

Figure 7.18.3 Multiple slit second order maximum enlargement

As well as increasing the brightness of the maxima, gratings with a large number of slits also produce very sharp maxima with well defined areas of cancellation between.

Angular separation

Figure 7.18.4 shows a picture of the type of pattern produced by a grating.

Figure 7.18.4

We can calculate the angle at which the first maximum is formed if we know the wavelength of the laser and the separation of the slits on the grating. One example uses laser light with a wavelength of 673 nm and a grating that is marked as having 100 lines per mm.

$$100 \text{ lines per mm} = 100,000 \text{ per m}$$

$$d = \frac{1}{100,000}$$

$$d = 1.0 \times 10^{-5} \text{ m}$$

Using the grating equation; $d \sin \theta = n\lambda$

$$\sin \theta = \frac{n\lambda}{d}$$

$$\sin \theta = \frac{1 \times 673 \times 10^{-9}}{1.0 \times 10^{-5}}$$

$$\theta = 3.86°$$

The screen is 1 m away from the grating. We can now calculate the distance between the central and first order maxima.

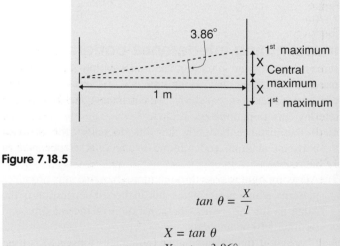

Figure 7.18.5

$$\tan \theta = \frac{X}{1}$$

$$X = \tan \theta$$

$$X = \tan 3.86°$$

$$X = 0.067 \text{ m}$$

The first maximum is therefore 6.7 cm away from the central maximum. This separation could be increased by moving the screen away from the grating or by using a grating with more closely spaced lines. If we used a monochromatic light source with a wavelength of 700 nm the first maximum would be formed on a screen placed 1 m away from the grating, at a distance of 7 cm from the central maximum.

If we used this grating to view a source which emitted light of both 673 nm and 700 nm it would be difficult to distinguish their maxima on the screen.

Measuring the wavelength of laser light

EXPERIMENT When monochromatic light from the laser shown in figure 7.18.6 falls on the grating, diffraction means that each slit acts as a coherent source so an interference pattern is formed on the screen. By taking measurements from this pattern we are able, knowing the separation of the slits on the grating, to calculate the wavelength of the laser light. In this experiment the laser allows us to use a double slit to form the interference pattern.

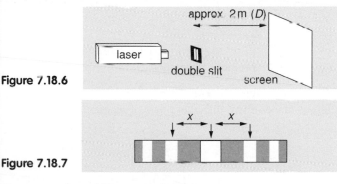

Figure 7.18.6

Figure 7.18.7

If a piece of graph paper is attached to the screen, the average distance, x, between the middle of the central maximum and the first order maxima on either side can be calculated. Knowing this value and the distance of the slit from the screen we can calculate $\tan \theta$. For $n - 1$ the angle θ and $\sin \theta$ can then be determined. The grating equation, $d \sin \theta = n\lambda$, can be used to find the wavelength of the light from the laser, since the separation of the slits is known.

Errors and uncertainties

When you measure the distance between the grating and the screen there will be a **reading uncertainty**. If you use a metre rule you must judge the exact distance. The reading uncertainty can be estimated as one half of the smallest division on the scale. If the metre rule is graduated only in centimetres the reading uncertainty would be 0.5 cm. If the distance between the grating and the screen is 1.98 m the reading is expressed as 1.98 ± 0.005 m where 0.005 m is the **absolute uncertainty**. The **percentage uncertainty** can also be calculated

$$Percentage\ uncertainty = \frac{0.005}{1.98} \times 100\% = 0.25\%$$

Although this uncertainty is small, its significance can be reduced even further by repeating the initial reading several times and calculating a mean. The mean is then the best estimate of the true length. Six repetitions of this reading might give; 1.98 m, 1.97 m, 1.99 m, 1.98 m, 1.98 m, 1.99 m and these would give a mean of 1.982 m.

The **approximate random uncertainty** in these results can be expressed as;

$$Uncertainty = \frac{Range\ of\ readings}{Number\ of\ readings} = \frac{1.99 - 1.97}{6}$$

$$Uncertainty = 0.0033\ m$$

The reading for the distance would be quoted as;

$$1.982 \pm 0.0033\ m$$

The separation of the maxima could also be measured and might be expressed as 6.7 ± 0.1 cm.

The size of the reading uncertainty in this measurement is less, only 1 mm, but the percentage error is greater;

$$Percentage\ uncertainty = \frac{0.1}{6.7} \times 100\%$$

$$\%\ uncertainty = 1.55\%$$

The uncertainty in the slit separation is much more difficult to measure but gratings are often supplied with the slit separation specified as, for example, $(1.0 \pm 0.05) \times 10^{-5}$ m. The % uncertainty in this reading is 5%. This source of uncertainty has a much greater effect than either of the others. When calculating the wavelength of the laser we combine quantities with 0.25%, 1.55% and 5% uncertainty.

Where one uncertainty is much larger than the others, this on its own gives a good estimate of the overall percentage uncertainty in the final numerical result of the experiment. If our calculated value for the wavelength of the laser light is 673 nm we would quote the result as 673 ± 34 nm, i.e. 673 nm ± 5%.

The whole area of error analysis is complicated and a wide variety of advanced mathematical techniques are used to increase the confidence with which experimental results are quoted.

Analysing grating patterns

Light sources which are not monochromatic can be analysed very accurately using the apparatus shown in figure 7.18.8.

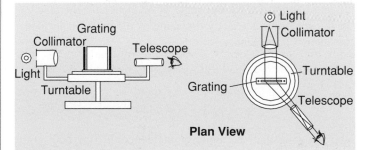

Figure 7.18.8 Spectrometer

Light from the source passes through the collimator onto the grating and the telescope is moved to find the positions of coloured maxima and minima. The wavelengths of the constituent colours of light can then be calculated to a high degree of accuracy.

7.19 Gratings, prisms and white light

Introduction

The path of a ray of light entering a glass block changes as a result of a change in the speed of the waves. Snell's Law allows us to quantify the change in direction and introduces the idea of the refractive index of a material. We saw that different materials have different refractive indices and also stated that the refractive index of any material differs slightly depending on the frequency of light passing through the material.

Typically a glass block will have a refractive index for blue light that is about 1% higher than the refractive index for red light.

Dispersion by a prism

White light is composed of many different frequencies of wave which fall within the visible part of the electromagnetic spectrum. As a glass block has a slightly different refractive index for each frequency of light, the paths followed by light of different frequencies through a prism will be slightly different. **This results in the white light passing through a prism being dispersed into its constituent colours.**

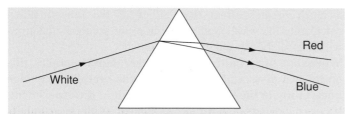

Figure 7.19.1 Colours from white light and a prism

From our work on refraction we know that;

$$n_{1 \to 2} = \frac{sin\ \theta_1}{sin\ \theta_2} = \frac{v_1}{v_2}$$

Since light at the blue end of the spectrum has a slightly higher refractive index than light at the red end, the blue light must travel through the glass slightly slower than the red.

Figure 7.19.1 shows a ray of white light incident on the surface of a glass prism. The white light is made up from each of the visible frequencies but the angle of incidence for the blue part of the spectrum is exactly the same as the angle of incidence for the red light. Since;

$$n_{blue} > n_{red}$$

$$\left[\frac{sin\ \theta_1}{sin\ \theta_2} \right]_{blue} > \left[\frac{sin\ \theta_1}{sin\ \theta_2} \right]_{red}$$

and knowing that the angle of incidence, θ_1, for both is the same, we can say;

$$\left[\frac{1}{sin\ \theta_2} \right]_{blue} > \left[\frac{1}{sin\ \theta_2} \right]_{red}$$

$$(sin\ \theta_2)_{blue} < (sin\ \theta_2)_{red}$$

$$(\theta_2)_{blue} < (\theta_2)_{red}$$

This means that the angle of refraction in the glass is smaller for blue than for red so the blue light is bent towards the normal more than the red.

The resulting spectrum can tell us something about the source of light being used. It is possible to analyse light from different sources in this way to find which frequencies of light they emit. However types of glass with very large refractive indices are required if the dispersion is to be observed easily.

A little further...

Chromatic aberration

Figure 7.19.2 shows that a lens can have slightly different focal lengths for each of the different colours which make up white light. This is called **chromatic aberration**.

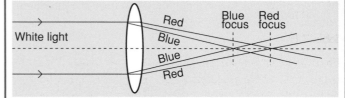

Figure 7.19.2 Chromatic aberration

High power lens systems correct this aberration by using lenses made of two different types of glass.

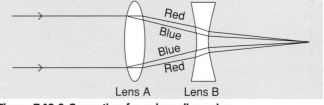

Figure 7.19.3 Correcting for colour dispersion

The overall action of this lens is to converge the rays of light because the convergence from lens A is greater than the divergence caused by lens B. However, if the materials from which the lenses are made are chosen carefully the dispersion from lens A as it converges the rays can be cancelled exactly by lens B which, when diverging the rays, disperses the colours in the opposite order. The resultant effect of this combination is to **focus both the red and blue constituents of the white light source at the same point.**

Dispersion by gratings

A grating will form interference maxima at angles, θ, given by;

$$d\ sin\ \theta = n\lambda$$

where d is the slit separation and n the order of the maxima. For any particular order in the spectrum where $n > 0$;

$$sin\ \theta \propto \lambda$$

This proportionality shows that light of larger wavelengths will form interference maxima at greater angles.

Since white light is a mixture of wavelengths, a grating will produce a series of different coloured interference maxima on a screen.

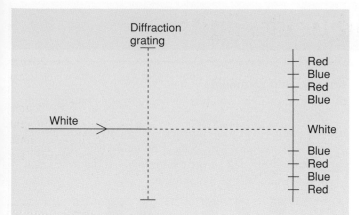

Figure 7.19.4 Grating dispersion

Typical values for the wavelengths of red, green, yellow and blue light are as shown in the table below.

Red	Yellow	Green	Blue
640 nm	590 nm	510 nm	480 nm

Table 7.19.1

From these values we can see that the wavelength of red light is greater than for blue light so the maximum for red light will be formed at a greater angle than the maximum for blue light. The pattern produced by the grating shows all parts of the white light spectrum. The white light has been **dispersed** into its constituent colours.

Calculating the dispersion of red and blue light

 Figure 7.19.5 shows white light from a bright source incident on a diffraction grating which is marked as having 300 lines per millimetre. At what angles will the first order blue and red maxima be formed?

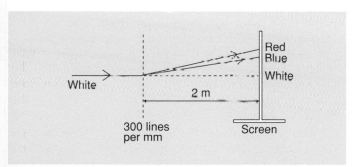

Figure 7.19.5

 We must firstly calculate the spacing of the slits in the grating.

> *300 lines per mm = 300,000 lines per m*

Therefore; the distance between adjacent slits, d, is given by;

$$d = \frac{1}{300,000}$$

$$d = 3.33 \times 10^{-6}\, m$$

So using $d\sin\theta = n\lambda$ and knowing that, since we are dealing with the first maxima, $n = 1$, we can use the value for the wavelength of blue light as given in table 7.19.1

For blue light;

$$3.33 \times 10^{-6} \sin\theta = 1 \times 480 \times 10^{-9}$$

$$\sin\theta = \frac{480 \times 10^{-9}}{3.33 \times 10^{-6}}$$

$$\sin\theta = 0.144$$
$$\theta = 8.29°$$

For red light

We can substitute the wavelength from table 7.19.1 to get the angle at which the first red maxima is formed

$$\sin\theta = \frac{640 \times 10^{-9}}{3.33 \times 10^{-6}}$$

$$\sin\theta = 0.192$$
$$\theta = 11.1°$$

The maximum for red light is formed at a greater angle than the maximum, of the same order, for blue light.

Comparing spectra from prisms and gratings

In this section we have considered two ways of splitting white light into its constituent colours. We have discovered;

1. The prism produces a single spectrum of colours in which red deviates least from the straight through position.
2. The grating produces more than one spectrum of colours.
3. The coloured bands caused by the grating are arranged symmetrically around a central white band.
4. With a grating the blue light deviates less than red from the straight through position.

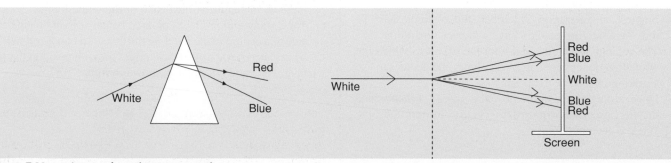

Figure 7.19.6 Prism and grating compared

7.20 Consolidation questions

Worked example 1

A grating is placed in front of a source of monochromatic light and a screen is positioned some distance from the grating.

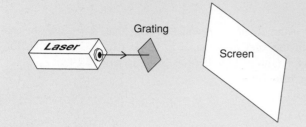

fig 7.20.1

a) Explain the meaning of the term monochromatic.
b) When the grating is removed a single red dot appears on the screen. Draw a diagram to show the type of pattern that is produced when the grating is placed in the path of the beam.
c) The grating is marked as having 300 lines per mm. What is the spacing of the lines?
d) The centre of the second order maximum is formed at an angle of 23°. What is the wavelength of monochromatic light from the laser?
e) Describe and explain what happens to the appearance of the pattern on the screen when the distance between the grating and the screen is increased.

Answers and comments

a) A beam of light which is monochromatic contains light of only one wavelength.
b) The grating will form a pattern of equally spaced bright and dark regions. The central region, where the single dot was positioned before the grating was put in the path of the beam, will be a maximum.

fig 7.20.2

c) $300 \ lines \ per \ mm = 300,000 \ lines \ per \ m$

$$d = \frac{1}{300,000}$$

$$d = 3.3 \times 10^{-6}$$

d) $$sin \ 23° = \frac{2 \times \lambda}{3.33 \times 10^{-6}}$$

$$\lambda = \frac{3.33 \times 10^{-6} \times sin \ 23°}{2}$$

$$\lambda = 651 \ nm$$

e) When the screen is moved back the centres of the maxima appear further apart but the angles at which they are formed remain the same.

Worked example 2

Figure 7.20.3 shows a ray of white light incident on one side of a glass prism. The prism disperses the white light into its constituent colours.

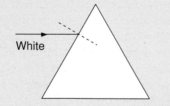

fig 7.20.3

a) Copy this diagram and sketch the path of the light as it passes through the glass.
b) Suggest one alteration which the student could make so that the emerging colours have a greater separation.
c) The student is not satisfied with the extent to which the white light is separated into its colours and decides to use a grating instead of the prism.
Draw a diagram to show how the grating will produce the first order maxima. Mark clearly the red and blue sides of each maxima.

Answers and comments

a) The glass prism will disperse the white light into its constituent colours because each different colour passes through the prism at a slightly different speed. The colours will emerge as shown in figure 7.20.4.

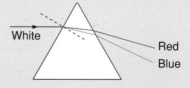

fig 7.20.4

b) The simplest alteration which would cause a slight increase in the separation of the colours would be to increase the angle of incidence of the white light.
c) The grating will produce coloured bands of light on either side of a central white band. In the coloured bands red will be deviated further from the straight through position than blue. The pattern will be as shown in figure 7.20.5.

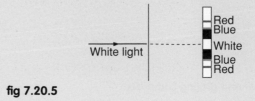

fig 7.20.5

d) The grating has 2×10^{-6} m between slits. What are the angles for the first order maxima for red light of wavelength 640 nm and blue light of wavelength 480 nm?

d)

For red	For blue
$\sin \theta = \dfrac{640 \times 10^{-9}}{2.0 \times 10^{-6}}$	$\sin \theta = \dfrac{480 \times 10^{-9}}{2.0 \times 10^{-6}}$
$\sin \theta = 0.32$	$\sin \theta = 0.24$
$\theta = 18.7°$	$\theta = 13.9°$

Questions

1 A pupil making revision notes on diffraction writes down the following statements about the pattern produced when light from a laser passes through a grating.
I   The grating will produce a series of well defined maxima positioned symmetrically about a central point.
II Increasing the number of slits illuminated by the laser light, without altering their separation will produce brighter, sharper maxima.
III Decreasing the space between the slits increases the separation of the maxima.
Which of these statements is/are correct?
A I only
B I and II only
C I and III only
D II and III only
E I, II and III.

2 The diagram shown below is drawn in many text books in the chapter on gratings.

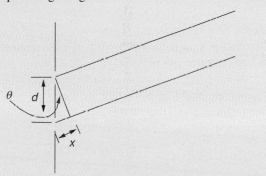

fig 7.20.6

The distance x is given by which of the following?
A $d \cos\theta$
B $d \tan\theta$
C $\dfrac{d}{(\tan\theta)}$
D $d \sin\theta$
E $d\,\theta$

3 A narrow beam of light from a sodium lamp which emits light of wavelength 589 nm is directed at a grating which has 200 lines per mm.
  a) Which order of maximum is detected in the interference pattern at an angle of 20.69°?
  b) At what angle will the sixth interference maximum be formed?
  c) Why are there only eight bright bands on either side of the central maximum for this grating?

4 A pupil sets up the following apparatus to measure the wavelength of laser light. She is told that the slit separation, d, is 1 ± 0.1 µm.

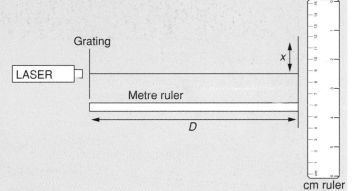

fig 7.20.7

She measures the distance, x, to the first bright fringe and the distance, D, and estimates the uncertainty in each. Her results are shown below.
  Measurement $x = 4.5 \pm 0.25$ mm
  Measurement $D = 2.95$ m ± 5 mm
She then makes the following statements about the accuracy of her measurements;
I   There is a 10% uncertainty in the slit separation measurement.
II The measurement uncertainty of 5 mm in D has least effect on the accuracy of the final numerical result.
III The overall uncertainty should be quoted as 10%.
Which of the following statements is/are true?
A I only
B I and II
C I and III
D II and III
E I, II and III.

5 A student investigating gratings, passes white light through a grating having 250 lines per mm. The student then uses colour filters to measure the angles at which different coloured maxima are formed.
  a) She uses a red filter allowing only light of wavelength 640 nm to pass through it. At what angle will the fourth order maximum be formed?
  b) Using a second filter the student notices that its fifth order maximum occurs at exactly the same position as the 4th order maximum for the red filter.
    (i) What wavelength of light does this second filter allow through?
    (ii) What is the colour of the light coming through this filter?

7.21 Exam style questions

Past paper question 1

Loudspeakers 1 and 2 are both connected in phase to the same signal generator which is set to produce a 1 kHz signal. Loudspeaker 1 is switched on but loudspeaker 2 is switched off.

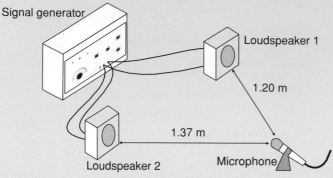

fig 7.21.1

State and explain what happens to the amplitude of the sound picked up by the microphone when loudspeaker 2 is switched on. Your explanation should include a calculation taking the speed of sound in air as 340 m s⁻¹. **SQA 1995**

Answers and comments

The amplitude detected by the microphone will be determined by the type of interference which takes place. This in turn depends on the wavelength of the sound and the path difference. If the frequency of the sound is 1 kHz, i.e. 1000 Hz, and the speed of sound in air is 340 m s⁻¹, then we can find the wavelength from;

$$v = f\lambda$$
$$v = 340 \ m \ s^{-1}$$

$$\lambda = \frac{v}{f} = \frac{340}{1 \times 10^3}$$

$$\lambda = 0.34 \ m$$

From loudspeaker 1 to the microphone is a path length of 1.20/0.34 = 3.53 wavelengths.
From loudspeaker 2 to the microphone is a path length of 1.37/0.34 = 4.03 wavelengths.
The path difference between sound arriving at the microphone from the loudspeakers is therefore 0.5 wavelengths. A path difference of 0.5 wavelengths results in destructive interference so the amplitude of the signal at the microphone will reduce when loudspeaker 2 is switched on.

Past paper question 2

a) The diagram of figure 7.21.2 shows the path of a monochromatic beam of light through a triangular plastic prism.

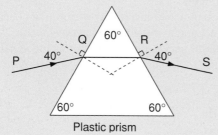

fig 7.21.2

(i) Calculate the refractive index of the plastic.
(ii) Sketch a copy of the above diagram with ray PQRS clearly labelled. (Sizes of angles need not be shown.) Add to your drawing the path which the ray PQ would take from Q if the prism were made of plastic with a *slightly higher* refractive index.

b) The original prism is now replaced with one of the same size and shape but made from glass of refractive index 1.80. Calculate the critical angle for this glass.

SQA 1994

Answers and comments

a) (i) By the symmetry of the diagram the angle between the ray from Q to R and the side of the prism is 60°. But the angle of incidence is measured to the normal so the angle of refraction in the glass is 30°. The refractive index is therefore;

$$n = \frac{sin \ i}{sin \ r} = \frac{sin \ 40°}{sin \ 30°}$$

$$n = 1.29$$

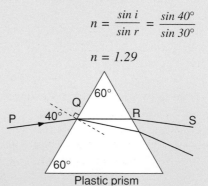

fig 7.21.3

b) To calculate the critical angle.

$$n_{material} = \frac{1}{sin \ \theta_c}$$

$$sin \ \theta_c = \frac{1}{n_{material}} = \frac{1}{1.80}$$

$$sin \ \theta_c = 0.555$$

$$\theta_c = 33.7°$$

Past paper question 3

The apparatus shown below is set up to determine the wavelength of laser light.

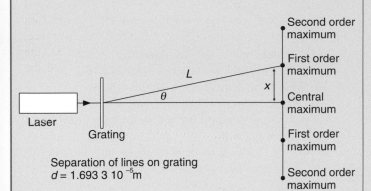

fig 7.21.4

The wavelength of the light is calculated using the equations;

$$d \sin \theta = n\lambda \quad and \quad \sin \theta = \frac{X}{L}$$

where angle θ and distances x and L are as shown in the diagram.

a) Seven students measure the distance L with a tape measure. Their results are as follows:
2.402 m, 2.399 m, 2.412 m, 2.408 m, 2.388 m, 2.383 m, 2.415 m.
Calculate the mean and the approximate random uncertainty in the mean.

b) The best estimate for the distance x is (91 ± 1) mm. Show by calculation whether x or L has the larger percentage uncertainty.

c) Calculate the wavelength, in nanometres, of the laser light. You must give your answer in the form:
Final value ± uncertainty.

d) Suggest an improvement which could be made so that a more accurate estimate of the wavelength could be made. You must use only the same equipment and make the same number of measurements. **SQA 1994**

Answers and comments

a) The mean distance L is given from

$$Mean\ L = \frac{2.402 + 2.399 + 2.412 + 2.408 + 2.388 + 2.383 + 2.415}{7}$$

$$Mean\ L = 2.401\ m$$

The approximate random uncertainty in this mean is given by;

$$Uncertainty = \frac{Range\ of\ readings}{Number\ of\ readings}$$

$$Uncertainty = \frac{2.415 - 2.383}{7}$$

$$Uncertainty = 0.0046\ m$$

b) To calculate the % uncertainty

$$\%\ uncertainty\ in\ x = \frac{1}{91} \times 100\% = 1.1\%$$

$$\%\ uncertainty\ in\ L = \frac{0.0046}{2.401} \times 100\% = 0.19\%$$

From this we can see that the uncertainty in x is more significant.

c) To calculate the wavelength of the laser we must use.

$$\sin \theta = \frac{X}{L} = \frac{0.091}{2.401}$$

$$d \sin \theta = n\ \lambda$$

$$1.693 \times 10^{5} \times \frac{0.091}{2.401} = 1 \times \lambda$$

$$\lambda = 6.41 \times 10^{-7}\ m$$

$$\lambda = 641\ nm$$

Since the uncertainty in x is much much larger than the error in L we can say that this is a good estimate of the uncertainty in the final answer. Therefore the wavelength of this laser is quoted as 641 nm ± 1.1% or 641 ± 7 nm.

d) The simplest alteration that could be made would be to move the screen further from the grating. This would make both x and L larger and so the uncertainty would be less significant.

Exam style questions – continued

Questions

1 Light of frequency 6×10^{14} Hz passes from air into glass of refractive index 1.5.
   a) Calculate the wavelength of this light in air. The speed of light in air is 3×10^8 m s^{-1}.
   b) What is the speed of the light in the glass?
   c) What is the frequency of the light in the glass?
   d) Calculate the wavelength of the light in the glass.

2 A ray of monochromatic light is directed at right angles into a rectangular glass block as shown in figure 7.21.5.

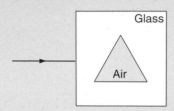

fig 7.21.5

The centre of the glass block is prism shaped and filled with air.
Copy figure 7.21.5 and continue the ray to show its passage until it emerges from the glass block.

3 Figure 7.21.6 shows a diagram copied from a page of a student's notebook.

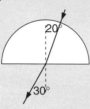

fig 7.21.6

A ray of monochromatic light passes through a semicircular glass block as shown in the diagram.
   a) What is meant by the term monochromatic?
   b) From the information given in the diagram calculate the refractive index of the glass relative to air.
   c) What angle of incidence in the block will produce an angle of 45° in the air?

4 A narrow beam of microwaves with a wavelength of 3 cm travelling through air with a speed of 3×10^8 m s^{-1} is directed towards a tank of oil. The waves meet the tank at an incident angle of 45° and are refracted so that the angle of refraction in the oil is 30°.
   a) What is the frequency of the microwaves in the oil?
   b) Calculate the speed of the microwaves in the oil.
   c) What is the wavelength of the microwaves in the oil?

5 A ray of monochromatic light, travelling in air, strikes the midpoint of one side of a glass block as shown in figure 7.21.7.

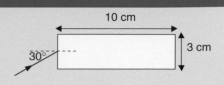

fig 7.21.7

   a) Calculate the angle of refraction in the glass. The refractive index of the glass block is 1.74.
   b) What is the critical angle for this type of glass?
   c) Draw an accurate copy of figure 7.21.7 and show the path of the ray until it emerges from the glass block.

6 A student wanting to disperse light into its constituent colours sets up the apparatus as shown in figure 7.21.8.

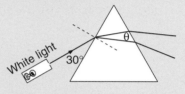

fig 7.21.8

   a) The refractive index of this material is 1.517 for red light, and 1.538 for blue light. Calculate the angles of refraction as the light enters the prism.
   b) What is the size of the angle marked θ on the diagram?
   c) Which colour of light travels fastest in the prism?
   d) All wavelengths in white light travel at 3×10^8 m s^{-1} in air. What is the speed of the red light in the block?
   e) What is the difference between the speed of the red light and the speed of the blue light in the prism?

7 The diagram of figure 7.21.9 shows an experimental arrangement used to produce an interference pattern. The sound from the loudspeaker has a frequency of 1 kHz.

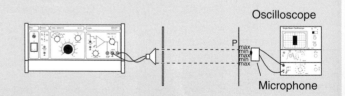

fig 7.21.9

   a) A student states that the gaps in the barriers are acting as coherent sources. Explain what is meant by coherent sources.
   b) Sketch a graph to show what the trace on the oscilloscope will look like when the microphone is placed at P and the lower gap is covered. Show a time scale on your graph.
   c) When both gaps are open, point P is found to be a point of constructive interference. Sketch on your graph the possible new trace recorded on the oscilloscope.

8 Figure 7.21.10 shows a ray of white light entering a block of crystal glass.

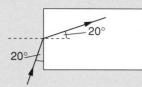

fig 7.21.10

   a) What is the refractive index for this type of glass?
   b) Explain why very powerful converging lenses made from materials with high refractive indices sometimes have "coloured edges".

9 A thin beam of monochromatic light is incident on a grating marked as having 400 lines per mm. The angle between the two second order maxima is 48° (as shown in figure 7.21.11).

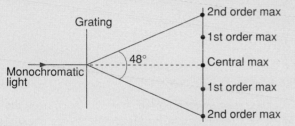

fig 7.21.11

   a) Calculate the wavelength of the monochromatic light.
   b) What colour would this monochromatic light appear?
   c) This monochromatic light source is replaced by another monochromatic source which is marked as having a frequency of 4.7×10^{14} Hz. Calculate the new angle between the second order maxima.

10 Microwaves of wavelength 2.8 cm pass through two narrow gaps, S_1 and S_2 as shown in figure 7.21.12.

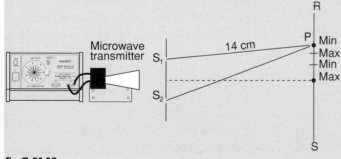

fig 7.21.12

When a microwave detector is moved along the line RS a series of maxima and minima are detected in the interference pattern.
   a) Explain in terms of waves how the minimum at point P is formed.
   b) How many more wavelengths would fit into the distance S_2P compared with the distance S_1P?
   c) The distance S_1P is 14 cm. How far is gap S_2 away from P?

11 Laser light with a wavelength 693 nm is passed through a grating and all the maxima produced are displayed on a semicircular screen.

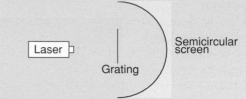

fig 7.21.13

The slit separation on the grating is 2.1×10^{-6} m.
   a) What is the angle between the second order maximum and the central maximum?
   b) Will there be a maximum or minimum of intensity at the straight through position? Explain your answer.
   c) How many maxima will there be on the screen?

12 The pattern produced by a diffraction grating is a result of both diffraction and interference.
   a) Explain, with the aid of a diagram, what is meant by the term diffraction.
   b) Explain, with the aid of a diagram, how coherent light from two sources can cause an interference pattern.

13 A grating marked as having 150 lines per mm is illuminated with light from a laser and the positions of some of the interference maxima are marked on a screen.

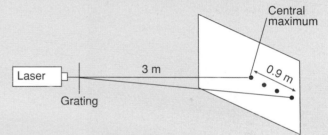

fig 7.21.14

The third interference maxima is 90 cm from the centre as shown.
   a) What is the angle at which this third maxima is formed?
   b) Calculate the wavelength of the laser light.
   c) What would be the effect on the appearance of the pattern on the screen if;
      (i) the grating is replaced by one having slits 5×10^{-6} m apart?
      (ii) the screen is moved further away from the grating?
   d) The original grating is now replaced and the laser is changed to a white light source. Describe the appearance of the pattern now produced on the screen.
   e) Describe how this pattern is different from that which would be produced by replacing the grating with a glass prism.

8.1 Irradiance

Introduction

Ever since the beginning of time mankind has been fascinated by the lights in the night sky. Over the centuries our understanding of what causes these lights has developed. We now know that the 'twinkling' light in the night sky comes from distant stars. Collections of stars are called galaxies and each galaxy is composed of many different types of star. The sun is our nearest star and fusion of the hydrogen nuclei at its core produces energy which radiates out into space.

Figure 8.1.1 Night sky

The sun is by no means the most energetic star in the sky but it has the ability, perhaps unique, to support life on one of its orbiting planets, earth.

Figure 8.1.2 The earth from space

Radiation and distance

The planets in orbit around the sun in our solar system can be represented as shown in figure 8.1.3.

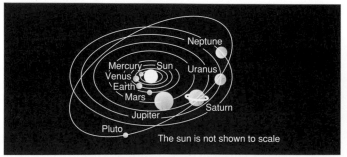

Figure 8.1.3 The solar system

Planets closer to the sun than the earth are too warm to sustain human life while those further away are too cold. The earth is special because it is the correct distance from the sun to receive just the right amount of energy.

From these considerations it would seem sensible to suggest that **the energy collected by each planet depends on its distance from the sun**. The energy collected also depends on the time for which a planet is irradiated. For a fair comparison of the energy received at different distances we must consider equal time intervals. For simplicity one second is chosen as the standard time interval. The energy arriving at any point in one second is called the power.

$$Power = \frac{Energy}{Time}$$

The lamp shown in figure 8.1.4 has a power rating of 60 W. This means that it uses up 60 joules of electrical energy per second. If the lamp is 50% efficient it turns half of its input energy into light, so it radiates 30 joules of light per second.

Figure 8.1.4 60 W light lamp

To calculate the power at a point some distance from the lamp we must first imagine that it radiates its energy equally in all directions. Such a source is called a **point source**.

The point source of light in figure 8.1.5 radiates equally in all directions. The radiated light travels away from the source at a speed of 3×10^8 m s^{-1}.

Figure 8.1.5 Point source

If we consider light emitted by the source at a certain time we can calculate how far the light travels from the source during 1 nanosecond.

$$Speed = \frac{Distance}{Time}$$

$$Distance = Speed \times Time$$
$$Distance = 3 \times 10^8 \times 1 \times 10^{-9}$$
$$Distance = 0.3 \ m$$

After the first nanosecond the energy produced by the source can be thought of as being spread evenly over a sphere of radius 0.3 m. The surface area of this sphere with radius, r, can be calculated from;

$$Sphere \ area = 4 \ \pi r^2$$
$$Sphere \ area = 4 \times 3.14 \times 0.3^2$$
$$Sphere \ area = 1.13 \ m^2$$

So in the first nanosecond after it leaves a point source the energy spreads out over an area of 1.13 m^2.

Irradiance

In common speech the word intensity is used to describe the loudness of a sound or the brightness of a lamp. **For scientific purposes irradiance is more rigidly defined as the energy per second per unit area or the power per unit area**.

$$Irradiance = \frac{Power}{Area}$$

If the point source of figure 8.1.5 radiates energy at a rate of 10 joules per second we can say that it has a power of 10 W. The irradiance at a point 0.3 m from the source can be found from;

$$Irradiance = \frac{Power}{Area}$$

$$Irradiance = \frac{10}{1.13}$$

$$Irradiance = 8.84 \; W m^{-2}$$

Two nanoseconds after leaving the source the radiation will have travelled twice as far, i.e. 0.6 m. The radiation energy will now have spread over an area given by;

$$Sphere \; area = 4 \; \pi r^2$$
$$Sphere \; area = 4 \times 3.14 \times 0.6^2$$
$$Sphere \; area = 4.52 \; m^2$$

The irradiance of the radiation at any point 0.6 m from the source in figure 8.1.5 is therefore;

$$Irradiance = \frac{Power}{Area}$$

$$Irradiance = \frac{10}{4.52}$$

$$Irradiance = 2.21 \; W m^{-2}$$

Similar analysis will show that for points 0.3 m, 0.4 m and 0.5 m away from the source the irradiance values are as shown in the table 8.1.1.

Distance/m	0.3	0.4	0.5	0.6
Irradiance/W m^{-2}	8.84	4.97	3.18	2.21

Table 8.1.1

The inverse square law

The irradiance of radiation at any point 0.3 m from the 10 W source has already been calculated as 8.84 W m^{-2}. When the distance from the source doubles to 0.6 m the irradiance falls to 2.21 W m^{-2}. **Doubling the distance from the source has quartered the irradiance**. This would suggest that irradiance, I, and distance, d, are connected by an inverse square law, therefore:

$$I \propto \frac{1}{d^2}$$

We can test the validity of this relationship for the data from the table 8.1.1. If the above equation is valid then;

$$Id^2 = constant$$

Multiplying the irradiance values by the square of the distances gives the results shown in table 8.1.2.

Distance/m	0.3	0.4	0.5	0.6
$I \times d^2$	0.80	0.80	0.80	0.80

Table 8.1.2

The Id^2 products are constant thus *confirming* the **inverse square relationship between the irradiance of radiation and distance from a point source**.

Measuring irradiance of light

EXPERIMENT The apparatus shown in figure 8.1.6 uses a light meter to measure the irradiance at various distances from a lamp.

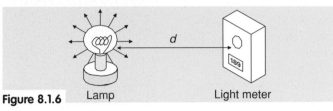

Figure 8.1.6   Lamp   Light meter

To confirm the relationship between the irradiance and distance values, a graph of irradiance versus the reciprocal of distance squared could be plotted.

Errors and uncertainty

This experiment is prone to a number of factors which can reduce the accuracy of the results. These factors are referred to as uncertainties or errors. If the classroom lights are left on while readings are taken the measured light irradiance will always be greater than it should be. This is a **systematic error or uncertainty** and could be eliminated by taking a reading with the lamp switched off and then with it on. The light irradiance due to the source alone is the *difference* between these readings.

The lamp has to be connected to a power supply so it cannot send much light into the area around the connection. The small quantity of light which would have gone into this area may be reflected in other directions. This means that the lamp is not a true point source but gives readings which are consistently greater than anticipated. This gives rise to another small systematic error. Although the lamp is not a true point source it is a close approximation to one for an observer a large distance away.

Stray reflections within the laboratory might also influence the readings. Some students might be wearing clothing which reflects light into the light meter, while others might be casting a shadow. Taking readings under these varying conditions gives rise to **random uncertainties**. This type of uncertainty will randomly cause readings higher or lower than the true value. **The influence of random uncertainties can be reduced by trying to improve the technique before repeating readings and calculating the mean.**

The approximate random uncertainty in a result which has been found by repeating readings can be calculated using the formula;

$$Uncertainty = \frac{Range \; of \; readings}{Number \; of \; repetitions}$$

To look for a trend in the results from this experiment you must consider both the readings and their associated uncertainties. You can quote a measured value of light intensity as;

$$Average \; value \pm uncertainty$$

For a trend to be identified from the results the uncertainty in each reading must be small enough so that the measurements with their associated uncertainties do not overlap.

8.2 Photoelectric effect

Introduction

What we now call the **photoelectric effect** was first observed by Henrich Hertz in 1887 during his work with radio waves. His experiment involved producing sparks by charging a metal sphere to a voltage high enough for sparks to jump across the gap to a neighbouring earthed sphere.

Figure 8.2.1 Photoelectric effect

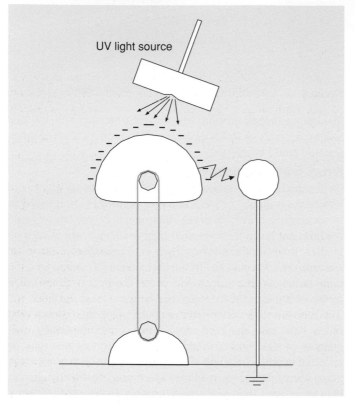

UV light source

The negatively charged sphere discharges when the electrons jump across the gap. **Hertz observed that stronger sparks were produced when the metal sphere was illuminated with ultraviolet light**.

Observing the photoelectric effect

We can confirm that **ultraviolet radiation causes certain negatively charged metal objects to discharge** using the apparatus shown in figure 8.2.2.

When the clean zinc foil is placed on the cap of the gold leaf electroscope and the flying lead from the negative output of the power supply, set to 3 kV, is touched momentarily against the zinc, the leaf rises. Negative charge has been transferred from the power supply to the electroscope. This charge spreads out evenly over the stem and the leaf. The repulsive forces between the charges on the fixed stem and those on the leaf are large enough to cause the leaf to diverge.

When the lamp emitting ultraviolet radiation with a wavelength of around 254 nm is switched on, the leaf collapses back towards the stem.

With this apparatus you can show that;

1 The leaf collapses when uv light is shone onto the zinc plate.
2 When the distance between the uv source and electroscope is increased the leaf still falls, only more slowly.
3 Replacing the uv source with a source of white light fails to collapse the leaf.
4 Using a laser with a wavelength of 639 nm also fails to collapse the leaf.
5 There is also no discharge if the zinc plate is replaced with steel or copper.
6 When the electroscope is charged *positively* the leaf *will not collapse* even when the zinc plate is irradiated with uv radiation.

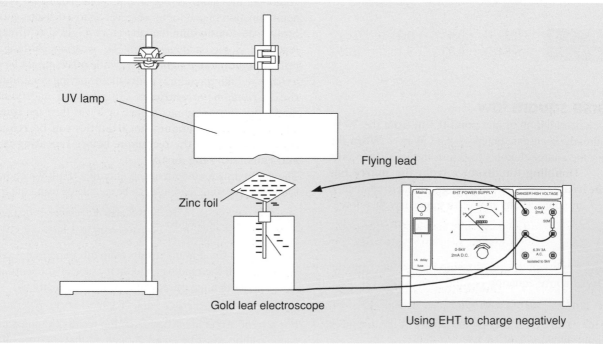

UV lamp

Flying lead

Zinc foil

Gold leaf electroscope

Using EHT to charge negatively

Figure 8.2.2

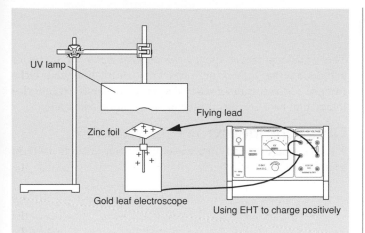

UV lamp

Flying lead

Zinc foil

Gold leaf electroscope

Using EHT to charge positively

Figure 8.2.3

Properties of the emitted electrons

When we consider that the electron was not discovered until 1897 by J.J. Thomson it is hardly surprising that Hertz, working ten years before Thomson, was unable to explain his observation of the photoelectric effect. Although unable to explain his observations, Hertz still reported them in his writings and others took up the search for an explanation.

Figure 8.2.4 J.J. Thomson

In 1899, Lenard established that the particles ejected from a negatively charged surface were electrons and went on to look at their kinetic energies. **He found that when photoelectric emission did take place the ejected electrons had a range of different kinetic energies**. He noticed that when each kind of material was irradiated with a particular incident frequency, a number of the ejected electrons had a maximum kinetic energy which was never exceeded.

This maximum kinetic energy of the ejected electrons depends on the frequency of the incident radiation and on the metallic surface irradiated. He also found that the maximum kinetic energy of the ejected electrons *did not* depend on the irradiance of the incident radiation.

Threshold frequency

When Lenard investigated the kinetic energies of electrons ejected from different metals he realised that for electrons to be ejected in the first place the frequency of the incident radiation had to be above a certain value. This critical value is called the **threshold frequency**. Different metals have different threshold frequencies.

	Threshold frequency
Gold	1.18×10^{15} Hz
Platinum	1.52×10^{15} Hz
Sodium	0.56×10^{15} Hz
Calcium	0.65×10^{15} Hz

Table 8.2.1

If the frequency of the radiation incident on a metal surface is below the threshold frequency no photoelectric emission occurs, even when the irradiance of the incident radiation is increased.

If the frequency of the incident radiation is above the threshold frequency the number of electrons ejected is directly proportional to the intensity of the radiation.

Lenard showed that *doubling* the irradiance *doubles* the number of electrons ejected but *does not* alter their maximum kinetic energy.

Summary

In the experiment shown in figure 8.2.2 the leaf collapsing indicates that the charge on the electroscope is reducing as a consequence of the interaction of the uv rays with the zinc metal. With the apparatus in figure 8.2.3 the leaf did not collapse when the electroscope was charged positively. Clearly then we can conclude that the uv rays are helping loosely held electrons to move off the charged electroscope.

Moving the uv source further away from the negatively charged electroscope reduces the irradiance of the radiation arriving at the cap of the electroscope. As a consequence the leaf falls more slowly. This observation seems quite understandable. At greater distances less of the radiated energy is gathered by the cap of the electroscope so fewer electrons are released. However, replacing the uv source with a more intense white light source or even with light from a laser did not make the leaf fall so we must conclude that no electrons were released.

The fact that a weak uv source can release electrons when other more intense sources fail must indicate that irradiance alone cannot explain the release of electrons from a metal surface.

It was not until 1905 and the work of Albert Einstein, who connected Lenard's work with the independent studies of Max Planck, that this apparent paradox was successfully explained. In 1887 when Henrich Hertz shone a uv source onto the charged conductors it made it easier for electron discharge to occur because the frequency of the uv source was above the threshold frequency of the metal conductors. Little did he realise that his observation, which he didn't understand, yet carefully recorded, would bring about such a radical rethink of the way we explain how electromagnetic waves interact with matter.

Planck and Einstein

Introduction

During the final part of the 1890s and the first few years of the 1900s, the phenomenon of photoelectric emission was thoroughly investigated. The fact that electromagnetic radiation, usually visible light or ultraviolet rays, incident on a metal surface could release electrons had been established beyond doubt. Furthermore it had been established that;

1 Photoelectric emission occurs only when the frequency of the incident radiation is above a minimum value called the threshold frequency.

2 For frequencies of radiation above the threshold frequency the number of photoelectrons emitted per second is directly proportional to the irradiance of the radiation.

3 Different metals have different threshold frequencies.

4 For frequencies of radiation greater than the threshold frequency photoelectric emission commences at the *very instant* the radiation starts to shine on the metal surface.

5 The electrons ejected from a particular energy level have a range of kinetic energies up to a maximum value which depends on the frequency of the incident radiation but not on the irradiance of the radiation.

Classical wave theory

What made the photoelectric effect of such interest to scientists was the fact that it could not be explained by the theories existing at the time.

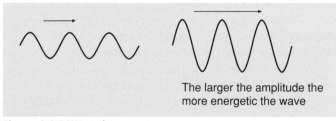

The larger the amplitude the more energetic the wave

Figure 8.3.1 Wave theory

In the early 1900s electromagnetic radiation was pictured as continuous waves where the energy depended on the amplitude. Larger amplitude vibrations caused more energetic waves and greater irradiances of radiation transferred more energy from a source to an irradiated object.

The classical wave theory suggested that light energy directed at a metal surface would be collected by the surface until *eventually* there was enough to release an electron. Clearly this was not what was observed in photoelectric experiments where low intensity sources of the correct frequency produced electrons *immediately*, but other high intensity sources *would not* release electrons even when the negatively charged surface was irradiated for long periods of time.

The classical wave theory was equally ineffective for explaining the threshold frequency and the dependence of the maximum kinetic energy of the emitted electrons on the frequency of the incident radiation.

The classical wave theory which had been so useful in explaining the reflection, refraction, diffraction and interference of electromagnetic waves could not explain the interaction between electromagnetic radiation and matter.

Quantum theory

The first step to modify the wave theory was taken in 1901 by Max Planck.

Figure 8.3.2 Max Planck

After investigating the infrared radiation emitted by a hot object, Planck suggested that his results could be explained by assuming that the radiation was emitted as separate, individual 'bundles' of energy rather than as a continuous wave. He called the smallest possible 'bundle' a **quantum** and stated that **the detected radiation was made up from 'bundles' having one, two, three, etc. quanta of energy but no fractional amounts.** Planck proposed that the energy, E, associated with any individual quantum was directly proportional to the frequency, f, of the radiation emitted by the source;

$$E \propto f$$
$$E = h f$$

The constant of proportionality in this equation, h, is now called **Planck's constant** and has a value of 6.63×10^{-34} Js. In 1905 Albert Einstein extended Planck's ideas by suggesting that **light and other electromagnetic waves could be considered to be streams of concentrated 'bundles' of energy.** In this case he called the 'bundles' **photons**. We can picture these photons as shown in figure 8.3.3.

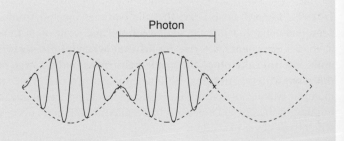

Photon

Figure 8.3.3 Photons

The energy, E, of a photon of electromagnetic radiation of frequency f, is given by;

$$E = hf$$

Einstein further proposed that **photons interact with electrons in an 'all or nothing way'**. When a photon interacts with an electron it gives either *all* or *none* of its energy to the electron. **The energy of a photon cannot be shared by more than one electron nor can a single electron absorb the energy of more than one photon**.

This helps to explain the threshold frequency observed in photoelectric experiments. Only above a certain frequency, would a photon have enough energy to release an electron from attractive forces within the metal.

Photoelectrons can be released from sodium by visible light with frequencies of 5.6×10^{14} Hz and above. We can therefore state that the threshold frequency for sodium is 5.6×10^{14} Hz. We can calculate the energy of photons with this frequency using;

$$E = hf$$
$$E = 6.63 \times 10^{-34} \times 5.6 \times 10^{14}$$
$$E = 3.7 \times 10^{-19} J$$

Therefore we can say that only photons with energies greater than 3.7×10^{-17} J have sufficient energy to release electrons from the surface of sodium metal.

Work function and electron energy

The minimum energy which a photon must have to release a photoelectron is called the work function. For the sodium used in the previous example the minimum energy needed to release an electron was calculated as 3.7×10^{-19} J. The work function of sodium is therefore 3.7×10^{-19} J. The threshold frequency, f_0, depends on the work function, W, of the metal, according to the equation;

$$W = hf_0$$

The threshold frequency for calcium can be shown by experiment to be 6.5×10^{14} Hz. The work function for calcium can therefore be calculated;

$$W = hf_0$$
$$W = 6.63 \times 10^{-34} \times 6.5 \times 10^{14}$$
$$W = 4.31 \times 10^{-19} J$$

So only photons with energies greater than 4.31×10^{-19} J will eject electrons from calcium.

The threshold frequencies for both sodium and calcium are in the visible part of the electromagnetic spectrum. We would therefore expect an ultraviolet source to be able to eject electrons from both sodium and calcium because frequencies in the uv part of the spectrum are *above* the threshold frequencies for both sodium and calcium.

One common uv source emits radiation with a wavelength of 254 nm. We calculate the frequency of these electromagnetic waves using the equation linking speed, frequency and wavelength;

$$c = f \times \lambda$$

Therefore;

$$f = \frac{c}{\lambda}$$

$$f = \frac{3 \times 10^8}{254 \times 10^{-9}}$$

$$f = 1.18 \times 10^{15} Hz$$

The energy of these photons of uv radiation can be calculated using;

$$E = hf$$
$$E = 6.63 \times 10^{-34} \times 1.18 \times 10^{15}$$
$$E = 7.82 \times 10^{-19} J$$

Photons of uv radiation having an energy of 7.82×10^{-19} J will have sufficient energy to emit electrons from a sodium surface where the work function is 3.7×10^{-19} J. Since each photon interacts with one electron in an 'all or nothing way' the uv photon will give all of its energy to an electron. Only part of the energy, 3.7×10^{-19} J, is needed to overcome the work function and release the electron. The remaining energy, 4.12×10^{-19} J, is given to the released electron as kinetic energy. The released electrons can never have more than this value of kinetic energy. So there is a limit to the maximum kinetic energy that a released electron can have.

Using this quantum model we have been able to explain another of the properties of photoelectric emission.

In more general terms we can say;

$$Energy\ of\ incident\ photon = Work\ function + KE_{max}$$

If the incident radiation has a frequency, f, and the threshold frequency for the metallic surface is f_0, we can say;

$$hf = hf_0 + \frac{1}{2}mv^2$$

where v is the maximum velocity with which the electrons can be released. This is known as Einstein's equation.

Figure 8.3.4 summarises how the maximum kinetic energy of the released electrons varies with the frequency of the incident radiation.

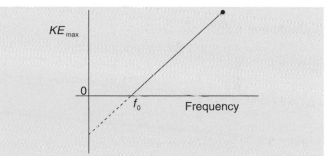

Figure 8.3.4

Quantum theory and irradiance

When the radiation incident on a surface has a frequency above the threshold frequency the number of electrons emitted is directly proportional to the irradiance of the radiation. Doubling the irradiance means that twice as many photons are arriving on unit area each second so it is not surprising that twice as many electrons are released. Thus again the quantum theory successfully explains another of the observations from photoelectric effect experiments.

If there are N photons each of energy, E, arriving on unit area each second, the irradiance, I, which can be stated as the energy per second per unit area, can be found from;

$$I = NE$$

If the photons are from a source of radiation of frequency, f, then;

$$I = Nhf$$

Worked example 1

A light sensor which gives an output voltage proportional to the irradiance of light falling on it is placed 20 cm away from a lamp in an otherwise darkened room. The lamp is acting as a point source.

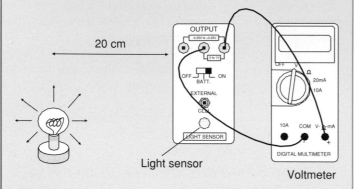

fig 8.4.1

The output voltage from the light sensor is 1 V when placed 20 cm from the lamp.
a) What is meant by calling the lamp a *point source*?
b) The light sensor is moved to a position 40 cm from the lamp. What reading will be shown on the voltmeter?
c) The lamp is radiating 24 W of light. What is the irradiance at a point 0.2 m from the lamp?

Answers and comments

a) By saying that the lamp is a point source we are assuming that the lamp is radiating energy equally in all directions.
b) The reading from the light sensor when it is placed 20 cm away from the lamp is 1 V. This reading is proportional to the irradiance. Moving the light sensor to a point 40 cm from the lamp doubles the distance from the lamp. Since irradiance is related to distance by the equation;

$$I \propto \frac{1}{d^2}$$

we can see that doubling the distance from the source will reduce the irradiance to one quarter of its former value. This means that the reading from the light sensor will also be one quarter of its former value. The new reading on the light meter will be 0.25 V.
c) The 24 W of energy are passing through an area given by $4\pi r^2$, where r is the distance from the point source. The irradiance 0.2 m from the source can be calculated from;

$$Irradiance = \frac{Power}{Area} = \frac{24}{4\pi r^2}$$

$$Irradiance = \frac{24}{4 \times \pi \times 0.2^2}$$

$$Irradiance = 47.75\ W\,m^{-2}$$

Worked example 2

A textbook states that gold has a work function of 7.84×10^{-19} J.
a) Explain what is meant by the term *work function*.
b) Calculate the threshold frequency of gold.
c) In what part of the electromagnetic spectrum does radiation of this frequency occur?
d) A gold foil is illuminated with radiation of frequency 1.7×10^{15} Hz. Calculate the maximum kinetic energy of the ejected electrons.

Answers and comments

a) The work function is the minimum energy which a photon must have to release a photoelectron.
b) The threshold frequency, f_o, is calculated from;

$$W = hf_o$$
$$7.84 \times 10^{-19} = 6.63 \times 10^{-34} \times f_o$$

$$f_o = \frac{7.84 \times 10^{-19}}{6.63 \times 10^{34}}$$

$$f_o = 1.18 \times 10^{15}\ Hz$$

c) This radiation is in the ultraviolet part of the electromagnetic spectrum.
d) The maximum kinetic energy which an electron can have is the difference between the photon energy and the work function.

$$Photon\ energy = hf = 6.63 \times 10^{-34} \times 1.7 \times 10^{15}$$
$$Photon\ energy = 11.3 \times 10^{-19}\ J$$
$$KE_{max} = Photon\ energy - Work\ function$$
$$KE_{max} = 11.3 \times 10^{-19} - 7.84 \times 10^{-19}$$
$$KE_{max} = 3.46 \times 10^{-17}\ J$$

The maximum energy that the released photoelectrons can have is 3.46×10^{-17} J.

Questions

1. The electronic systems on a space probe which is 300 million km from the sun are powered from solar panels with an area of 4 m².
   a) Explain why guidance systems are needed to keep these solar panels pointing directly at the sun.
   b) The space probe moves closer to the sun. What happens to the power output from the solar panels? Explain your answer.
   c) What area of solar panels will be required to power the probe's electronic systems when it is positioned 1.5×10^{11} m from the sun?

2. In an experiment to investigate the relationship between light irradiance, I, and distance, d, from a point source the following measurements are obtained.

Distance/m	0.2	0.3	0.4	0.5	0.6
Light irradiance (arbitrary units)	160	70	39	25	18

Table 8.4.1

   a) Draw a graph to show how the light irradiance changes with distance.
   b) From your graph determine the reading on the light irradiance meter when it is placed 25 cm from the lamp.
   c) Draw another graph to confirm the relationship between light irradiance and distance.

3. At a point just outside the earth's atmosphere the irradiance of radiation arriving from the sun is 1360 W m⁻².
   a) Rays leaving the sun take 8 minutes to arrive at the point. Show that the point is 1.44×10^{11} m from the sun.
   b) The surface area of a sphere of radius, r, can be calculated from the equation $4\pi r^2$. What is the area of a sphere of radius 1.44×10^{11} m?
   c) Assuming that the sun is a point source, estimate its power output.

4. A student studying the photoelectric effect directs electromagnetic radiation from different sources onto a charged zinc plate placed on top of an electroscope.
   a) Explain the meaning of the terms threshold frequency and work function.
   b) (i) When the plate is negatively charged, ultraviolet radiation will discharge the electroscope. Explain this observation.
      (ii) Describe and explain the effect of increasing the irradiance of the uv radiation.
   c) Explain why, when the plate is negatively charged, light from a sodium lamp will not discharge the electroscope.

5.
   a) Calculate the energy of a photon of radiation X which has a frequency of 7×10^{14} Hz.
   b) Calculate the energy of a photon of radiation Y which has a wavelength of 3×10^{-7} m.
   c) Which of these radiations will release electrons from a metal whose work function is 5×10^{-19} J? Justify your answer.
   d) What is the maximum kinetic energy of the electrons released when both types of radiation are shone on a sample of metal at the same time?

6. Radiation with a wavelength of 0.4 μm incident on a metal surface ejects electrons with kinetic energies up to a maximum value of 2×10^{-19} J.
   a) What is the frequency of the incident radiation?
   b) What is the value of the work function for this metal?
   c) What is the threshold frequency for the metal being irradiated?
   d) Describe and explain what will happen when the metal is irradiated with radiation having a wavelength of 700 nm.

7. In an experiment to investigate the photoelectric effect a caesium coated electrode is placed in an evacuated glass tube. This is connected to a power supply as shown in figure 8.4.2.

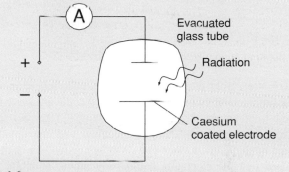

fig 8.4.2

   a) What is the threshold frequency for caesium whose work function is quoted as 3.04×10^{-19} J?
   b) What is the maximum kinetic energy of the electrons released when the caesium electrode is irradiated with radiation having a frequency of 6.5×10^{14} Hz?
   c) This source of radiation directs 3.2×10^{13} photons per second onto the caesium electrode.
      (i) What is the total energy per second collected by the electrode?
      (ii) The area of the electrode is 1 cm². What is the irradiance of the radiation arriving at the electrode?
      (iii) Each photon releases one electron. What is the current in the circuit?

8.5 Electrons and emission spectra

Introduction

While Ernest Rutherford and his collaborators laboured for a detailed knowledge of the structure of the atomic nucleus (see section 9.1), they left it to others to unravel the nature of the electrons in the atom.

Thomson's plum pudding model had pictured electrons moving in orbits within a sphere of positive charge. Following the discovery of the atomic nucleus it was suggested that the electrons orbited this nucleus. Initially the idea was not well received as it seemed to contradict one of the fundamental parts of classical electromagnetic wave theory.

Niels Bohr

**Figure 8.5.1
Niels Bohr**

In 1913 Niels Bohr proposed that electrons did indeed move in circular orbits centred on the nucleus and that the attractive force needed to maintain these orbits was provided by the forces between the negatively charged electrons and the positively charged nucleus. What made Bohr's theory different was that he proposed that the **electrons could only move in orbits with certain radii.**

Since the electrons can only exist at certain distances from the nucleus they must only have certain energy values; called **energy levels.** Bohr, who had been inspired by Einstein's explanation of the photoelectric effect using a quantum model, went on to suggest that electron transitions between energy levels could only involve certain fixed amounts (or quanta) of energy. Bohr correctly suggested that when an electron in an outer energy level with energy W_2 falls to an inner level with energy W_1 radiation of frequency, f, is released according to the equation;

$$W_2 - W_1 = hf$$

Many of Bohr's early ideas have been superseded but his basic concept of electron levels and of the quantised jumps between them still remains.

Energy levels

The energy levels within an atom which can be occupied by the electrons are often represented by horizontal lines arranged one above the other like a ladder with unequal rungs. The type of diagram shown in Figure 8.5.2 is called an **energy level diagram**.

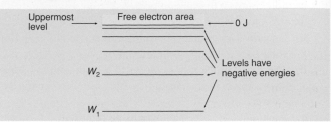

Figure 8.5.2 Energy level diagram

The uppermost level indicates the energy of an electron which is just, and no more, free from the influence of a nucleus. Any electron with more than this uppermost level of energy is considered free from the nucleus and will have some kinetic energy of its own. It will have a positive quantity of energy, consequently the uppermost energy level in any energy level diagram is labelled zero. To be consistent therefore, electrons in lower levels have to be labelled as having negative energies.

Figure 8.5.3 shows the energy level diagram for an electron in an atom of hydrogen.

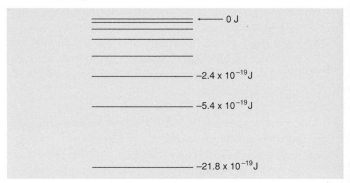

Figure 8.5.3 Energy level diagram for hydrogen

In the case of hydrogen atoms the lowest energy level which an electron can occupy has an energy of -21.8×10^{-19} J. This is the energy level normally occupied by the single electron in a hydrogen atom. An atom with its electrons in their lowest possible levels is said to be in its **ground state**.

To raise the electron in a hydrogen atom from its lowest energy level into the next highest energy level requires an input of energy.

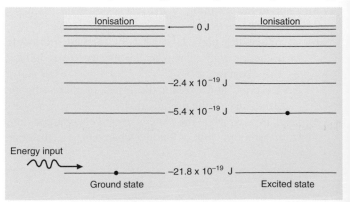

Figure 8.5.4 Promoting electrons

To promote electrons between levels requires the atoms to absorb an amount of energy *exactly* equal to the difference

between levels in the energy level diagram. Therefore to raise the electron in a hydrogen atom from the lowest level to the next highest level requires an input energy given by;

$$Energy\ input = 21.8 \times 10^{-19} - 5.4 \times 10^{-19}\ J$$
$$Energy\ input = 16.4 \times 10^{-19}\ J$$

If a hydrogen atom with its electron in the lowest possible level absorbs 21.8×10^{-19} J of energy the electron will now have enough energy to be just outside the influence of the nucleus. Such atoms are said to be ionised. The zero energy level in an energy level diagram is called the **ionisation level**.

Evidence for energy levels

Atoms with their electrons in the lowest available energy levels are said to be in their ground states. If the electrons are promoted to other levels the atoms are in an **excited state**. An atom in an excited state can revert to a ground state if the electrons revert to the lowest energy levels available.

When the hydrogen atom in the excited state shown in figure 8.5.5 reverts to the ground state, energy must be released.

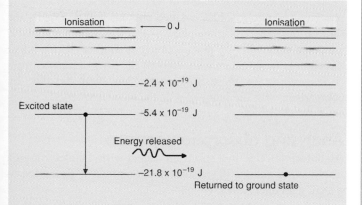

Figure 8.5.5 Excited atom reverts to ground state

The quantity of energy released will exactly equal the difference in the levels within the energy level diagram.

$$Energy\ input = 21.8 \times 10^{-19} - 5.4 \times 10^{-19}\ J$$
$$Energy\ input = 16.4 \times 10^{-19}\ J$$

When the electron moves to a lower energy level, a photon of electromagnetic radiation is released. The frequency of this photon can be calculated using;

$$E = hf$$
$$16.4 \times 10^{-19} = 6.63 \times 10^{-34} \times f$$
$$f = \frac{16.4 \times 10^{-19}}{6.63 \times 10^{-34}}$$
$$f = 2.47 \times 10^{15}\ Hz$$

The wavelength of this radiation can also be found from;

$$c = f\lambda$$
$$\lambda = \frac{c}{f}$$
$$\lambda = \frac{3 \times 10^{8}}{2.47 \times 10^{15}}$$
$$\lambda = 1.21 \times 10^{-7}\ m$$

The radiation emitted when an electron makes a transition between the first excited level and the ground state of atoms of hydrogen, has a single wavelength. This radiation can be detected in the ultraviolet part of the electromagnetic spectrum. An electron transition from the second level into the first produces visible light with a wavelength of 660 nm.

Spectroscopy

For the idealised atom of figure 8.5.6 there are six possible transitions between the four energy levels.

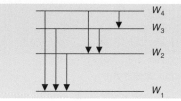

Figure 8.5.6

Electron transitions from a level with energy W_2 to a lower level having energy W_1 will produce radiation with a frequency given by;

$$W_2 - W_1 = hf$$

The wavelength of the radiation emitted can be calculated from;

$$W_2 - W_1 = \frac{hc}{\lambda}$$

Since each of the six transitions between the levels in figure 8.5.6 involves a different quantity of energy the radiation emitted from an excited sample of these atoms will contain six different wavelengths.

These wavelengths can be separated using a grating which disperses different wavelengths to different angles according to the equation;

$$n\lambda = d sin\theta$$

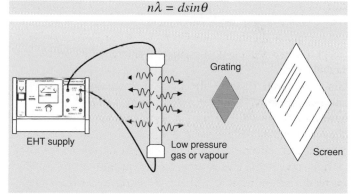

Figure 8.5.7

Each of the different wavelengths will appear at a different angle θ. The electron transitions within many atoms produce photons with wavelengths in the visible region of the electromagnetic spectrum.

The patterns produced by the different gratings are often sets of coloured lines. These are called optical line spectra or line emission spectra. By measuring the wavelengths of the lines in an element's emission spectrum, the energies of the permitted electron transitions can be determined, giving evidence for the arrangement of the electrons within the atoms of that element.

8.6 Absorption spectra

Introduction

In the apparatus shown in figure 8.5.7, the element producing a line spectrum is kept at low pressure in a discharge tube to ensure that each atom is as free as possible from the influence of other atoms.

Figure 8.6.1 Emission spectrum

When electrons in excited atoms return to their lowest level, photons of radiation are emitted. The grating splits the emitted light into its constituent wavelengths producing coloured line spectra. The hydrogen line spectrum consists of a red, a blue-green, and a violet line while the spectrum of sodium vapour contains bright yellow lines with wavelengths of 589 and 589.6 nm.

The study of spectra, known as spectroscopy, provides much information about the arrangement of electrons within the atoms. No two elements give the same line spectrum and spectroscopic methods are so sensitive that they can identify the presence of even the most minute quantities of materials.

Producing absorption spectra

With the apparatus shown in figure 8.6.2 the element to be analysed is again kept at very low pressure in a tube. However this time no high voltage supply is connected across the tube. Instead a source of white light is shone through the tube.

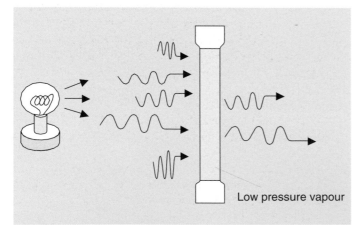

Figure 8.6.2 Absorption apparatus

White light, as we already know, is made up from every possible frequency of radiation within the visible part of the electromagnetic spectrum. Therefore, using quantum considerations we can say that a white light source emits photons with a continuous range of energies.

When the photons of white light pass through the tube the vast majority of the incident photons emerge. **However a small number of incident photons will have energies which match *exactly* the energies needed to move the electrons in ground states into higher energy levels. The atoms in the low pressure gas absorb these photons and use their energy to promote electrons into higher levels. Consequently, photons of these precise energies are missing from the radiation emerging from the tube.** When the radiation emerging from the tube is analysed with a grating, the coloured spectrum of the white light is *almost* but not *quite* complete. Black lines can be seen at certain points in the spectrum. These are caused by the absorption of photons which have the precise energy required to cause upward transitions within the energy levels of the atom.

The atoms excited by absorbing photons of energy from the incident white light can return to their ground state by emitting photons. The energy of the emitted photons will be exactly the same as the energy of those photons absorbed in the first place. However only *part* of this radiation is emitted in the same direction as the incident white light so the parts of the spectrum corresponding to these frequencies are seen as dark by comparison with the intensities of the other frequencies which were not absorbed.

Analysing absorption spectra

An absorption line in a spectrum occurs when an electron in a level having energy W_1 absorbs a photon and is promoted to a level with energy W_2.

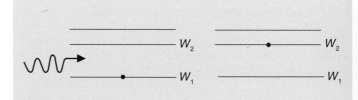

Figure 8.6.3 Absorption of photon

Considering the energies involved we can say that;

$$W_2 = W_1 + Photon\ energy$$

Since the energy levels are spaced very precisely, only those photons having an energy exactly equal to the difference between two levels will be absorbed.

$$W_2 - W_1 = Photon\ energy$$
$$W_2 - W_1 = hf$$

The **emission** spectrum for an element contains coloured lines corresponding to electron transitions from higher to lower energy levels within the atoms of the element. The **absorption** spectrum for the same element will contain black lines because photons of particular energies have been absorbed to promote electrons from lower to higher levels within the atom. Since the energy transitions in both the emission and absorption spectra involve the same energy levels it is not surprising that the coloured lines in an emission spectrum correspond to exactly the same frequencies as the dark lines in an absorption spectrum.

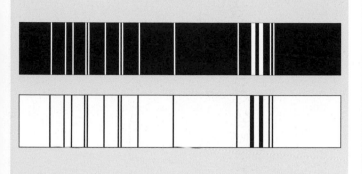

Figure 8.6.4 Correspondence of emission (top) and absorption (bottom) spectra

The absorption spectrum of hydrogen consists of thin black lines in the red, blue-green and violet parts of the spectrum. Similarly the absorption spectrum of sodium vapour has two black lines in the yellow region of the spectrum. These black lines correspond to wavelengths of 589 nm and 589.6 nm.

Observing absorption of light

EXPERIMENT The apparatus shown in figure 8.6.5 uses a sodium vapour lamp which emits light with a wavelength of approximately 589 nm.

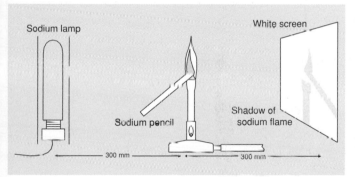

Figure 8.6.5

The energy of the photons emerging from this source can be calculated from;

$$c = f\lambda$$

$$f = \frac{c}{\lambda} = \frac{3 \times 10^{-8}}{589 \times 10^{-9}}$$

$$f = 5.1 \times 10^{14}\ Hz$$

Therefore;

$$E = hf$$
$$E = 6.63 \times 10^{-34} \times 5.1 \times 10^{14}$$
$$E = 3.4 \times 10^{-19}\ J$$

Almost all of these photons will pass through a simple gas flame because there is nothing to absorb photons of this particular energy. The orange light from the sodium lamp will illuminate the entire screen. If a sodium pencil is placed in the flame this produces a concentration of sodium vapour in the flame. The photons of light from the lamp must now travel through a region where they encounter many atoms capable of absorbing photons having this exact amount of energy. Many of the photons passing through the beam are absorbed and consequently cannot travel on towards the screen. **This means that there is a dark area on the screen due to the sodium vapour in the flame. The screen appears to show a shadow of the flame because of the absorption of the photons from the sodium lamp by the sodium vapour.**

Fraunhöfer lines

In the demonstration using the apparatus of figure 8.6.5, the dark shadow indicates the presence of sodium vapour in the path of light from a sodium lamp.

If normal daylight is examined with a spectroscope, absorption lines in the coloured spectrum can be clearly seen. (With dull or cloudy conditions the absorption lines are still visible and you should remember that **sunlight must not be viewed directly**.) The dark lines in the absorption spectrum of sunlight, called Fraunhöfer lines, are due to photons of certain energies being absorbed by atoms in the outer, cooler layer of gases which surround the sun's hot core. The atoms in this cooler layer are in less excited states than those in the hot core so they can absorb certain photon energies which try to pass through.

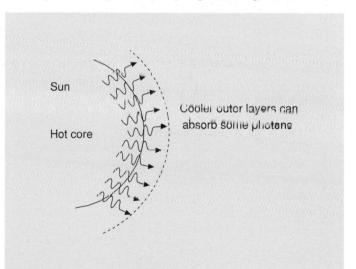

Figure 8.6.6 The sun's two layers

The Fraunhöfer lines indicate the presence of hydrogen, helium, sodium and other elements in the sun's atmosphere. From these observations scientists have been able to devise theories describing the composition and activity of the sun and other stars.

Worked example 1

Figure 8.7.1 shows a diagram taken from a student's notebook. This diagram represents the energy levels in a free atom.

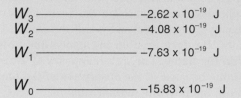

W_3 ——————— -2.62×10^{-19} J
W_2 ——————— -4.08×10^{-19} J

W_1 ——————— -7.63×10^{-19} J

W_0 ——————— -15.83×10^{-19} J

fig 8.7.1

a) What is meant by the term *free atom*?
b) Explain why all the energy levels are marked as having negative energies.
c) Calculate the energy required to promote an electron from level W_0 to W_3.

Answers and comments

a) Atoms are described as free atoms when they are well away from the influence of surrounding atoms. This means that the electrons in a free atom are only affected by the nucleus of that atom. The atoms in gases at low pressures are considered free atoms.
b) An electron which has just, and no more, been removed from its atom is considered as having zero energy. To promote the electron to this state, energy must be given to the electron. Since energy must be added to reach zero, the electron initially has a negative level of energy.
c) To promote an electron between these two levels requires an input of energy exactly equal to the difference in energy between the two levels.

$$Energy\ input = (15.83 - 2.62) \times 10^{-17}\ J$$
$$Energy\ input = 13.21 \times 10^{-17}\ J.$$

Worked example 2

A textbook states that the electrons in hydrogen atoms can exist in the following discrete energy levels.

Level name	Energy ($\times 10^{-19}$ J)
O	0.0
P	−0.61
Q	−0.86
R	−1.36
S	−2.42
T	−5.4
U	−21.8

Table 8.7.1

a) What level will the electron in a hydrogen atom occupy if the atom is in its ground state?
b) Draw a diagram which represents the energy levels that electrons can occupy within the hydrogen atom.
c) Calculate the energy of the photons released when electrons move from level P to level T.
d) What frequency of radiation will be radiated when electrons move from level P to level T?

Answers and comments

a) If an atom is in its ground state the electrons must be in their lowest possible energy levels. Since a hydrogen atom has only one electron this will be in the lowest level. This is the level labelled U in the table.
b) Translating the information from the table into a diagram results in figure 8.7.2.

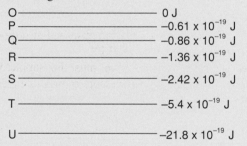

O ——————— 0 J
P ——————— -0.61×10^{-19} J
Q ——————— -0.86×10^{-19} J
R ——————— -1.36×10^{-19} J

S ——————— -2.42×10^{-19} J

T ——————— -5.4×10^{-19} J

U ——————— -21.8×10^{-19} J

fig 8.7.2

c) The energy released is equal to the difference in energy between the two levels.

$$Energy\ released = Energy\ of\ P - Energy\ of\ T$$
$$Energy\ released = (-0.61 - (-5.4)) \times 10^{-19}\ J$$
$$Energy\ released = 4.79 \times 10^{-19}\ J$$

d) In part (c) we calculate the energy that the photons will have when electrons fall from level P to level T. We can now use Planck's equation to calculate the frequency of radiation for these photons.

$$E = hf$$

$$f = \frac{4.79 \times 10^{-19}}{6.63 \times 10^{-34}}$$

$$f = 7.2 \times 10^{14}\ Hz$$

The radiated photons will have a frequency of 7.2×10^{14} Hz.

Questions

1 The diagram of figure 8.7.3 shows part of the energy level diagram for the electrons in an atom.

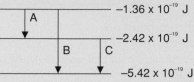

fig 8.7.3

a) For atoms to emit light they must be in an excited state. Explain the meaning of the term excited state.

b) Which of the transitions in the diagram above would emit photons of the highest frequency? Explain your answer.

c) What wavelength of photons would be emitted by electrons making the transition labelled C in figure 8.7.3?

2 Figure 8.7.4 shows part of the energy level diagram for electrons in a certain atom.

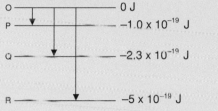

fig 8.7.4

a) The 0 J energy level is sometimes called the ionisation level. Explain the term ionisation level.

b) Calculate the energies released in each of the transitions shown on the diagram.

c) What is the energy of a photon absorbed to promote an electron from level R to level P?

d) What frequency of radiation must be supplied to a sample of this gas at low pressure to promote electrons from level R to level O?

3 When visible light from the sun is examined with a spectrometer some of the frequencies are found to be missing.

a) Explain why some frequencies and not others are absorbed.

b) How can analysis of the missing frequencies help determine the composition of the outer layers of the sun.

c) Light of wavelength 589 nm is missing from the sun's spectrum. What is the energy of the absorbed photons?

4 The apparatus shown in figure 8.7.5 is used to demonstrate the absorption of photons of light by free sodium atoms produced by placing a sodium pencil in a flame.

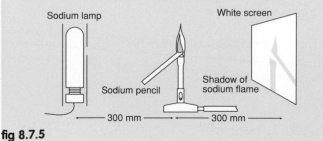

fig 8.7.5

a) With a pencil in the flame a dark shadow is seen on the screen because light is prevented from passing through the flame. Explain what prevents some of the light from passing through the flame.

b) Analysing light from the sodium lamp with a grating shows that it emits light with a wavelength of 589 nm. What is the energy of a photon of this light?

c) What is the energy of a photon absorbed by the flame when the sodium pencil is being burned?

5 With the apparatus shown in figure 8.7.6 a high voltage supply is connected across a discharge tube which contains gas at low pressure. The emission spectrum of the discharge tube is examined with a spectrometer.

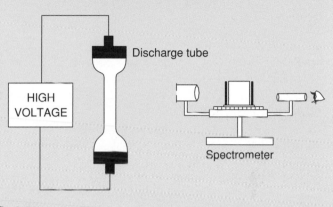

fig 8.7.6

a) Draw a diagram to represent the electron energy levels within an atom which would emit photons having three different energies.

b) A student identifies the following three wavelengths of radiation in the emission spectrum. She calls them, P, Q and R

Name	Wavelength/nm
P	400
Q	934
R	700

Table 8.7.2

(i) What are the energies of photons of radiation with these wavelengths?

(ii) The student tries to draw an energy level diagram to explain how these wavelengths could be produced.

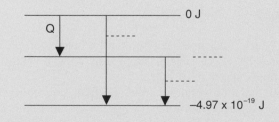

fig 8.7.7

Copy and complete the diagram by marking on the missing energy level and identifying which electron transitions emit radiations P and R.

Past paper question 1

a) Laser light is *monochromatic* and *coherent*. Briefly explain the meaning of these terms.

b) A laser radiates energy when electrons are stimulated to fall from energy level E_2 to energy level E_1 as shown in the diagram.

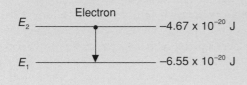

fig 8.8.1

  (i) What is the frequency and wavelength of the radiation emitted?

  (ii) Name the section of the electromagnetic spectrum in which the radiation occurs.   **SQA 1993**

Answers and comments

a) Monochromatic light is light of a single wavelength. It is created by a transition of electrons between two specific energy levels.

Coherent means that all parts of the beam, an equal distance from the source, are at the same stage of their oscillation.

b) i) To calculate the frequency and wavelength of the radiation we must first calculate the energy change involved.

$$Energy\ released = (-4.67 - (-6.55)) \times 10^{-20}\ J$$
$$Energy\ released = 1.88 \times 10^{-20}\ J$$
$$The\ frequency\ is\ calculated\ using\ E = hf$$

$$f = \frac{E}{h} = \frac{1.88 \times 10^{-20}}{6.65 \times 10^{-34}}$$

$$f = 2.84 \times 10^{13}\ Hz$$

ii) The wavelength for this frequency is given by;

$$\lambda = \frac{c}{f} = \frac{3 \times 10^8}{2.84 \times 10^{13}}$$

$$\lambda = 1.06 \times 10^{-5}\ m$$

By referring to the data sheet on an exam paper we can see that a wavelength of 10,000 nm is in the infrared region of the spectrum.

Worked example

Figure 8.8.2 represents four possible energy levels in mercury atoms.

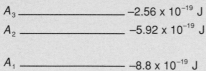

fig 8.8.2

a) Which transitions will produce photons of radiation capable of releasing electrons from a metal whose work function is 9×10^{-19} J?

b) If radiation resulting from the transition between A_3 and A_0 irradiates a metal whose work function is 8.5×10^{-19} J what will be the maximum possible kinetic energy of the ejected electrons?

Answers and comments

a) Photoelectrons are ejected from metals by incident radiation made up from photons with energies greater than the work function. In this case the work function is 9×10^{-19} J so we must look for transitions releasing photons with this amount of energy or more. The ones which will release photoelectrons are;

$$A_3\ to\ A_0\ (energy\ released = 14.08 \times 10^{-19}\ J)$$
$$A_2\ to\ A_0\ (energy\ released = 10.72 \times 10^{-19}\ J)$$

b) The maximum kinetic energy of the released electrons is the difference between the energy of the incident photons and the work function. Therefore;

$$KE_{max} = Photon\ energy - Work\ function$$
$$KE_{max} = 14.08 \times 10^{-19} - 8.5 \times 10^{-19}\ J$$
$$KE_{max} = 5.58 \times 10^{-19}\ J$$

The maximum kinetic energy of the released electrons is 5.58×10^{-19} J.

Questions

1   A point source of monochromatic light with a wavelength of 589 nm is used to irradiate a piece of zinc which is placed on top of a negatively charged metal object.

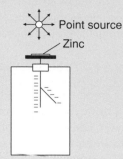

Point source
Zinc

fig 8.8.3

   a) Explain why the charge held on the object decreases.
   b) State and explain the effect of doubling the distance between the source and the piece of zinc.
   c) Explain, in terms of electron transitions, why the source can produce light of wavelength 589 nm.
   d) The zinc is now replaced by a piece of platinum of the same size. Will the charged object discharge? Justify your answer. (The work function of platinum is 1×10^{-18} J.)

2   At a point P the irradiance of light from a point source of monochromatic light is 5 W m^{-2}. The light has a frequency of 7.5×10^{14} Hz.
   a) What is the light irradiance at point Q which is twice as far as P from the point source?
   b) What is the energy of the photons radiated by this monochromatic light source?
   c) How many photons of this light fall on an area of 1 m^2 each second?

3   In the apparatus shown in figure 8.8.4 ultraviolet light with a wavelength of 254 nm falls on a zinc plate.

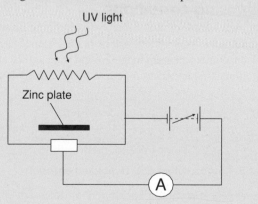

UV light

Zinc plate

A

fig 8.8.4

   Each photon arriving at the zinc plate releases a photoelectron.
   a) How many photons per second must be arriving for the current in the circuit to be 6 mA?
   b) What happens to the reading shown on the ammeter when the power supply connections are reversed?

4   In an experiment to investigate the photoelectric effect, the maximum kinetic energy of photoelectrons was determined for a range of different frequencies of incident radiation.

Frequency ($\times 10^{14}$ Hz)	10	8	6
Max KE of photoelectrons ($\times 10^{-19}$ J)	3.25	1.92	0.58

Table 8.8.1

   a) Draw a graph of maximum kinetic energy versus frequency.
   b) From your graph determine the threshold frequency for the metal being investigated.
   c) Calculate the work function for this metal.

5   The emission spectrum of a particular gas contains a line of frequency 5×10^{14} Hz.
   a) Calculate the energy of the photons causing this line in the spectrum.
   b) A student decides to examine this radiation with a very accurate grating and a spectrometer.

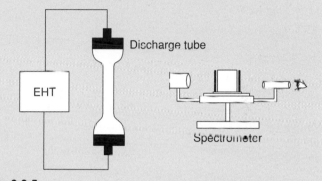

Discharge tube

EHT

Spectrometer

fig 8.8.5

   The grating is marked as having 500 lines per mm. At what angle will the first order maximum in the interference pattern be seen?

6   X-rays are produced by allowing very fast moving electrons to strike a target made from a metal such as molybdenum. These fast moving electrons force atoms in the target into an excited state.
   a) Explain what is meant by the term excited state.
   b) When the excited atoms return to the ground state photons of radiation are emitted.
      Explain why these photons have certain precise energies rather than a continuous range of different values.
   c) The two different wavelengths of X-rays characteristic of one particular target are a result of the electron transitions shown in figure 8.8.6.

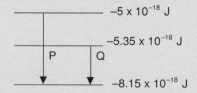

-5×10^{-18} J

-5.35×10^{-18} J

P   Q

-8.15×10^{-18} J

fig 8.8.6

What is the wavelength of X-rays radiated as a result of the transitions labelled P and Q?

 **Lasers**

Introduction

Some of the experiments and demonstrations studied earlier in this book were enhanced by the special properties of the light from a laser. The demonstrations of the diffraction and interference of light relied on the fact that the laser was a source of coherent light with a single wavelength. Sources with a single wavelength are called **monochromatic**. The coherent nature of laser light means that each point right across the entire width of the beam is composed of waves which are in phase or where there is a constant phase difference. Consequently when a beam of laser light falls on a grating the slits act as coherent secondary sources.

Lasers have been used in industry and in medicine for some time. Modern laser printers produce images with very high resolution. In these printers a narrow beam of laser light scans across the drum, electrostatically imprinting an image of the text or graphic.

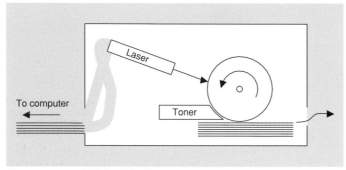

Figure 8.9.1 Principles of a laser printer

In the rest of the printing process the drum collects fine particles of toner and deposits them onto paper.

Spontaneous emissions

A laser is monochromatic because the photons of laser light are created as a result of electrons making one particular transition between two specific energy levels within excited atoms. When atoms with electrons in their ground states absorb specific quantities of energy the electrons move into higher levels. These excited atoms can then return to their ground state by emitting photons of a particular frequency.

Within any sample of material it is extremely unlikely that all of the atoms will naturally be in the ground state. Some will have their electrons in excited states. Atoms in excited states naturally return to the ground state, emitting a photon of radiation in the process. Such emissions are called **spontaneous emissions**.

Excited atoms returning to their ground states will not all do so simultaneously. Some atoms will spend longer in their excited states than others. Spontaneous emission, like radioactive decay, is a **random process**. We cannot predict the exact moment when a particular atom will return to its ground state but since there are so many atoms in even a small sample we can predict what fraction of the excited atoms will return to their ground state in a particular time.

Stimulated emissions

Excited atoms will return to their ground state by radiating photons with a certain energy. If the electron in the higher level has energy E_1 and in the ground state has energy E_0 then we can say;

$$E_1 - E_0 = hf$$

where f is the frequency of the radiation.

Excited atoms can be encouraged or stimulated to return to their ground state. **The presence of photons with energies exactly equal to those which the excited atoms will radiate when returning to their ground state, stimulates an excited atom to return to its ground state.** As it does so it releases a stimulated photon of exactly the correct energy to stimulate other excited atoms to return to their ground states. If the proportion of excited atoms in a sample is high enough a **chain reaction** can occur.

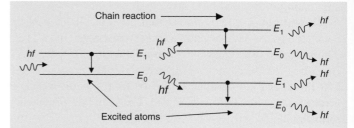

Figure 8.9.2 Chain reaction

Stimulated photons not only have the same frequency as the photons causing the stimulation but they are always in phase and travel in the same direction. If the conditions inside a laser are suitable, the number of photons in phase and travelling in the same direction rapidly increases so that a **concentrated beam** is produced.

Optimising conditions

In a normal sample of atoms there will always be some that are in an excited state while others will be in the ground state. A photon which encounters an atom in the ground state will be absorbed by the atom making an electron go to a higher energy level.

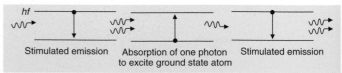

Stimulated emission   Absorption of one photon   Stimulated emission
to excite ground state atom

Figure 8.9.3 Absorption and stimulated emission

Alternatively if a photon encounters an atom which is *already* in an excited state it can cause a **stimulated emission**. For the laser to produce a concentrated beam of light the conditions must be such that the beam gains *more* energy by stimulated emissions than it *loses* by absorption. This will be the case only if the proportion of the atoms in an excited state is relatively high. Different types of laser adopt different techniques for ensuring that the proportion of excited atoms is high.

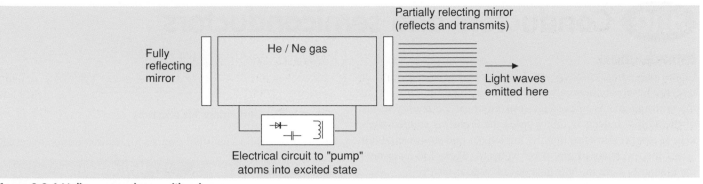

Figure 8.9.4 Helium-neon laser with mirrors

The laser's reflective mirrors

We have already seen the need for a laser to have a high proportion of its constituent atoms in an excited state. Only when this is so can photons cause even more photons to be produced. It is therefore important that not all of the photons leave the laser immediately after they are produced. They must spend some time increasing the irradiance of the beam by stimulating further emissions.

In the laser commonly used in schools a mixture of helium and neon atoms is contained in a tube with mirrors at either end. **The mirror at one end of the tube is totally reflecting while the mirror at the other end is partially reflecting and allows 1% of the light intensity in the tube to be emitted**. This arrangement ensures that each photon undergoes many reflections before escaping. Therefore the radiation is trapped in the tube long enough to stimulate sufficient further photons to maintain the irradiance of the beam.

The amplification of the beam using these stimulated emissions of radiation gives the laser process the name; **Light Amplification by Stimulated Emission of Radiation**.

Irradiance of laser beam

The light which leaves the laser is monochromatic, coherent across the full width of the beam and is all travelling in the same direction. The beam emerging from a laser is therefore almost parallel along the entire length of its travel.

Figure 8.9.5 Parallel laser beam

The fact that the beam is not divergent means that its irradiance is constant all along its length. The irradiance is defined as;

$$Irradiance = \frac{Power}{Area}$$

The He-Ne laser used in schools radiates energy at a rate of 0.1 millijoules per second and has a beam of circular cross-section, typically 2 mm in diameter. The irradiance can be calculated as;

$$Irradiance = \frac{Power}{Area}$$

$$Irradiance = \frac{0.1 \times 10^{-3}}{\pi \times (1.0 \times 10^{-3})^2}$$

$$Irradiance = 31.8 \, W m^{-2}$$

This value is much higher than the light irradiance 1 m away from a 100 W light bulb. Even though the laser is radiating only one millionth of the bulb's energy the fact that its beam does not diverge makes it considerably more intense. The high intensity of the laser makes it useful in surgery and engraving but also means that it can be hazardous if used incorrectly.

Holograms

These 3D images are produced by splitting a laser beam into two parts. One beam shines on the object and forms an image on a photographic plate. The second beam directly illuminates the plate. When the photographic plate is developed and viewed under the correct conditions the hologram is revealed.

Hologram images are used widely in security applications like the credit cards shown below and researchers would like to produce 3D moving pictures – perhaps using moving holograms.

Figure 8.9.6

8.10 Conductors and semiconductors

Introduction

In the preceding chapters we have looked at spectra and at lasers. These phenomena have been explained by considering the rearrangements of the electrons amongst the energy levels of the atoms. Indeed in many cases the elements studied were kept at low pressures so that the nucleus of one atom would not affect the electrons of any other atom. At these low pressures the elements were in their gaseous or vapour states.

When elements are in their solid or liquid states their properties are different. Nevertheless we can explain some of their properties by considering the behaviour of the electrons surrounding the nuclei.

Conductors

We can explain the conduction properties of metals by considering that the outer electrons in a metal are only loosely bound within the metal rather than being under the influence of any one specific nucleus. Using this model insulators are pictured as materials where all of the electrons are involved in the bonding process and are consequently bonded tightly to an atom.

Air is regarded as a good insulator but under the correct conditions it can be made to conduct.

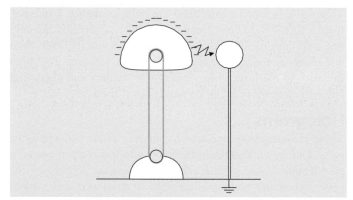

Figure 8.10.1 Van der Graaf spark

When the dome of the Van der Graaf generator is charged to a very high voltage and a grounded sphere is brought up close a spark will jump across the gap. The very high voltage makes the air conduct.

Quantifying conduction

Good conductors have a low resistance so a small potential difference will cause an appreciable current. Poor conductors have high resistances. We might therefore imagine that the resistance of a sample of material is a measurement of its ability to conduct. Although *resistance* gives an indication of ability to conduct, the resistance of a sample depends upon its dimensions. The **resistivity** of a material takes account of the size and shape of a sample and allows the conducting properties of similar materials to be compared.

A material's resistivity is numerically the same as the resistance of a sample of the material which is 1 m long and has a cross-sectional area of 1 m^2.

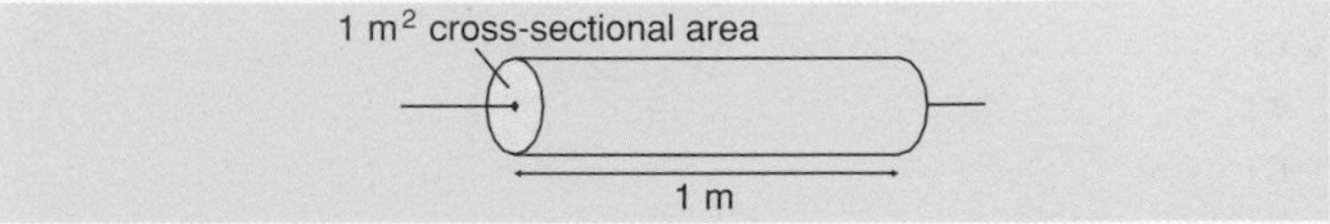

Figure 8.10.2 Dimensions for resistivity

Table 8.10.1 shows resistivity values for some common materials.

Material	Resistivity/Ω m
Silver	1.6×10^{-8}
Copper	1.7×10^{-8}
Iron	1.0×10^{-8}
Glass	1.0×10^{12}
Quartz	1.0×10^{17}

Table 8.10.1

Clearly the materials listed in the table fall into two main categories; conductors with low values of resistivity and insulators with high resistivity values.

Altering resistance

The resistance of a sample depends on its dimensions. In addition, **the resistance of certain materials can be altered by adding atoms of other materials. This process is called doping.**

The resistance of a doped sample of previously pure silicon or germanium depends on the concentration of the doping element. For similar sized samples the resistance of doped silicon or germanium will fall midway between the resistances typical of conductors and insulators. Doped samples of silicon or germanium are called **extrinsic semiconductors**.

Conductors $\longrightarrow$ Semiconductors $\longrightarrow$ Insulators
resistivity ~ 10^{-9} Ωm   resistivity ~ 10^{1} Ω m   resistivity ~ 10^{9} Ωm

Figure 8.10.3

In a sample of semiconducting material some of the atoms have electrons which are tightly bound while other atoms have some electrons which can contribute to the conduction process.

Doping silicon: n-type semiconductors

Silicon, like carbon, is in group 4 of the periodic table and has four electrons in its outer shell available for bonding.

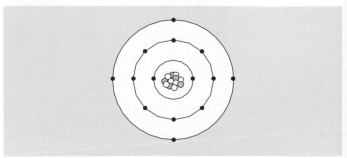

Figure 8.10.4 Electron structure of silicon

To achieve a stable arrangement of eight bonding electrons a silicon atom will share its electrons with four neighbouring silicon atoms. The bonds formed by the sharing of electrons between atoms are called **covalent bonds**. In silicon, as in the diamond form of carbon, these bonds make samples of pure silicon very strong and chemically unreactive. The 3D arrangement adopted by the bonded atoms in a silicon lattice can be represented as shown in figure 8.10.5.

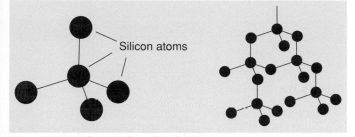

Figure 8.10.5 Silicon in bonding lattice

The sharing of electrons can be pictured as shown in figure 8.10.6.

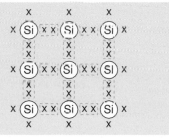

Figure 8.10.6

Doping introduces small but carefully controlled quantities of impurities such as arsenic into crystalline silicon. Arsenic is a group 5 element and has five electrons in its outer shell available for bonding. Four of these are used by neighbouring silicon atoms to form covalent bonds but the *fifth* helps improve conduction.

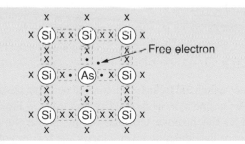

Figure 8.10.7

It would be easy to think that the 'free' electron gives the sample of doped silicon an overall negative charge. This is however not correct. Yes, there are electrons which are not involved in the bonding process but the sample is electrically neutral because every electron in both the arsenic and silicon atoms has a corresponding proton in its atomic nucleus.

Doping pure silicon with arsenic produces what is called an **n-type semiconductor** which can be represented as shown in Figure 8.10.8.

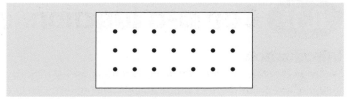

Figure 8.10.8 n-type semiconductor

The dots represent the free electrons responsible for conduction in n-type semiconductors. These are called the **majority charge carriers**.

Making p-type semiconductors

If group 3 elements such as indium are used to dope pure silicon there are insufficient electrons to form four covalent bonds.

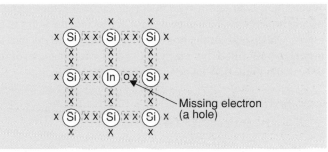

Figure 8.10.9

The 'hole' in the bonding structure can be thought of as a site in search of an electron. The sample illustrated in figure 8.10.9 is electrically neutral due to the presence of equal numbers of protons and electrons.

Within the material some of the electrons from other silicon atoms will move from their bonded positions into this hole leaving behind another hole and an atom having a positive charge.

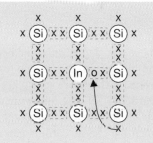

Figure 8.10.10 Si atom becomes positive with "hole"

Since other electrons can move to fill this hole and as a consequence create further 'positive holes' this type of semiconductor is called **p-type**. **The majority charge** carriers in a p-type semiconductor are positive holes.

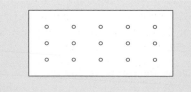

Figure 8.10.11

8.11 The p-n junction

Introduction

An n-type semiconductor has free electrons as its majority charge carriers. In p-type semiconductors the majority charge carriers are positive holes. Both p-type and n-type semiconducting materials are electrically neutral because they contain equal numbers of protons and electrons.

When samples of p-type and n-type semiconductor are together on a single piece of silicon the combined material has special properties. The junction where the p-type and n-type meet is called a **p-n junction**.

At the junction

When the p-type and n-type materials come together some of the electrons from the n-type diffuse across into the holes in the p-type while some of the holes from the p-type diffuse into the n-type to be filled by electrons. As a consequence of the diffusion a **depletion layer** is formed at the junction.

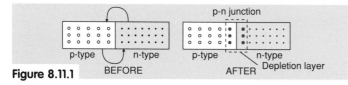

Figure 8.11.1

When the electrons leave the electrically neutral n-type material it becomes positively charged. Similarly if positive holes diffuse away from the p-type it must become negatively charged.

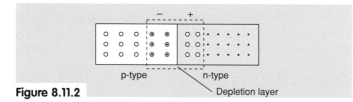

Figure 8.11.2

The p-type region of the depletion layer becomes negatively charged while the region in the n-type charges positively. **A potential difference is therefore set up across the depletion layer as a consequence of the movement of the charge carriers**. The exchange at the junction is only momentary as a time is reached when the charge carriers can no longer migrate across the depletion layer because of the potential barrier. In silicon the potential barrier across the depletion layer is typically around 0.6 V while for germanium the barrier is around 0.1 V. Consider a single charge carrier in the n-type region of figure 8.11.3 attempting to migrate.

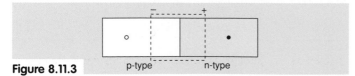

Figure 8.11.3

When it tries to enter the p-type region the positive side of the potential barrier will try to keep it where it is while the negative side of the potential barrier will prevent it from entering the p-type. Similarly a majority charge carrier in the p-type will find it equally difficult to migrate across the depletion layer.

The p-n junction as a diode

Charge carriers can be made to pass across the junction barrier if a power supply is connected across the p-n junction. The positive of the power supply must be connected to the p-type region and the negative to the n-type region.

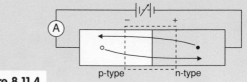

Figure 8.11.4

If the output from the power supply is increased from 0 V, a voltage is reached where the potential difference across the p-n junction is high enough to overcome the potential barrier and make charge carriers migrate through the semiconducting material.

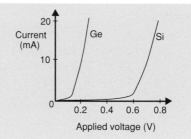

Figure 8.11.5

Figure 8.11.5 shows how the current through a p-n junction changes as the voltage from the supply increases.

If the power supply is connected across the junction with the opposite polarity the effect is quite different.

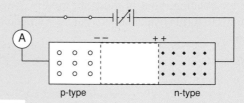

Figure 8.11.6

When the switch in figure 8.11.6 is closed the positive holes congregate at one side while the free electrons in the n-type are attracted to the positive terminal of the power supply. As a consequence the depletion layer widens, the barrier voltage rises and this prevents conduction. **The p-n junction is behaving as a diode. Conduction will only take place when the p-type is connected to the positive terminal of a power supply capable of overcoming the potential barrier at the junction.**

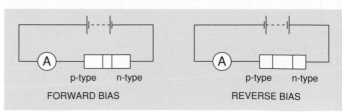

Figure 8.11.7

The p-n junction diode will only conduct when it is **forward biased** as shown in figure 8.11.7. If the positive of the supply is joined to the n-type the diode does not conduct and this connection is called **reverse bias**.

The symbol for a diode is as shown in figure 8.11.8 and forward and reverse connections are made as shown in figure 8.11.9.

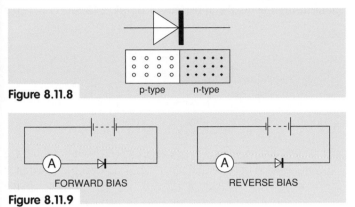

Figure 8.11.8

p-type   n-type

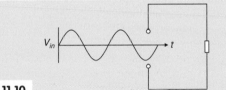

FORWARD BIAS   REVERSE BIAS

Figure 8.11.9

Using diodes

Diodes are often described as components which will only allow charge to move one way around a circuit. The alternating voltage of figure 8.11.10 applied to the resistor in the circuit will cause a changing current in the resistor.

V_{in}

Figure 8.11.10

In the first half of the cycle the current will be in one direction in the resistor but when the polarity of the voltage changes the current will be in the other direction. With the diode of figure 8.11.11 in the circuit the current will be blocked when the polarity of the voltage changes so the current in the resistor and indeed the voltage across it can be represented as shown in the figure 8.11.11.

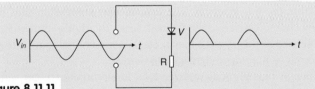

Figure 8.11.11

A little further...

The output voltage of this circuit has only positive values although the input voltage has both positive and negative values. The output is described as being **half wave rectified**. With the diode bridge circuit of figure 8.11.12 **full wave rectification** can be achieved.

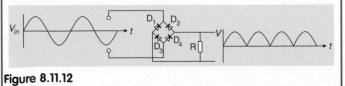

Figure 8.11.12

Light emitting diodes

In chapter 5 we looked at how LEDs and LED arrays could be used in basic electronic circuits, without explaining how they work. The holes and electrons in a forward biased diode can be pictured as shown in figure 8.11.13.

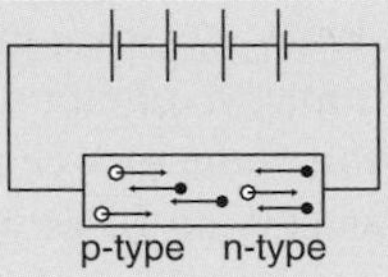

Figure 8.11.13

p-type   n-type

In an earlier section we described holes as *sites in search of electrons*. When current flows in a forward biased diode some of the electrons in the n-type side **recombine** with holes from the p-type at the junction. The result is a more stable arrangement so some energy has to be given up. In most diodes this results in heating. **However in a few diodes where the junction is close to the surface the energy is radiated as photons of light. Such diodes are called light emitting diodes or LEDs.**

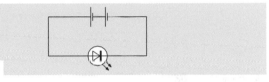

Figure 8.11.14

Semiconducting material fabricated with different compositions of the elements gallium, arsenic and phosphorus can provide visible light with red, yellow, green and blue wavelengths. The photons emitted from an LED all have the same frequency so the light from an LED is monochromatic. Modern developments have even seen LEDs emitting coherent, monochromatic light in a single direction. These laser LEDs have many applications and are widely used in processing data stored on CDs.

Recombination energy

In LEDs the recombination of electrons and holes produces photons of visible light. The energy released as a result of the recombination can be found by examining the light from the LED with a grating.

Q For an LED a first order line is observed at an angle of 10° when using a grating with 300 lines per millimetre. Calculate the recombination energy for the electrons and holes in the LED.

A The wavelength of the light emitted from the LED can be found from;

$$d \sin \theta = n\lambda$$

$$\frac{1}{300,000} \sin 10° = 1 \times \lambda$$

$$\lambda = 579 \ mm$$

The energy of photons with this wavelength can be calculated from;

$$E = \frac{hc}{\lambda} = \frac{6.63 \times 10^{-34} \times 3 \times 10^8}{579 \times 10^{-9}}$$

$$E = 3.38 \times 10^{-19} \ J$$

8.12 Light on p-n junctions

Introduction

Light incident upon metal surfaces, in certain situations, can cause the photoelectric effect while white light passing through various gases or vapours at low pressures can produce absorption spectra. Similarly, photons of light falling upon specially fabricated semiconductors can alter their conducting properties. Light incident on an LDR lowers its resistance by producing electrons and holes which take part in the conduction process.

A solid state device, where positive and negative charges are produced by the action of light on a p-n junction is called a photodiode. When reverse biased, photodiodes can respond quickly to changes in surrounding light levels.

Photoconductors

In the circuit of figure 8.12.1 the diode connected to the power supply is reverse biased.

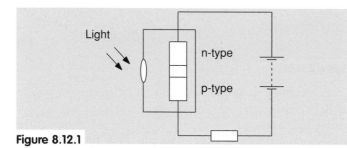

Figure 8.12.1

In theory no current should flow in the resistor because the p-n junction is reverse biased. In practice, however, there will be a very small current. This is caused by bonded electrons within the semiconductor being released from their normal positions by heat absorbed from the surroundings. The current through a reverse biased photodiode is called the **leakage current**. **Photons of light shining on the p-n junction can liberate electrons leaving behind holes. These electron-hole pairs increase the leakage current.** When used in this mode the photodiode is said to be operating in **photoconductive mode**.

The symbol for a photodiode is shown in Figure 8.12.2.

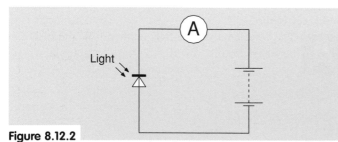

Figure 8.12.2

Light incident upon the transparent window of the photodiode falls on the p-n junction and the charge carriers created will flow, in opposite directions, through the diode to the power supply terminals. More photons mean more electron-hole pairs so more current. Therefore, supplying more photons by increasing the intensity of the light falling on a p-n junction increases the leakage current.

Leakage current and light irradiance

We can show that the leakage current in the reverse biased photodiode is directly proportional to the irradiance of the incident light using the apparatus of figure 8.12.3.

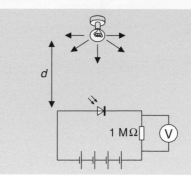

Figure 8.12.3

By Ohm's Law the current in the circuit is directly proportional to the potential difference across the 1 MΩ resistor. Doubling the distance between the light source and the photodiode will quarter the current in the photodiode. This is as expected since doubling the distance quarters the irradiance of the incident light. If we assume that N photons per unit area, each of frequency, f, fall on the photodiode per second, we can say that the irradiance, I, is given by;

$$I = Nhf$$

where h is Planck's constant.
If we assume that h and f are constant;

$$I \propto N$$

The irradiance of light incident on the photodiode is directly proportional to the number of photons. The leakage current is proportional to the number of free charge carriers so if all of the incident photons produce free charges, the leakage current is directly proportional to the irradiance of the light falling on the photodiode.

This analysis assumes that there is no build up of electrons within the p-n junction. The reverse bias voltage ensures that any charge carriers once formed, leave the diode immediately. The size of this reverse bias voltage is not crucial, but it should not be so high as to damage the diode.

Reverse bias photodiodes are used in circuits to trigger timers for mechanics experiments. In figure 8.12.4 the light shining on the photodiode causes a high leakage current in the resistor so the output voltage is high.

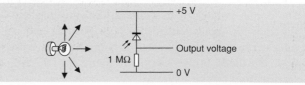

Figure 8.12.4

Interrupting the light beam cuts off the supply of photons to the photodiode and almost immediately the leakage current drops so the output voltage falls. The speed with which electron-hole pairs recombine once the light is cut off makes the photodiode very responsive to changes in light level. By contrast LDRs respond relatively slowly to changes in light level so they can cause inaccuracies if used to trigger timers in mechanics experiments.

Photodiodes as solar cells

The circuit shown in figure 8.12.5 also has **photons of light incident on the transparent window of a photodiode**. This time however there is *no* battery connected in the circuit.

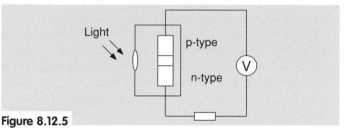

Figure 8.12.5

If the photodiode is covered the reading on the voltmeter reduces to zero. **When light shines on the p-n junction in the photodiode a voltage is produced.** The photodiode is now said to be working in the **photovoltaic mode**.

The photons of light are again separating some electrons from the holes to form electron-hole pairs. The holes will congregate on one side of the potential barrier and the electrons on the other. As the number of pairs builds up it becomes more difficult for electrons and holes to separate so the build up eventually stops.

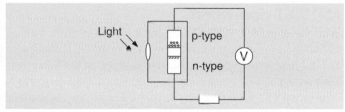

Figure 8.12.6

The charge separation means that the photodiode operating in photovoltaic mode can act as a **source of electrical energy**. Photodiodes can supply small quantities of current to resistors or other transducers connected in an external circuit. The photodiode in figure 8.12.6 is transforming some of the light from the sun into electrical energy and is sometimes called **a solar cell**.

**Figure 8.12.7
Solar panels in the Mir Space Station**

Large areas of panels are required if solar cells are to produce high power outputs. Spacecraft are fitted with orientation systems which keep their solar panels pointing directly at the sun. Since solar panels fitted to spacecraft are closer to the sun

than those used in ground based applications the intensity of the light they receive is greater. Similarly, solar panels on satellites in space also receive more of the sun's energy because the rays do not have to pass through the earth's atmosphere. The power output from a particular area of panels in space is therefore higher than the power output from the same area of panels used in ground based applications.

Solar power is a very clean source of energy and much research is taking place into developing semiconducting materials which will efficiently and cost effectively harness the sun's energy.

Demonstrations with photodiodes

EXPERIMENT The speed with which photodiodes respond to changes in light level can be shown with the apparatus illustrated in figure 8.12.8.

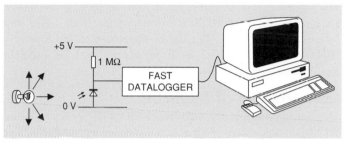

Figure 8.12.8

With light shining on the reverse biased photodiode, the fast datalogger is started and a piece of card is quickly passed through the gap between the light and the photodiode. When the recorded voltages are displayed the changes are seen as sharp, almost vertical, 'edges'. If the photodiode is replaced with an LDR and the experiment repeated the changes are more gradual because the LDR's response is slower.

In the circuit of figure 8.12.9 the photodiode, operating in photovoltaic mode is connected to an oscilloscope and a lamp, connected to a signal generator, is held near the photodiode. When the frequency of the a.c. supplied to the lamp is increased and the timebase of the oscilloscope suitably adjusted, the flickering of the lamp is still shown on the oscilloscope long after the eye has detected continuous illumination.

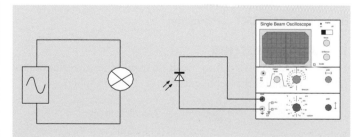

Figure 8.12.9

While doing this experiment with the lamp switched off you might still detect a small signal if the photodiode is picking up the flicker of the lights in the classroom.

As well as being used in light meters, the speed and sensitivity of semiconducting photodiodes means that they have many applications in practical electronic sensing and timing circuits.

8.13 MOSFETs

Introduction

In section 5.2 you were introduced to the transistor symbols shown in figure 8.13.1.

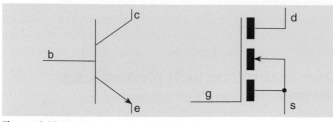

Figure 8.13.1

The bipolar npn transistor is made from layers of n and p type semiconductor. The MOSFET is also made from n-type and p-type semiconductors but the arrangement of the layers is slightly more complex. Both transistors can be used in amplifier or switching circuits.

Composition of MOSFETs

The starting material for the type of MOSFET shown in figure 8.13.1 is p-type silicon. This material is called the **substrate**. The majority charge carriers in the p-type substrate are positive holes but there are still electrons present as minority carriers. Some areas on the surface of the substrate are coated with a thin layer of **insulating silicon oxide**. Other regions in the substrate have atoms of n-type semiconductor added. These regions are called **implants**. Metallic connections are made to the three different areas on the surface of the substrate which is itself usually connected either externally or internally to one of the implants. The connections to the different areas on the MOSFET are called the **source, s, gate, g,** and **drain, d**.

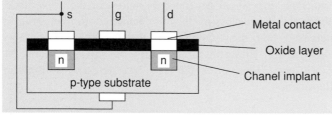

Figure 8.13.2 MOSFET construction

Induced charges

Figure 8.13.3 shows a thin piece of insulating material placed on top of a metal disc connected to a coulomb meter.

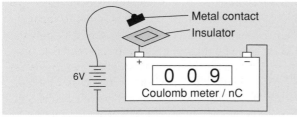

Figure 8.13.3 Inducing charges

When the metal contact connected to the positive of the battery is touched against the top of the insulator the coulomb meter shows that the metal disc becomes charged. When the contact is removed from the insulator the reading shown by the coulomb meter returns to zero. The battery causes negative charge on the metal disc even though there is no charge transferred from the battery to the metal disc.

When the positive terminal of the battery is held against the top of the insulator the tightly bound positive and negative charges in the insulator polarise (figure 8.13.4). This produces an electric field and leaves a net positive charge on the lower surface of the insulator. This charge induces electrons in the stem of the metal disc to move onto the disc.

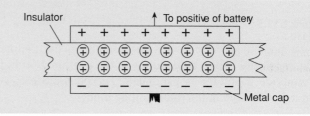

Figure 8.13.4

When the positive contact is removed from the top of the insulator the electric field inside the insulator collapses and the electrons in the metal disc return to their normal position so the coulomb meter reading returns to zero.

MOSFET operation

With the apparatus shown in figure 8.13.3, a positive voltage connected to one side of an insulating material produced an electric field which induces negative charges on the other side. We can use similar ideas to explain how a MOSFET operates. With no positive voltage applied to the gate the p-type and n-type regions in the MOSFET behave as 'back to back' diodes neither of which conduct, so the transistor is switched off.

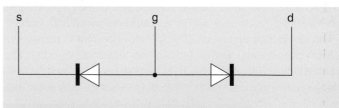

Figure 8.13.5a MOSFET switched off

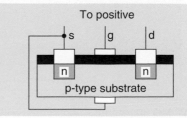

Figure 8.13.5b Positive voltage applied to gate

Applying a positive voltage to the gate enhances the concentration of electrons in the substrate under the insulating oxide layer in the region beneath the gate (figure 8.13.5b). This forms a conduction channel, called an **n-channel**, connecting the source and drain. When a p.d. is applied between the source and drain there is a current in the channel causing the MOSFET to switch on. MOSFETs which are normally switched off but can be switched on by applying a positive voltage to the gate are said to be working in **enhancement** mode. The type of MOSFET described so far is referred to as an **n-channel enhancement mode MOSFET**.

A little further...

Manufacturing MOSFETS

From what you have read already you are perhaps beginning to view the MOSFET as a fairly complex device. If we consider that many tens of millions of MOSFETs can be placed on a single integrated circuit we begin to appreciate something of the skill of those involved in the design and manufacture of integrated circuits. The p-type silicon substrate is made by slowly pulling a crystal of pure silicon out of a bath of suitably doped molten silicon. As the crystal is pulled out, other crystals form around it to such an extent that ingots approximately 50 cm wide and 3 m long can be produced. The ingots are then cut into wafers less than 1 mm thick.

Figure 8.13.6
A silicon ingot

The wafers are polished and the layered structure of the MOSFET is built up in different stages by coating the clean surface with different materials. Etching away some of the deposited materials then leaves behind the desired pattern. The shape of the patterns on the surface is defined using **photolithography** and many different masks are used to build up the layers of insulator, implants and interconnections between transistors within the **integrated circuits**.

Why use MOSFETS?

In section 5.2 you saw how a small current in the base of the bipolar transistor in figure 8.13.7 can switch on a much larger current in the lamp. The bipolar transistor is a current controlled current switch.

The MOSFET is a voltage controlled current switch. It is switched on when a voltage greater than the switching threshold is applied to the gate. There is no *transfer* of charge

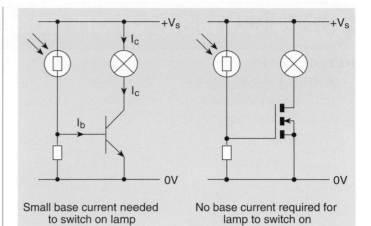

Small base current needed to switch on lamp

No base current required for lamp to switch on

Figure 8.13.7

from the gate to the substrate so there is no current from the gate into the MOSFET. **The MOSFET therefore has an infinite input resistance.** This enables low current integrated circuits involving many millions of MOSFETs to be constructed.

A little further...

CMOS

In this chapter we have outlined how a MOSFET with a p-type substrate and n-type implants can be constructed. This transistor switches on when a positive voltage is applied to the gate. In modern terminology this is called an NMOS transistor. Another type of MOSFET, the PMOS, is manufactured from an n-type substrate and has p-type implants. The PMOS transistor shown in figure 8.13.8 switches on when its gate is at zero volts.

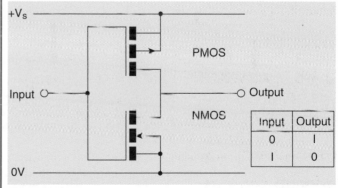

Figure 8.13.8

Figure 8.13.8 shows a circuit made from complementary NMOS and PMOS transistors. When the input is made high only the NMOS transistor is switched on, effectively connecting the output to 0V. When the input is zero, only the PMOS is on so the output is effectively connected to $+V_s$. Figure 8.13.8 shows the type of CMOS NOT gate that is the basic building brick in logic, memory and processor circuits. Considering the extent to which these type of circuits have become a part of modern life it is not difficult to realise why MOSFETs are the most commonly used component in modern electronics.

Worked example 1

The diagram shown in figure 8.14.1 represents the internal operation of one type of laser.

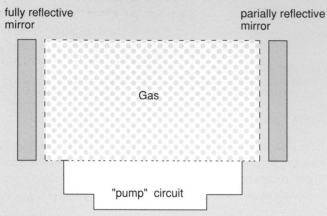

fig 8.14.1

a) Explain the function of the totally and partially reflecting mirrors.
b) Explain why a laser beam having a power even as low as 0.1 mW may cause eye damage.
c) Calculate the irradiance of the light from the 0.1 mW laser if the beam has a diameter of 2 mm.

Answers and comments

a) The mirrors 'trap' the stimulated photons within the gas mixture for a certain time. By continually reflecting them back and forth through the gas mixture they can stimulate further photons. The partially reflecting mirror allows some of the photons to be emitted as the laser beam.

b) Light intensity is defined as;

$$Intenstity = \frac{Power}{Area}$$

Even though the power of 0.1 mW is low, the narrowness of the laser beam and the fact that it does not spread out as it gets further away from the source means that the cross-sectional area of the beam, expressed in square metres, is very small. A small power on a very small area can give a high irradiance which could supply enough energy to damage cells in the retina.

c) If the beam has a diameter of 2 mm its area is only;

$$\pi \times (1 \times 10^{-3})^2 . = 3.14 \times 10^{-6} \, m^2.$$

$$Irradiance = \frac{Power}{Area} = \frac{0.1 \times 10^{-3}}{3.14 \times 10^{-6}}$$

$$Irradiance = 31.85 \, Wm^{-2}$$

Worked example 2

Figure 8.14.2 shows a representation of a p-n junction diode.

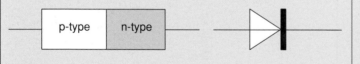

fig 8.14.2

a) Explain how a sample of pure silicon can be made into a n-type semiconductor.
b) Draw a diagram showing this p-n junction connected in series with a 6 V battery and a lamp so that the lamp lights.
c) When a p-n junction is forward biased it conducts. Describe the movement of the charge carriers which produce the current.

Answers and comments

a) In a sample of pure silicon each of the four outer electrons are bonded to other silicon atoms. Adding a doping material, such as arsenic, which has five electrons in its outer shell produces an n-type semiconductor. Where the arsenic atoms replace silicon in the bonding structure an electron not used in bonding with silicon atoms is left over. This electron is free to take part in the conduction process.

b) For the p-n junction to conduct it must be connected in the forward biased mode. This happens when the p-type is joined to the positive of the battery. The necessary circuit is as shown in figure 8.14.3.

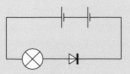

fig 8.14.3

c) The p-type material will have an excess of positive holes which will be attracted across the junction by the negative of the battery.
In the n-type material electrons are the majority charge carriers and when the diode is forward biased these are attracted across the junction to the positive of the battery.

Questions

1 A student, making revision notes, writes down the following statements about the operation of lasers.

I Spontaneous emission is a random process analogous to radioactive decay.

II Photons of energy *hf* incident on excited atoms may stimulate these atoms to emit photons also of energy *hf*.

III In stimulated emission the incident and emitted radiations are in phase and travel in the same direction.

Which of these statements is/are correct?

a) I only
b) II only
c) I and II only
d) II and III only
e) I, II and III.

2 A 1 mW He-Ne laser of the type used in schools produces light with a wavelength of 633 nm.

a) What is the energy of a single photon of this light?
b) How many photons per second are emitted by the laser?
c) The beam emerging from the laser has a diameter of 2 mm and is passed through a converging lens. What is the irradiance at a point beyond the lens where the beam diameter is 0.1 mm?

3 Table 8.14.1 shows the number of outer electrons in the atoms of certain elements.

Silicon	4
Arsenic	5
Indium	3

Table 8.14.1

a) Draw a diagram to show how the electrons in a sample of pure silicon will bond the atoms together.
b) What type of semiconductor will be formed when pure silicon is doped with indium? Explain your answer.
c) Draw a diagram to show how a small number of arsenic atoms added to pure silicon can change its conducting properties.
d) Does doping pure silicon with arsenic or indium atoms produce semiconducting samples with overall positive or negative charges? Explain your answer.

4 Many digital readouts are made from seven segment displays such as that shown in figure 8.14.4. Each of the segments is a separate LED which can be switched on by itself or along with others.

fig 8.14.4

a) Describe what occurs at the junction when one of the segments is forward biased.

b) Explain what happens at the junction if the LED is reverse biased.
c) In a component catalogue one LED is described as having a recombination energy of 3.14×10^{-19} J.
   (i) What is the wavelength of the light emitted by this LED?
   (ii) What is the colour of this display?

5 Figure 8.14.5 shows a circuit where a photodiode is acting as a switch.

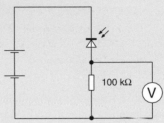

fig 8.14.5

a) Is the photodiode forward or reverse biased? Explain your answer.
b) Is the photodiode operating in the photovoltaic or photoconductive mode? Explain your answer.
c) Explain why light shining on the photodiode lowers its resistance.
d) The photodiode is carefully shielded so that only a small quantity of light from a point source falls on the p-n junction.
   (i) Explain what happens to the charge carriers in the photodiode as it is moved away from the point source.
   (ii) Explain why the reading on the voltmeter reduces as the photodiode is moved away from a point source of light.

8.15 Exam style questions

Past paper question 1

a) The word laser is an acronym for 'light amplification by the stimulated emission of radiation'.

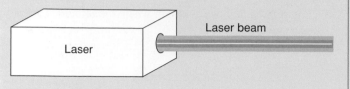

fig 8.15.1

Describe what is meant by 'stimulated emission' and describe how amplification is produced in a laser.

b) Radiation from a laser is directed at a small solid cylinder of copper as shown below. The cylinder has a cross sectional area of 1.25×10^{-5} m^2. The intensity of the laser beam at the surface of the cylinder is 4.00×10^5 W m^{-2}.

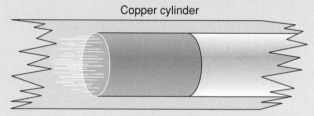

Copper cylinder

fig 8.15.2

Show that the energy delivered to the cylinder in 100 seconds is 500 J.   **SQA 1996**

Answers and comments

a) Within any gas at room temperature there will always be some atoms with electrons in excited energy levels. When an electron returns to a lower level it emits a photon of radiation. This is called a spontaneous emission. A spontaneous photon can cause other excited atoms to return to their ground state by emitting photons. The photons induced from excited atoms by the spontaneous emission are called stimulated emissions.

b) Irradiance is defined from the equation;

$$Irradiance = \frac{Power}{Area}$$

$$Power = 4.00 \times 10^5 \times 1.25 \times 10^{-5}$$
$$= 5\ W$$
$$Energy = Power \times Time$$
$$= 5 \times 100$$
$$= 500\ J$$

Past paper question 2

The following graph shows the variation of current with voltage for a diode when it is forward biased.

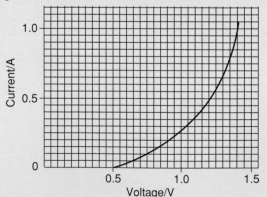

fig 8.15.3

a) What is the minimum voltage required for this diode to conduct?

b) What happens to the resistance of the diode as the voltage is increased above this minimum value?
Use information from the graph to justify your answer.   **SQA 1995**

Answers and comments

a) The diode begins to conduct when the voltage is large enough to cause a current in the diode. From the graph we can see that the current starts when the voltage is 0.5 V. Therefore a minimum of 0.5 V is required to make this diode conduct.

b) In the region where the diode is conducting the slope of the line is increasing. In this region as the voltage increases we can see that equal changes in voltage are causing progressively larger changes in current. We can explain this by saying that as the voltage increases the resistance of the diode reduces.

Questions

1

a) Explain the function of the fully and partially reflecting mirrors in a laser.

b) Explain why a beam of light from a 0.1 W laser can cause more damage to the human eye than the light from a 100 W lamp.

2 An npn transistor is one example of a bipolar transistor. This transistor is made from regions of n-type and p-type silicon.

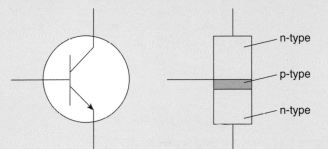

fig 8.15.4

a) Is a sample of pure silicon, a conductor, a semi-conductor or an insulator? Explain your answer.

b) Explain how a sample of pure silicon can be made into an n-type semiconductor.

c) Explain how a sample of pure silicon can be made into a p-type semiconductor.

d) What are the majority charge carriers in an n-type semiconductor?

3 Figure 8.15.5 represents the different parts of a MOSFET.

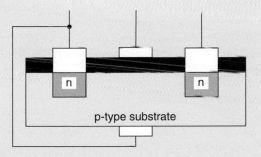

fig 8.15.5

a) Copy figure 8.15.5 and label the gate, source, drain, substrate and oxide layer.

b) Draw the circuit symbol for an n-channel enhancement mode MOSFET.

c) Describe how the n-channel enhancement mode MOSFET can be used as a switch.

d) State another use for a MOSFET.

4 Figure 8.15.6 shows an LED connected to a 6 V battery and resistor, R. The LED is of the type that has a p.d. of 2 V across its terminals when the current in it is 10 mA.

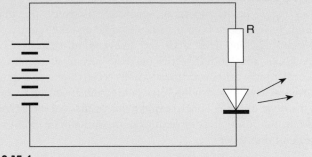

fig 8.15.6

a) Resistor R is chosen so that the current in the LED is 10 mA.
  (i) What is the p.d. across the resistor?
  (ii) Show that the resistor has a value of 400 ohms.
  (iii) Show that the LED dissipates 200 mW of power.

b) The LED emits green light with a wavelength of 500 nm.
  (i) Calculate the energy of a photon of this green light.
  (ii) Assuming that all of the energy used by the LED is converted to light calculate the number of photons being radiated each second.

5

a) Describe what you would observe when the power supply in figure 8.15.7 is switched on.

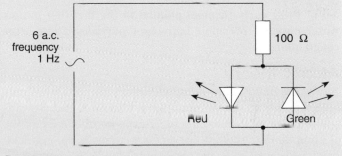

fig 8.15.7

b) Using the terms forward bias and reverse bias explain your observation.

9.1 Atomic theory

Introduction

Ever since the beginning of scientific discovery, thoughtful minds have probed both the macroscopic and the microscopic worlds. Astronomers scan the heavens in search of the mysteries they hold while particle physicists probe the microscopic world inside the atom in search of new sub-atomic particles. The great driving force behind much current research is the hope of being able to explain the vastness of the universe and the microscopic complexities of the sub-atomic world with one unifying theory.

Early ideas about matter

The study of matter and its behaviour is by no means a new subject. The ancient Greek philosophers had theories on the composition of matter and on how different substances reacted with each other.

The origins of our modern atomic theory can be traced back to the nineteenth century when the development of experimental chemistry brought about the need for a greater knowledge of how elements and compounds combine. In the early 1800s John Dalton suggested that the atoms of one element could only combine with the atoms of another element in simple fixed proportions. Following on from Dalton's theory and further research by Joseph Louis Gay-Lussac, Amedeo Avogadro proposed that under the same conditions of temperature and pressure, equal volumes of all gases contain an identical number of molecules.

All of these early theories pointed to the fact that matter was composed of small individual particles. These were called atoms.

Atomic theory

By the beginning of the twentieth century the then commonly accepted idea that atoms were hard, indivisible particles was beginning to be challenged. J.J. Thomson's discovery of the electron in 1897 showed that atoms were made up of at least two fundamental parts; the negatively charged electrons and the positively charged particles needed for the electrical neutrality of the atom. Other researchers showed that 'heavy' atoms like uranium and thorium were unstable and changed into other atoms by emitting radiation or particles. The first two types of radioactive particle to be detected were called alpha and beta rays.

Following his discovery of the electron J.J. Thomson proposed that electrons moved in concentric shells within a sphere of uniform positive electric charge. This model is often called **Thomson's plum pudding model**.

Ernest Rutherford

Ernest Rutherford was born in New Zealand in 1871 and is best remembered for the work carried out under his direction at Manchester and Cambridge Universities. During this time, he collaborated with many researchers who went on to achieve fame in their own right but, as the guiding hand behind the research into atomic structure, it is Rutherford himself who is given most of the credit for the success of his research teams.

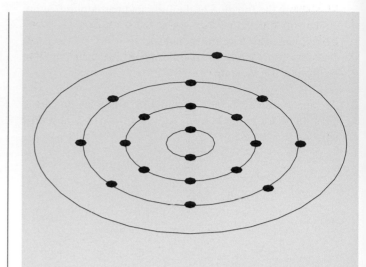

Figure 9.1.1 Plum pudding model

Figure 9.1.2 Ernest Rutherford

In 1902 Rutherford and Snoddy showed that the alpha rays emitted from the decay of unstable radioactive materials were electrically charged helium nuclei travelling at high speed. In experiments with Geiger and Marsden in 1909, Rutherford used alpha particles to investigate the composition of thin gold foils.

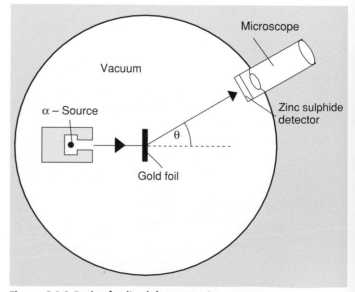

Figure 9.1.3 Rutherford's alpha scattering apparatus

Rutherford's investigations found that;
1 Most of the alpha particles passed through the foil undeviated.
2 A few alpha particles were deflected from their incoming path but continued through the foil.
3 A very small number (about 1 in 8000) were deviated by more than 90° i.e. they rebounded.

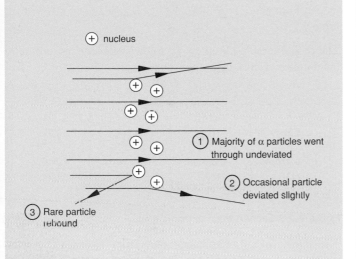

Figure 9.1.4 Rutherford's results

Since most alpha particles pass through the atom undeviated, Rutherford concluded that most of the atom is actually empty space.

He was able to explain the second observation by assuming that the positively charged alpha particles were deflected from their original path by other positive charges within the foil.

He explained the small number that bounced back by suggesting that the alpha particles had collided with something much heavier containing a very concentrated region of positive charge. As a consequence of his observations Rutherford proposed that an atom has a positively charged centre containing most of the mass. He called the heavy charged centre of the atom the **nucleus** and went on to suggest that the positively charged nucleus was surrounded by the orbiting electrons needed for electrical neutrality.

Modern measurements show that an average nucleus has a radius of the order of 10^{-15} m. This is some 100,000 times smaller than the radius of a typical atom. Bearing this in mind it is not surprising then that most of the alpha particles passed through the thin gold foil.

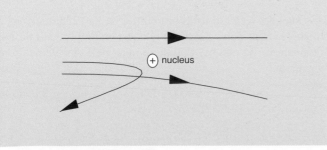

Figure 9.1.5 Rutherford's results

Even if the foil is many thousands of atomic radii thick, there is still lots of space for the alpha particles to pass through. Any alpha particles which pass close to a nucleus will be deflected from their path by the force of repulsion between the positively charged alpha particle and the positively charged atomic nucleus. Alpha particles scattered back along their incoming path are a result of direct encounters between alpha particles and the much more massive gold nuclei.

Protons, neutrons and electrons

Rutherford's work had established that atoms had heavy positively charged nuclei while J.J. Thomson had established the electron as one of the fundamental particles. It was not surprising therefore that researchers began to probe the atomic nucleus for a positive counterpart for the electron. The simplest positively charged particle came from hydrogen atoms and is called a **proton**. Its charge is equal in magnitude and opposite in sign to that of an electron but its mass is some 1836 times greater than the mass of the electron.

Neutrons were not discovered until 1932 when James Chadwick isolated a particle which has approximately the same mass as a proton but no electrical charge. It quickly became apparent that Chadwick's neutrons were another basic building block of the atomic nucleus.

Modern advances in the theory of the atom

Rutherford and his contemporaries were the first to probe the atomic nucleus by bombarding it with particles. Modern subatomic research continues to use this technique and over 300 sub-atomic particles have been found. Researchers today use more and more energetic particles as their tools for probing the nucleus. Many different designs of particle accelerators have been built for this purpose. The scale and the expense involved in probing the most minute parts of matter has resulted in collaboration amongst scientists from many nations at establishments such as CERN near Geneva.

Figure 9.1.6 This photograph of CERN shows the circumference of the Large Electron Positron ring accelerator.

9.2 Nuclear model

Introduction

In earlier chapters we reviewed some of the major developments which have shaped modern thinking on the proton, neutrons and electrons making up the atom.

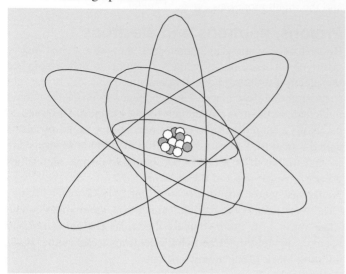

Figure 9.2.1 An atom

The work of Ernest Rutherford and his research teams in the early years of the twentieth century led to the idea of the atom having a nucleus.

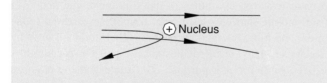

Figure 9.2.2 Rutherford's results explained

Most of the alpha particles bombarding the thin gold foil, in Rutherford's famous experiment, passed through without changing direction. Rutherford therefore concluded that most of the atom was empty space. A few of the positively charged alpha particles were deflected so he said that the repulsion was caused by centres of positive charge within the atom. The small number of nuclei reflected back led Rutherford to the conclusion that the vast majority of the atom's mass is concentrated in a very small volume. He called the dense centre of mass which contained the positive charge a **nucleus**.

This model of the atom pictures the nucleus as being made up from protons and neutrons surrounded by orbiting electrons as shown in figure 9.2.1. **The protons and neutrons in the nucleus of an atom are collectively called the nucleons.**

Nomenclature

All matter is made up from **elements** such as carbon, hydrogen, oxygen, sodium etc. The smallest part of each element which is chemically identical to all other parts, is called an **atom**. The atoms of different elements contain different numbers of protons. The number of protons in each atom of a particular element is called the **atomic number** or the **proton number, Z**.

Name	No. of protons	Symbol
Hydrogen	1	H
Helium	2	He
Carbon	6	C
Silicon	14	Si
Uranium	92	U

Table 9.2.1

The periodic table classifies all known elements according to the number of protons in their nucleus.

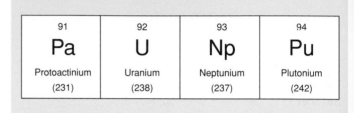

Figure 9.2.3 Section from periodic table

The nuclei of all elements, apart from hydrogen, contain **neutrons** as well as protons. The **mass of an atom is almost entirely due to the protons and neutrons**. The **mass number, A**, is the total number of protons plus neutrons.

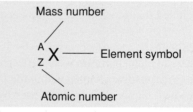

Figure 9.2.4

Using the notation shown in figure 9.2.4 the element identified by the symbol X has an atomic number Z and a mass number A. Therefore, we can say that element X has A nucleons of which Z are protons.

One particular form of uranium is identified as shown in figure 9.2.5.

$$^{238}_{92}\text{U}$$

Figure 9.2.5 Uranium-238

This nuclide, referred to as uranium-238, has 238 nucleons of which 92 are protons. Since the nucleus is composed of protons and neutrons we can say that there must be 238 − 92 neutrons. In general;

> *A = Number of protons + Number of neutrons*
> *Z = Number of protons*
> *A − Z = Number of neutrons*
> *Number of neutrons = Mass Number − Atomic Number*

Therefore each nuclide of uranium represented by the symbols shown in figure 9.2.5 has 146 neutrons in its nucleus.

Isotopes

Another form of uranium can be identified by the symbol shown in figure 9.2.6.

$$^{235}_{92}U$$

Figure 9.2.6 Uranium-235

Uranium-235 still has 92 protons in each nucleus. (Indeed it is this fact that identifies the element *as* uranium.)
However in this case;

Number of neutrons = Mass Number – Atomic Number
Number of neutrons = 235 – 92
Number of neutrons = 143

Figures 9.2.5 and 9.2.6 represent different isotopes of uranium. **Isotopes of any particular element contain the same number of protons but different numbers of neutrons**.

	$^{16}_{8}O$	$^{17}_{8}O$	$^{18}_{8}O$
No. of protons	8	8	8
No. of neutrons	8	9	10

Table 9.2.2

Table 9.2.2. shows three different isotopes of oxygen. Each isotope has eight protons per nucleus but the number of neutrons differs from one isotope to the next.

Radioisotopes

Most of the isotopes which occur in nature are stable. The forces within their nuclei are sufficient to keep the protons and neutrons in the same arrangement indefinitely. However, a few naturally occurring isotopes and almost all of the man made ones are unstable. Unstable nuclei can change into more stable arrangements by releasing different kinds of particles. A rearrangement within a nucleus resulting in the emission of a particle or energy is called **radioactive decay**. The particular isotopes of elements with nuclei which can undergo radioactive decay are called **radioisotopes** or **radionuclides**.
Carbon-14 is a naturally occurring radioisotope of carbon. Each nucleus contains six protons and eight neutrons. Tritium is a man made radioisotope of hydrogen where each nucleus contains one proton and two neutrons.

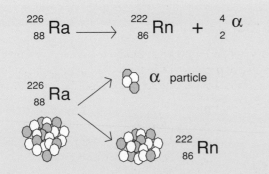

Figure 9.2.7

Alpha and beta particles

Radioactive decays result in the emission of either an alpha particle, a beta particle or a gamma ray.
In the alpha decay shown in figure 9.2.7 radium-226 (Ra) changes into radon-222 (Rn) by releasing a particle which has two protons and two neutrons from its nucleus. This is called an **alpha particle. An alpha particle is identical to the nucleus of a helium atom**.
This decay can be written as shown in figure 9.2.8.

Figure 9.2.8

When a radium atom releases an alpha particle its mass number (226) must decrease by four because each nucleus is losing two protons and two neutrons. Consequently the resulting radon atoms have a mass number of 222. Similarly the atomic number of radon must be two less than that of radium because of the two protons in the emitted alpha particle.
The radon atoms formed by the alpha decay are themselves radioactive. They too emit alpha particles to form nuclei of the element polonium (Po) which has an atomic number of 84 and a mass number of 218.

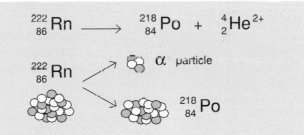

Figure 9.2.9

Polonium-218 is another radioisotope and decays by emitting a beta particle. **The beta particle is a very fast moving electron released from the nucleus because a neutron has changed into a proton and an electron**. As a result the nucleus has one *less* neutron but one *extra* proton. The atomic number increases by one but the mass number remains the same.

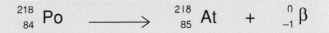

Figure 9.2.10

The sum of the atomic numbers and the sum of the mass numbers on either side of a decay equation must balance so the beta particle must have a mass number of 0 and an atomic number of – 1!

Gamma rays

Gamma rays are not charged particles like alpha and beta particles. They are electromagnetic radiation with a high frequency.
When atoms of one element change into atoms of a different element by emitting alpha or beta particles the nuclei formed may still have too much energy to be completely stable. The excess energy is emitted as gamma rays. Typical photons of gamma radiation have energies of around 1×10^{-12} joules.

9.3 Nuclear reactions

Introduction

An alpha emission can be summarised by the equation shown in figure 9.3.1. The polonium nucleus splits into two particles one of which is much bigger than the other.

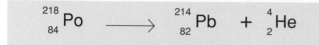

$$^{218}_{84}\text{Po} \longrightarrow {}^{214}_{82}\text{Pb} + {}^{4}_{2}\text{He}$$

Figure 9.3.1 Radioactive decay equation

The discovery of the neutron by James Chadwick in 1932 gave the early particle physicists a new tool for probing the nature of the nucleus. Before Chadwick's discovery, Rutherford and his contemporaries had to use alpha particles as projectiles. But since both the projected alpha particles and the target nuclei are positively charged, electrostatic repulsion stops alpha particles from penetrating the nucleus.

In 1939 Enrico Fermi found that bombarding uranium atoms with neutrons caused their nuclei to split into two parts roughly equal in size. **The splitting of a nucleus with a high mass number into two nuclei, of roughly equal mass numbers, is called nuclear fission**.

Nuclear fission

Some artificial radioisotopes have nuclei which undergo **spontaneous fission**. When they are created their nuclei are highly unstable and decay spontaneously by splitting into two smaller nuclei. Generally, these spontaneous decays are accompanied by the release of neutrons.

The fission of uranium which Fermi discovered was not spontaneous but was initiated by neutrons bombarding uranium atoms. This type of fission is called **induced fission**.

The equation in figure 9.3.2 illustrates one possible way in which the fission of uranium-235 can occur.

$$^{235}_{92}\text{U} + {}^{1}_{0}\text{n} \longrightarrow {}^{144}_{56}\text{Ba} + {}^{90}_{36}\text{Kr} + 2{}^{1}_{0}\text{n}$$

Figure 9.3.2

The bombarding neutron causes the uranium-235 nucleus to split into two parts called **fission fragments**. One fission fragment has 56 protons while the other has 36. By referring to a periodic table we can see that these nuclei will have the same chemical properties as other isotopes of barium and krypton.

Fission of uranium into barium and krypton nuclei is by no means the only way in which the uranium-235 nuclei can be split. There are many pairs of fission products but in the vast majority of cases the heavier nuclides formed will have mass numbers between 130 and 149. The particular pair of fission fragments formed depends largely on the **energy of the bombarding neutron**.

The induced fission of uranium-235 is of particular importance because of the potential it has as a source of energy. Of the three nuclides which can be induced to fission (uranium-235, uranium-233 and plutonium-239), only the uranium-235 is naturally occurring. It occurs (along with the more common isotope uranium-238) in sufficient quantities for power stations generating energy from its fission to be considered *commercially* viable.

What happens during a fission?

When a neutron, travelling at an appropriate speed, strikes a uranium-235 nucleus it is captured.

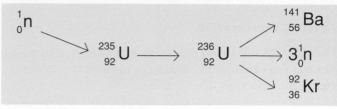

Figure 9.3.3

The uranium-236 nucleus formed by the capture is very unstable. It quickly transforms into an elongated shape before splitting completely into two fission fragments and releasing neutrons.

Narrow neck – little short-range attraction between lobes

Figure 9.3.4 Uranium-236 nucleus before splitting

Energy from nuclear fission

Both the products of the fission reaction and the released neutrons are moving very fast. **The total kinetic energy of the products exceeds that of the bombarding neutron and uranium-235 nucleus.** Clearly, then, **energy is being released as a result of the fission reaction**.

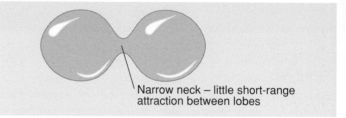

$$^{235}_{92}\text{U} + {}^{1}_{0}\text{n} \longrightarrow {}^{138}_{55}\text{Cs} + {}^{96}_{37}\text{Rb} + 2{}^{1}_{0}\text{n}$$

Figure 9.3.5

A very careful examination of the atomic masses on either side of the fission equation of figure 9.3.5 will reveal the source of this energy. Accurate values for the masses of all particles concerned in this equation are shown in table 9.3.1.

	Atomic mass/kg
${}^{235}_{92}\text{U}$	3.9014×10^{-25}
${}^{138}_{55}\text{Cs}$	2.2895×10^{-25}
${}^{92}_{37}\text{Rb}$	1.5925×10^{-25}
${}^{1}_{0}\text{n}$	1.6750×10^{-27}

Table 9.3.1

The total mass on the left hand side of this equation is;

$3.9014 \times 10^{-25} + 1.675 \times 10^{-27} = 3.91815 \times 10^{-25}\ kg$

The total mass on the right hand side of this equation is;

$2.2895 \times 10^{-25} + 1.5925 \times 10^{-25} + 2 \times 1.675 \times 10^{-27}$
$= 3.9155 \times 10^{-25}\ kg$

The total mass of the matter before a fission is greater than the total mass after the fission. This difference in mass is given the symbol, m. In this example;

Mass difference, $m = (3.91815 - 3.9155) \times 10^{-25}\ kg$

$m = 2.65 \times 10^{-28}\ kg$

Einstein's theory of special relativity showed that a mass, m, is equivalent to an energy, E, given by;

$$E = mc^2$$

where c is the speed of light in a vacuum ($3 \times 10^8\ m\,s^{-1}$). Using this equation we can calculate the energy equivalent to the mass difference found for the fission reaction considered previously.

$E = 2.65 \times 10^{-28} \times (3 \times 10^8)^2$
$E = 2.385 \times 10^{-11}\ J$

At first sight, this may not seem a particularly large amount of energy. However, considering that it is produced from the fission of a single nucleus we can show that nuclear fission has considerable potential as a source of power on a large scale.

Energy on a large scale

Each uranium-235 atom has a mass of 3.9014×10^{-25} kg. The number of atoms in 1 kg of uranium-235 can be found from:

Number of atoms in 1 kg $= \dfrac{1}{3.9014 \times 10^{-25}}$

Number of atoms in 1 kg $= 2.56 \times 10^{24}$

If each fission produces 2.385×10^{-11} J then the total possible energy available from the fission of 1 kg of uranium-235 is given by;

Total possible energy = Energy per fission × Number of atoms
Total possible energy $= 2.385 \times 10^{-11} \times 2.56 \times 10^{24}$
Total possible energy $= 6.11 \times 10^{13}\ J$

If we assume that burning 1 kg of coal releases about 30 MJ of energy we can see that each 1 kg of uranium-235 can produce the same energy as more than 2 million kg of coal.

If each 1 kg of uranium-235 has the potential to produce as much energy as approximately 2000 tonnes of coal it would seem, at first sight, that nuclear power has many advantages over coal fired power production. Less mining would be required and there would be no ash discharged into the atmosphere. However, the by-products of nuclear fission are themselves radioactive substances. The handling and storage of even small quantities over long periods of time raises issues which must be considered alongside the potential of nuclear fission to produce large quantities of energy from small amounts of raw materials.

Some of the issues associated with this nuclear debate are outlined in section 9.12.

Nuclear fusion

The calculations concerning nuclear fission have already shown the potential of nuclear reactions to release vast quantities of energy. In a **nuclear fusion** two nuclei with low mass numbers combine to produce a single nucleus with higher mass number.

$$^2_1H + {}^3_1H \longrightarrow {}^4_2He + {}^1_0n + \text{ENERGY}$$

Deuterium + Tritium   Helium + Neutron + Energy

Figure 9.3.6 Deuterium and tritium fusion

The fusion reaction shown in figure 9.3.6 forces together two different isotopes of hydrogen. Deuterium nuclei have one proton, like hydrogen, but also have an additional neutron while the tritium nuclei have a proton and two neutrons. This gives deuterium an atomic number of 1 and a mass number of 2 while tritium still has an atomic number of 1 but a mass number of 3. The mass of the deuterium and tritium nuclei is greater than the mass of the helium and the neutron produced. The accurate atomic masses of the nuclei involved in this fusion are shown in table 9.3.2. From these values we can calculate the mass differance and hence the energy released by each fusion.

Isotope	Atomic mass/kg
$^2_1$H	3.345×10^{-27}
$^3_1$H	5.008×10^{-27}
$^4_2$He	6.647×10^{-27}
$^1_0$n	1.675×10^{-27}

Table 9.3.2 Atomic masses for fusion

The total mass on the left hand side of this equation =

$5.008 \times 10^{-27} + 3.345 \times 10^{-27} = 8.353 \times 10^{-27}\ kg$

The total mass on the right hand side of this equation =

$6.647 \times 10^{-27} + 1.675 \times 10^{-27} = 8.322 \times 10^{-27}\ kg$

Therefore;

Mass difference, $m = (8.353 - 8.322) \times 10^{-27}\ kg$

$m = 3.1 \times 10^{-29}\ kg$

So the energy released per fusion is calculated from;

$E = mc^2$
$E = 3.1 \times 10^{-29} \times (3 \times 10^8)^2$
$E = 2.79 \times 10^{-12}\ J$

Nuclear fusion, like fission, has the potential to provide energy on a large scale and has the additional benefit of fewer harmful by-products. However, since both the deuterium and tritium nuclei are positively charged, their nuclei do not fuse together easily. Current attempts at **thermonuclear fusion** require a plasma containing hydrogen to be heated to about 100,000,000°C! Clearly such high temperatures present problems. Stars, such as our sun, produce their energy by thermonuclear fusion and scientists are striving towards the goal of **cold fusion**: the controlled release of energy by fusing nuclei together at manageable temperatures.

9.4 Harnessing nuclear power

Nuclear fission

The particular pair of products from any nuclear fission depends on the conditions under which fission takes place. Irrespective of how the nucleus splits, the fission produces particles with kinetic energy. **When the released particles move through a material which slows them down, the material becomes warmer as the kinetic energy of the particles is transferred to heat**.

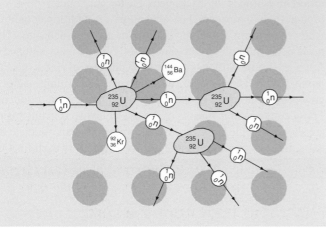

Figure 9.4.1 Fission fragments warming up the medium by collision

The challenge for the designers of nuclear power stations is to use this heat to produce electricity.

Nuclear fuel

Fissionable uranium-235 is not naturally abundant. In mined uranium only about one atom in 140 is uranium-235. In order to make mined uranium suitable for use as a nuclear fuel, the proportion of uranium-235 must be increased to about 2–3%. Suitably **enriched uranium** is fabricated into fuel rods ready to be placed in the core of a nuclear reactor.

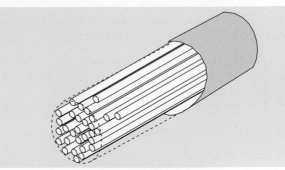

Figure 9.4.2 Cut away diagram of a fuel rod

Chain reactions

Maintaining steady supplies of electrical power from nuclear fission is by no means a simple process. Induced nuclear fissions must occur at a pre-determined rate in order to maintain steady power production.

In the fission represented by figure 9.4.3 a single neutron causes an individual uranium-235 nucleus to split.

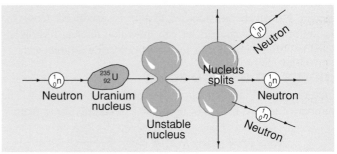

Figure 9.4.3 Chain reaction

As well as the energetic fission fragments, three further neutrons are released and if these were to go on to cause further fissions within the sample of uranium, there would be a continuous release of energy.

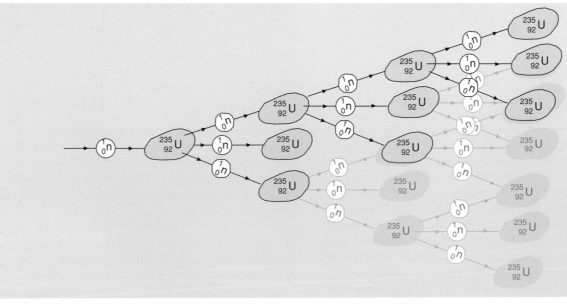

Figure 9.4.4 Explosion

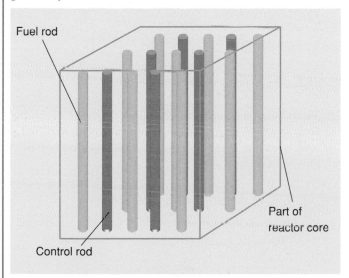

The process represented by figure 9.4.4 is called a **chain reaction** because the neutrons from one fission go on to cause further fissions. Such a chain reaction would produce energy at an ever increasing rate. **This uncontrolled release of energy would result in a nuclear explosion**.

Controlling chain reactions

Figure 9.4.5 represents a controlled chain reaction. Three neutrons are released by a fission but only one goes on to cause further fissions. The other two neutrons released are captured by **control rods** made from materials such as boron which have neutron absorbing properties.

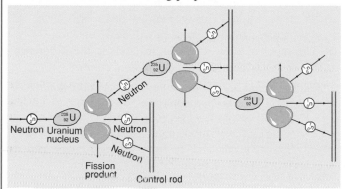

Figure 9.4.5 Controlled chain reaction

The control rods are arranged within the reactor core in close proximity to the fuel rods.

Figure 9.4.6 Core with fuel rods and control rods

When the rods are fully inserted into the core there are not enough neutrons available to sustain the chain reaction. Partially inserting the control rods and arranging their position allows the density of the neutrons within the core to be maintained at a level sufficient to sustain the chain reaction and to control the rate of energy production. If the temperature of the reactor core goes above a specified safety limit the control rods are automatically inserted into the core to shut down the reactor.

Moderation

Although the progress of the chain reaction can be controlled by using boron rods, the reaction as represented in figure 9.4.3 will not be self sustaining. The neutrons needed to cause fission

are called **slow** or **thermal neutrons** because of the quantities of kinetic energy they require in order to cause fission in uranium-235. **The neutrons released as a result of the fission have more kinetic energy and are called fast neutrons**.

When a fast neutron is captured by a uranium-235 nucleus fission does not occur. Instead, the unstable uranium-236 decays by emitting a beta particle.

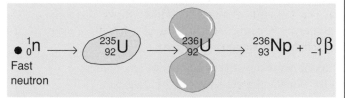

Figure 9.4.7 Fast neutron and uranium nucleus

To maintain a chain reaction the fast neutrons produced in each fission must be **slowed down** so they can cause further fissions. This process is known as **moderation**. The fuel rods and the control rods are surrounded by a moderating material such as **graphite**.

As the fast neutrons pass through the graphite moderator they collide with the carbon atoms. This warms the graphite and slows the neutrons to speeds where they can cause further fissions.

Extracting heat from the core

The movement of fission fragments within the fuel rods, the capture of neutrons by the control rods and the slowing of fast neutrons by the moderator all cause the reactor core to warm up. This heat is extracted by pumping carbon dioxide gas through the core and the hot gas is used to generate electricity.

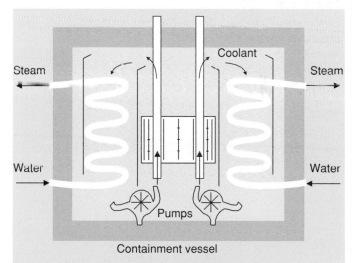

Figure 9.4.8 Carbon dioxide heat extraction

The warm carbon dioxide is passed through a heat exchanger where it converts water into steam. This turns the turbines to drive the generators in a power station.

Heat must be transferred very quickly from the reactor core to the heat exchanger so the carbon dioxide coolant needs to be pumped through the core at very high pressures. This means that the **containment vessel** around the reactor has to stop the release of pressurised gases as well as preventing radiation from escaping.

Worked example 1

Figure 9.5.1 shows the periodic table listing for the radioactive source americium which is kept in many schools.

$$^{241}_{95}\text{Am}$$

fig 9.5.1

a) What is the atomic number of americium?
b) What is the mass number of americium?
c) How many electrons will there be in an electrically neutral americium atom?
d) How many neutrons are there in an atom of this isotope of americium?

Answers and comments

a) The number at the lower left hand side of the symbol is the atomic number so the atomic number of americium is 95.
b) The mass number is the number at the upper left hand side of the symbol so the mass number of this americium nuclide is 241.
c) For an atom to be electrically neutral there must be equal numbers of protons and electrons. The atomic number, 95, tells us how many protons there are so in an electrically neutral atom there will be 95 electrons.
d) The number of neutrons in an atom is the difference between the mass number and atomic number.

Number of neutrons = Mass number – Atomic number
Number of neutrons = 241 – 95
Number of neutrons = 146

Worked example 2

The diagram of figure 9.5.2 shows two different ways that bismuth (Bi) can decay into lead (Pb).

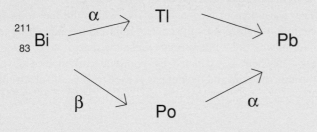

fig 9.5.2

a) Write an equation for the decay of bismuth into polonium (Po).
b) What are the atomic and mass numbers of thallium (Tl) and lead (Pb)?
c) Write an equation for the decay of thallium into lead.

Answers and comments

a) This is the standard equation for a beta decay. The mass number of the atom produced is the same as the mass number of the decaying atom. The atomic number of the produced atom is one more than the atomic number of the decaying atom.

$$^{211}_{83}\text{Bi} \longrightarrow {}^{211}_{84}\text{Po} + {}^{0}_{-1}\beta$$

fig 9.5.3

b) Thallium is formed from the alpha decay of bismuth. This means that the mass number of thallium will be four less than the mass number of the bismuth while thallium's atomic number will be two less.

Mass number of Tl = 207
Atomic number of Tl = 81

Lead is produced from bismuth by a beta decay followed by an alpha decay. The decay equation is as shown;

$$^{211}_{83}\text{Bi} \xrightarrow{\beta} {}^{211}_{84}\text{Po} \xrightarrow{\alpha} {}^{207}_{82}\text{Pb}$$

fig 9.5.4

Mass number of lead = 207
Atomic number of lead = 82

c) The equation for the decay of thallium into lead is shown in figure 9.5.5.

$$^{207}_{81}\text{Tl} \longrightarrow {}^{207}_{82}\text{Pb} + {}^{0}_{-1}\beta$$

fig 9.5.5

Questions

1 The following are incomplete statements representing the emission of different types of particle.

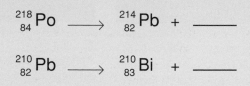

$$^{218}_{84}Po \longrightarrow \ ^{214}_{82}Pb \ + \ \underline{\hspace{2cm}}$$

$$^{210}_{82}Pb \longrightarrow \ ^{210}_{83}Bi \ + \ \underline{\hspace{2cm}}$$

fig 9.5.6

a) Complete each equation by adding the symbol for the particle emitted with its appropriate mass and atomic numbers.

b) Explain the origin of the beta particle emitted during the beta decay.

c) Explain why the element Pb which appears in both equations can have different mass numbers.

2 Figure 9.5.7 shows an outline of the apparatus used by Rutherford and his collaborators which led to the discovery of the atomic nucleus.

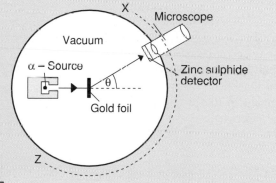

fig 9.5.7

a) Explain why a source of alpha radiation is used rather than a beta or gamma source.

b) What is observed as the microscope and screen are moved from X to Z?

c) Explain how these observations account for the nucleus being massive, small, and positively charged.

d) Why must the gold foil be thin?

3 A nucleus of uranium-238 is unstable and decays into lead-206. Part of this decay series is shown in figure 9.5.8.

$$^{238}_{92}U \xrightarrow{\text{(i)}} \ ^{234}_{90}Th \xrightarrow{\text{(ii)}} \ ^{234}_{91}Pa \xrightarrow{\text{(iii)}} \ ^{234}_{92}U \xrightarrow{\alpha} \ \underline{\hspace{1cm}} \xrightarrow{\alpha} \ ^{226}_{88}Ra \xrightarrow{\text{(iv)}} \ ^{222}_{86}Rn$$

fig 9.5.8

a) How many protons are there in the nucleus of a uranium-238 atom?

b) Some uranium atoms occur as the isotope uranium-235. Explain how these atoms differ from those in the uranium-238 isotope.

c) What kind of particle is emitted at each of the decay stages labelled (i), (ii), (iii) and (iv)?

4 A sequence of radioactive decays of radioisotope P into T can be summarised as shown in figure 9.5.9.

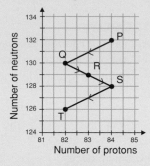

fig 9.5.9

a) What is the atomic number and mass number of element P?

b) What is the atomic number and mass number of element Q?

c) Write an equation for the decay of element P into element Q including information about the particle emitted during the decay.

d) Write an equation for the decay of element R into element S including information about the particle emitted during the decay.

e) Which letter represents an isotope of the same element as T? Explain your answer.

5 The nuclear equation for the beta decay of gallium-70 (Ga) into germanium-70 (Ge) is shown in figure 9.5.10. The masses of the nuclides involved are as shown.

$$^{70}_{31}Ga \longrightarrow \ ^{70}_{32}Ge \ + \ ^{0}_{-1}\beta$$

fig 9.5.10

Isotope	Atomic mass/ $\times 10^{-25}$ kg
$^{70}_{31}Ga$	1.16077
$^{70}_{32}Ge$	1.16074

Table 9.5.1

Calculate the energy released during each beta decay.

6 When beryllium (Be) is bombarded with alpha particles of a certain energy, carbon-13 is formed and gamma radiation is released. The masses of the nuclides involved are as shown in table 9.5.2.

Isotope	Atomic mass/ $\times 10^{-27}$ kg
$^{9}_{4}Be$	14.9649
$^{4}_{2}He$	6.6466
$^{13}_{6}C$	21.5925

Table 9.5.2

a) Write an equation for this fusion reaction.

b) Explain what is meant by the term fusion reaction.

c) How many neutrons are there in the carbon atoms formed?

d) Calculate the energy released as a result of this fusion.

9.6 Activity and half-life – revisited

Random decay

Radioactive decay is a **random** process. This means that within a sample of radioactive material each atom has an equal probability of emitting a particle and changing into another element.

$$^{238}_{92}\text{U} \longrightarrow {}^{234}_{90}\text{Th} + {}^{4}_{2}\text{He}$$

Figure 9.6.1 Uranium decay

One gram of uranium-238 contains approximately 2.5×10^{21} atoms. Of these an average of 12,000 decay per second. **Like any random process it is impossible to predict which nucleus will decay next or when a specific nucleus will decay. However since there are so many nuclei in even a small sample we can predict the average number that will decay during any particular time interval.**

The average activity, A, of a sample is defined as the **number of nuclei decaying per second**. The unit of activity is the **becquerel (Bq)**, where one becquerel is one decay per second.

$$Average\ activity = \frac{Number\ of\ nuclei\ decayed}{Time}$$

$$A = \frac{N}{t}$$

The average activity of a 1 gram sample of uranium-238 is stated as 12,000 becquerels or 12 kBq.

Changing activity

The activity of a sample depends upon the number of nuclei which are available to undergo decay. Once a radioactive nucleus has decayed the number of nuclei still available for decay decreases. However, the probability that a nucleus will change remains the same, so as more and more nuclei decay, we would expect the activity of the sample to reduce because fewer and fewer undecayed nuclei remain. For radioisotopes which decay quickly the activity will change quickly.

If a sample of radioactive material has 10,000 nuclei and a probability that 1 in 10 will decay each second, then during the first second we would expect 1000 nuclei to decay. The average activity over the first second will be 1000 Bq.

After the first second, only 9000 undecayed nuclei remain, but the probability of decay is still 1 in 10, so 900 nuclei will decay during the second interval of 1 second. The average activity during this time will be 900 Bq. At the start of the third second only 8,100 undecayed nuclei remain, so 810 decay, giving an average activity during the third second of 810 Bq. Using similar reasoning we can calculate the average activity for each of the first 10 seconds. The values are shown in table 9.6.1.

The principles governing the decay of real radioisotopes are identical to the simplified example in figure 9.6.2.

Thus the activity of the radiation emitted by radioisotopes changes with time.

Time	No. of undecayed nuclei	Second	Activity
Start	10000	1st	1000
1	9000	2nd	900
2	8100	3rd	810
3	7290	4th	729
4	6561	5th	656
5	5905	6th	590
6	5314	7th	531
7	4783	8th	478
8	4305	9th	430
9	3874	10th	387
10	3487		

Table 9.6.1

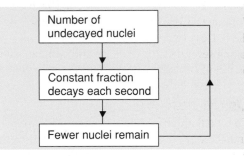

Figure 9.6.2

How quickly do radioisotopes decay?

In our simplified example 1 in 10 of the undecayed nuclei decays per second. This gives rise to an activity–time graph as shown in figure 9.6.3. This graph is often called a **decay curve**. From this graph we can see that

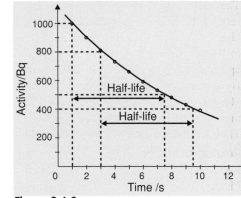

Figure 9.6.3

the activity falls from 1 kBq at $t = 1$ to 0.5 kBq at $t = 7.5$ seconds. Therefore we can say that the activity halves in 6.5 seconds. **The time taken for the activity emitted by a particular sample to halve is called the half-life.** You will also notice that the time taken for the activity to fall from 800 Bq to 400 Bq is also 6.5 seconds. By taking a few accurate measurements from the graph we could show that the half-life of this sample is 6.5 seconds.

A little further…

Measuring the half-life of a radioactive substance

EXPERIMENT An aqueous solution of proactinium-234 is placed in a bottle with an organic solvent. By shaking the solutions in the bottle, protactinium-234 (Pa-234) is

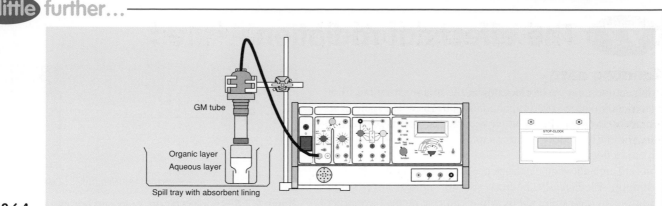

Figure 9.6.4

separated from the aqueous layer into the solvent. After shaking, the liquid is allowed to settle and when two separate layers are visible the number of beta particles emitted by the organic layer is measured by the GM tube. Recording the counts for 10 seconds at 30 second intervals for a period of 7 minutes gives data for a graph of **count rate** versus time for the β decay of protactinium-234. By drawing a best fit line and taking several starting points various values for the half-life can be found and an average calculated.

Using half-life

The half-life of a radioisotope is a very useful tool in archaeological research. The time taken for half of the radioactive material in any sample to decay is believed to be constant over many half-lives.

The age of moon rocks was first estimated by looking at the relative amounts of uranium-238 and lead-206 in a sample. Uranium-238 decays into lead-206 with a half-life of 4500 million years. The sample of moon rock contains approximately equal proportions of uranium-238 and lead. By assuming that, when created the rock contained no lead-206, we can say that approximately half of the original uranium-238 must have decayed to form the lead in the rock. Therefore moon rock is thought to be 4500 million years old.

Radiocarbon dating

All living things take in carbon from their surroundings. Plants are able to photosynthesise both the radioactive form of carbon, carbon-14, and its more common non-radioactive counterpart, carbon-12. In living things the proportions of both isotopes will be constant but when dead the radioactive carbon-14 starts to decay and will not be replaced by fresh supplies from the surroundings.

A certain mass of living bone produces an activity of 60 counts per minute. A bone of the same mass that has been dead for some time will give a lower count rate because some of its carbon-14 will have decayed. If the dead sample produces only 15 counts per minute we can say the activity has quartered. As the activity quarters in two half-lives, scientists will say that the dead bone is 11,400 years old since the half-life of carbon-14 is 5700 years.

It is worth keeping in mind that this technique makes some fundamental assumptions. It assumes:

1. that 11,400 years ago the bone had exactly the same relative proportions of carbon-12 and carbon-14 as is common today.
2. that the half-life, and consequently the decay rate, has been constant throughout this time.

Correcting count rates

When monitoring radioactive samples, particularly those producing very low count rates, it is worthwhile asking if all of the counts recorded are due to the sample. The answer to this question is almost certainly no.

If we move the sample away from the GM tube some counts are still detected. These are caused by **background radiation**, present at all times in the atmosphere. **To calculate the counts caused by the sample alone we must first measure the level of background radiation.** By subtracting background levels from the measured counts we obtain a value called the **corrected count rate**.

Corrected counts = Counts from sample − Background counts

A typical value for the background level of radiation is 30 counts per minute. The average activity caused by background radiation is therefore 0.5 Bq. The actual level of background activity in any environment depends on a number of factors, some of which are discussed in a later chapter.

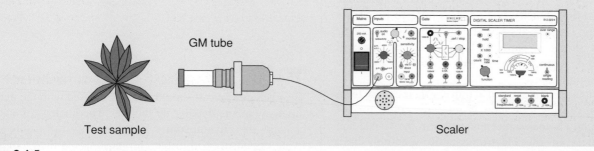

Figure 9.6.5

9.7 The effects of radiation

Creating ions

Altering the number of electrons in the atomic structure of an electrically neutral atom is called **ionisation**. When some of its electrons are removed, an atom becomes a **positively charged ion**. Similarly, adding electrons to an electrically neutral atom creates a **negative ion**.

Ions are created when electrons are transferred during chemical bonding. Electrically neutral sodium atoms have single electrons in their outer shells while chlorine atoms have seven outer electrons; one short of the desired stable configuration. When sodium chloride is formed, the sodium atoms donate their outer electron to chlorine atoms and the compound contains both sodium and chloride ions.

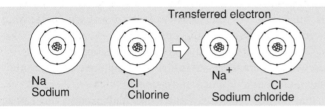

Figure 9.7.1 Sodium chloride formulation

Ionisation also occurs when electrons are stripped away from atoms by what we call **ionising radiation**. Alpha particles, because of their size, speed and double positive charge, are particularly good at dragging electrons away from the atoms of materials through which they pass.

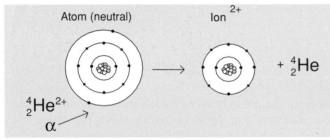

Figure 9.7.2 Alpha particle stripping electrons

Beta particles will also interact with electrons when passing through matter. However, beta particles are only singly charged, are smaller than alpha particles and although they move considerably faster, they cause less ionisation.

As ionising particles pass through matter, their kinetic energy is transferred to the electrons in the matter. Since alpha particles cause more ionisation than beta particles, their energies are used up more quickly. **So alpha particles produce a greater ionisation density but are less able to penetrate dense materials**. They are absorbed by a few millimetres of air or a couple of sheets of paper. A typical beta particle will penetrate about 1 metre of air but will be stopped by a few millimetres of aluminium.

Photons of gamma radiation can also transfer their energy to an atom's electrons causing ionisation. X-rays are another type of ionising radiation. Photons with frequencies in the visible and radio wave parts of the electromagnetic spectrum have insufficient energy to cause ionisation.

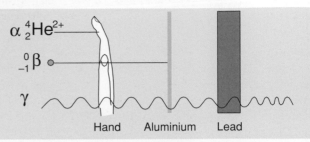

Figure 9.7.3 Stopping alpha, beta and gamma radiations

The effects of radiation on a cell

When ionising radiations eject electrons from the inner orbitals in an atom, the vacancy is quickly filled by other electrons falling to lower energy levels.

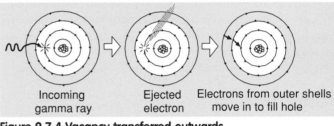

Figure 9.7.4 Vacancy transferred outwards

In this way any vacancy is transferred to the outer shell of the atom. Since it is the outer electrons which bond atoms together into molecules, the loss of any electron might cause the molecule to rearrange or even to break up. **When ionising radiations cause molecular changes in living cells various kinds of damage can occur**. The effect of large doses of ionising radiation may be immediately obvious but the damage caused to cells by smaller doses may take longer to become apparent.

Cells are the basic unit of every living organism. They have a nucleus surrounded by an aqueous material and enclosed within a membrane.

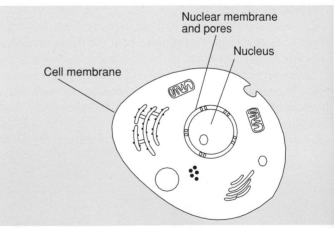

Figure 9.7.5 A cell

The nucleus of a cell should not be confused with the nuclei of the many atoms making up the cell. The nucleus of each cell contains the **chromosomes** which specify the nature of the living organism. Within the chromosomes is the DNA which holds the genetic information needed for the cell to reproduce.

When ionising radiations remove electrons from atoms within a chromosome, the DNA chains can change structure or may even break. This damage caused by the radiation can impair the normal functions of the cell or prevent it from reproducing properly.

Cancer

In a normal fully grown animal, new cells are produced at just the right rate to replace cells which die. The mechanism which governs the degeneration and subsequent replacement of cells is controlled by the chromosomes within the nucleus of the cell. If the DNA of the chromosomes is altered by ionising radiation, a single cell or group of cells may multiply uncontrollably. This results in a growth which develops and feeds at the expense of the healthy neighbouring cells. This growth is a type of malignant tumour, or cancer. Much research is still being done into the alterations in cell behaviour which produce cancer. A few of the factors which cause these alterations, such as exposure to ionising radiation, inhaling asbestos dust or tobacco smoke are known but many are still not proven.

Radiotherapy

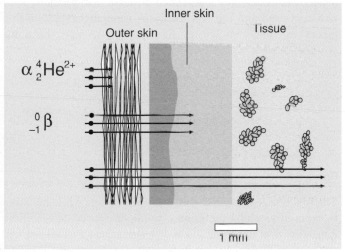

Figure 9.7.6 Penetration of skin

Various types of ionising radiation can be used in the treatment of cancer because cancer cells are more susceptible to damage by radiation than their normal healthy neighbours. For tumours lying deep within the body X-rays or gamma rays are used because they can penetrate beyond the outer skin and tissue. Although alpha and beta particles cannot penetrate far into the body they can be used along with beams of protons and neutrons to treat some types of tumours. **The aim of any treatment regime is to give as high a radiation dose as possible to the tumour, with as little effect as possible on the surrounding healthy cells.**

Using radioisotopes

Iodine-131 is a radioisotope which emits beta particles and gamma rays. It is used in nuclear medicine to monitor the activity of the thyroid gland. This gland is located in the neck and is responsible for the release of naturally occurring iodine into the blood stream. To check the activity of the thyroid, a small quantity of radioactive iodine is injected into the body. After a period of time this is taken up by the thyroid and, like the naturally occurring iodine, is released into the blood stream. Doctors can check the function of the thyroid by monitoring the radiation emitted. This technique has been used for more than 40 years to check for underactive or overactive thyroid glands. A radioisotope used in this way is called a **tracer**.

Detecting radiation

Using radioisotopes as tracers depends upon the availability of a detector which can sense the presence of the emitted radiation. The most common radiation detector is the **Geiger–Muller (GM) tube**.

When ionising particles enter into the GM tube through a 'window' they ionise molecules of a gas. The high voltage between the central wire and its surroundings results in a small pulse of current as the ions are collected. By counting the pulses the activity of a particular radioactive source can be found.

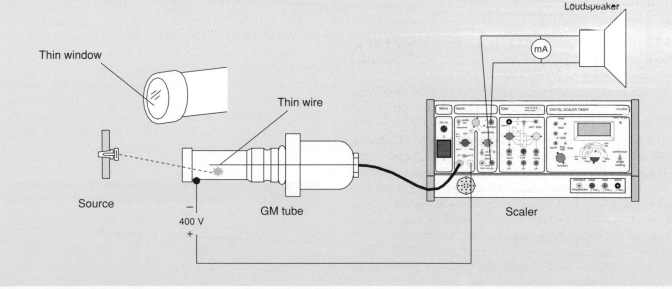

Figure 9.7.7 Geiger–Muller tube

9.8 Radiation and dose

Introduction

In section 9.5 we learned that ionising radiation can have harmful effects on human tissue. Alpha radiation is the most ionising but will be stopped from entering the body by the outer layer of the skin. However, radioactive gases emitting alpha radiation are very dangerous, because inhaled alpha particles will come into contact with the lungs and other sensitive parts of the body. Similarly, beta sources which are able to penetrate about 1 cm of flesh, are most dangerous if swallowed or inhaled.

X-rays and gamma rays produce little ionisation but can pass from outside the body through some of the most sensitive organs. Indeed high doses of gamma rays are used to sterilise prepacked syringes and needles for the very reason that they can kill any living bacteria present.

The harm that ionising radiation can do to living cells depends on a number of factors:

1. *the energy of the absorbed particles*
2. *the duration of the exposure*
3. *the mass of matter absorbing the radiation*
4. *the type of radiation absorbed*
5. *the type of tissue irradiated.*

Absorbed dose

The term **absorbed dose** is a scientific term and is defined as the energy absorbed per unit mass.

$$Absorbed\ dose = \frac{Energy\ absorbed}{Mass\ of\ absorbing\ material}$$

$$D = \frac{E}{m}$$

The unit of absorbed dose is the **gray (Gy)** where an absorbed dose of 1 Gy is equivalent to one joule of energy being absorbed per kilogram of material irradiated.

Q If a man of mass 80 kg is irradiated with a total of 30 J of radiation from a gamma source, calculate the absorbed dose.

A
$$Absorbed\ dose = \frac{Energy\ absorbed}{Mass\ of\ absorbing\ material}$$

$$D = \frac{30}{80}$$

$$D = 0.375\ Gy$$

In this example, we have assumed that the gamma radiation is absorbed evenly over the whole 80 kg of the man's body. If an alpha emitter is swallowed or inhaled the radiation does not penetrate the entire body. Instead the radiation is absorbed by a small mass of material immediately surrounding the source. The absorbed dose in this region will be high because the mass of material absorbing the radiation is low.

The gray is too large a unit for most practical purposes. Table 9.8.1 shows typical absorbed doses for diagnostic techniques commonly used in medicine.

X-ray examination	Typical absorbed dose
Chest X-ray	0.27 mGy
Barium Meal	17 mGy
Dental X-ray	0.1 mGy
Leg X-ray	~0.12 mGy

Table 9.8.1

These figures are averages and the absorbed doses depend on the size of the patient being examined and on the diagnostic technique used. The aim of any technique is to use the minimum dose required to complete the medical procedure.

In a typical dental X-ray the average dose of X-rays is 0.1 mGy. The mass being irradiated in this case would include a portion of gum, cheek and jaw as well as the tooth under investigation.

Equivalent dose

The damage that ionising radiation can do to living cells depends not only on the absorbed dose but also on the type of radiation used and the type of tissue irradiated. The two thin pieces of tissue shown in figure 9.8.1 receive the same absorbed dose of radiation.

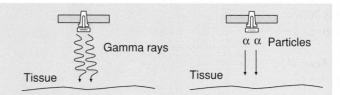

Figure 9.8.1

The first one is irradiated with a gamma source while the other is exposed to alpha radiation. The alpha source causes more ionisation than the gamma source so the tissue it irradiates will suffer more biological damage.

The biological effect of the radiation is quantified in a term called the equivalent dose, *H*, where;

Equivalent dose = Absorbed dose × Radiation weighting factor
$$H = DW_R$$

The **radiation weighting factor, W_R**, assigned to each type of radiation is a comparative measurement of the biological effect that the radiation has on irradiated tissue.

X-rays and gamma rays have a weighting factor of 1, while alpha particles have a weighting factor of 20. Therefore for the example illustrated in figure 9.8.1 the tissue irradiated by the alpha source is 20 times more likely to be damaged. Weighting factor values for other types of ionising radiation are shown in table 9.8.2.

Type of radiation	Weighting factor
α	20
β	1
γ and X rays	1
Slow neutrons	2.3
Fast neutrons	10

Table 9.8.2

Equivalent dose is measured in units called **sieverts, Sv**, but again this is too large a unit to be practical in many cases so equivalent dose values are frequently quoted in mSv or even μSv.

 Following an incident in a laboratory, a researcher estimates that her finger of mass 10 g was irradiated by 1 mGy of alpha radiation and 100 μGy of gamma radiation. What is the total equivalent dose received by her finger?

 Equivalent dose = Absorbed dose × Weighting factor
$$H = DW_R$$
For the alpha exposure;
$$H = 1 \times 10^{-3} \times 20$$
$$H = 20 \ mSv$$
For the gamma exposure;
$$H = 100 \times 10^{-6} \times 1$$
$$H = 100 \ \mu Sv$$
Total equivalent dose = 20 mSv + 100 μSv
Total equivalent dose = 20.1 mSv

By assigning each type of radiation its appropriate W_R value we have been able to calculate the total equivalent dose from the individual absorbed doses.

If the finger received the equivalent dose of 20.1 mSv over a period of 4 seconds we can say that the **equivalent dose rate** is 20.1/4 or approximately 5 mSv s⁻¹. The equivalent dose rate, given the symbol $\dot{H}$, (pronounced H dot) is defined as;

$$Equivalent \ dose \ rate = \frac{Equivalent \ dose \ received \ in \ a \ certain \ time \ interval}{Length \ of \ the \ time \ interval}$$

Equivalent dose rates are useful for calculating the equivalent dose received from low activity sources over long periods of time. For example a worker in a uranium mine may be exposed to an equivalent dose rate of approximately 1 mSv per month. It is therefore easy to calculate the annual equivalent dose for a worker in such an occupation.

Many organisations now employ health physicists to monitor levels of radiation in working environments where there is a risk of increased levels of exposure.

The effective equivalent dose

In the previous example it was a finger which was irradiated. Clearly there is less potential for major biological damage to be done in a finger than in many other parts of the body. To take account of the different susceptibilities to harm when tissues are irradiated we must consider a new quantity called the **effective equivalent dose**.

The effective equivalent dose value is found by multiplying the equivalent dose value by an internationally agreed **factor**. This factor will be largest for parts of the body with cells *most* at risk of damage from radiation which include the reproductive organs, eyes and the lungs.

Summary

In the previous sections we have introduced a number of new terms. Each has a symbol and some are defined by equations. Table 9.8.3 summarises the main points about these terms.

Term	Symbol	Equation	Unit
Average activity	A	$A = \dfrac{N}{t}$	Bq
Absorbed dose	D	$D = \dfrac{E}{m}$	Gy
Weighting factor	W_R	–	no unit
Equivalent dose	H	$H = DW_R$	Sv
Equivalent dose rate	$\dot{H}$	$\dot{H} = \dfrac{H}{t}$	Sv h⁻¹ or Sv s⁻¹

Table 9.8.3

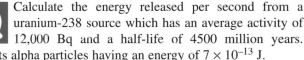 Calculate the energy released per second from a uranium-238 source which has an average activity of 12,000 Bq and a half-life of 4500 million years. It emits alpha particles having an energy of 7×10^{-13} J.

An activity of 12,000 Bq means that 12,000 alpha particles are being released per second and each has an energy of 7×10^{-13} J.

Therefore energy released in 1 second = 12,000 × 7 × 10⁻¹³ J
Energy released per second = 8.4 × 10⁻⁹ J

 If this source is used to irradiate 1 g of tissue for 5 minutes what is the absorbed dose?

$$Absorbed \ Dose = \frac{Energy \ absorbed}{Mass \ of \ absorbing \ material}$$

$$D = \frac{8.4 \times 10^{-9} \times (5 \times 60)}{1 \times 10^{-3}}$$

$$D = 2.52 \ mGy$$

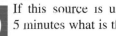 What is the equivalent dose received by the tissue while it is being irradiated? (The radiation weighting factor for the alpha particles is 20.)

Equivalent dose = Absorbed dose × Weighting factor
$$H = DW_R = 2.52 \times 20$$
$$H = 50.4 \ mSv$$

 What is the equivalent dose rate at which the tissue receives radiation?

 Equivalent dose rate =
$$\frac{Equivalent \ dose \ received \ in \ a \ certain \ time \ interval}{Length \ of \ the \ time \ interval}$$

$$\dot{H} = \frac{50.4 \ mSv}{5 \times 60 \ s}$$

$$\dot{H} = 0.168 \ mSv \ s^{-1}$$

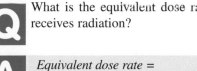

9.9 Background radiation and dose limits

Introduction

From earlier sections it should be clear that some sources of ionising radiation occur naturally while others are man-made. The combined effect of the naturally occurring and man-made sources, which find their way into the environment, is called the **background radiation**. If a GM counter is switched on it will record the arrival of radioactive particles even if there is no obvious radioactive source. By measuring these background counts over an extended period we can calculate the average background count rate. Indeed when doing accurate activity calculations we need to subtract the background count rate from the measured count rate to determine the corrected count rate for a source.

Corrected counts/s = Measured counts/s – Background counts/s

Natural sources of radiation

Even if we live in what we would consider a healthy environment we are still exposed to natural sources of radiation. Carbon-14 and uranium-238 are just two examples of radioisotopes which occur naturally in the earth's crust.

High energy electromagnetic radiations called **cosmic rays** are constantly bombarding the earth from outer space. They have higher energies than gamma rays and are an ionising radiation. Our bodies are exposed to gamma radiation from soil and rocks and many foodstuffs contain very small quantities of radioisotopes such as potassium-40 or carbon-14.

Radon-222 gas released into the air following the decay of small quantities of uranium, which occurs naturally in granite rocks, is a major contributor to natural radiation levels in some areas. Radon-222 atoms and their daughter products attach themselves to dust particles which can be inhaled into the lungs. Since these are alpha emitters the mean equivalent dose levels for people living in granite localities is higher than the national average.

	Average dose/μSv
Internal radiation $^{12}_{6}C$ $^{40}_{19}K$	370
Gamma rays from ground	400
Cosmic rays	300
Radon and daughter products	800
Total	**1870**

Table 9.9.1

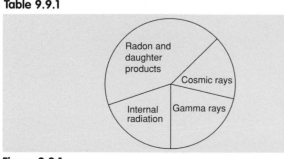

Figure 9.9.1

Figure 9.9.1 shows how these different sources contribute to the overall annual level of natural radiation. **The annual effective equivalent dose that a person in the UK receives from natural sources is about 2 mSv.** These are of course average figures and will vary from region to region.

Artificial sources of radiation

Natural radiation is not the only category of ionising radiation in our environment. There are a number of artificial sources to which we are all exposed.

We are exposed to sources of radiation during medical examinations using X-rays. This exposure varies from individual to individual but, on average, medical examination contributes some 250 μSv to our annual effective equivalent dose.

For the most part, large scale testing of nuclear weapons has ceased. However, some radioactive atoms formed during high altitude tests in the 1960s are still falling to earth. This type of radiation is called fallout and contributes slightly to our annual exposure.

Disposal of the radioactive materials used in domestic applications such as smoke detectors or industry, scientific research and medicine, as well as the by-products of nuclear power generation, further increases the level of radiation in the environment. Grouped together these are referred to as nuclear or radioactive waste.

	Average dose/μSv
Medicine	250
Fallout	10
Radioactive waste	2
Others	15
Total	**277**

Table 9.9.2

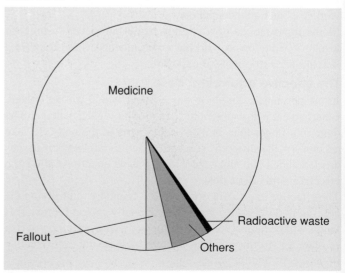

Figure 9.9.2

Figure 9.9.2 shows how the different artificial sources contribute to the overall annual effective equivalent dose from artificial sources. By comparing the data in tables 9.9.1 and 9.9.2 we can see that the vast proportion of background radiation is from natural sources.

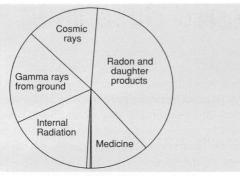

Figure 9.9.3 The composition of background radiation

Monitoring radiation levels

Radiation levels in the UK are monitored by the **National Radiological Protection Board (NRPB)**. The NRPB has been responsible for measuring the levels of natural and artificial radiations in different environments and for setting effective equivalent dose limits for members of the public.

While the average annual effective equivalent dose for members of the public can be found by adding together the values from natural and artificial sources, some people have occupations where increased doses are unavoidable.

Workers in the nuclear industry, radiographers and others working near radioisotopes will be at increased risk of exposure. **Higher annual effective equivalent dose limits are permitted for workers in these occupations**. Workers most at risk are required to wear a specially designed badge which records their exposure. These badges are called dosimeters.

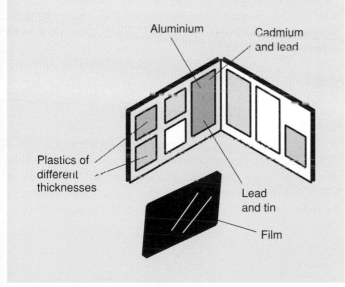

Figure 9.9.4 Dosimeter badge

When visible radiation falls on a photographic film a reaction occurs and the film surface darkens or 'fogs' when developed. One type of dosimeter uses the fact that fogging occurs when ionising radiation strikes a photographic film. The dosimeter badge contains a piece of photographic film, parts of which are shielded by different materials. Some radiations will penetrate one part of the badge but not others, so at the end of the monitoring period the fogging on the developed film indicates the total exposure to radiation as well as specific exposures to

particles of particular energies. In this way workers in 'at risk' occupations can be kept informed of their cumulative exposure and can reduce the risk of biological harm by staying within the annual permitted effective equivalent dose limit currently set at 50 mSv.

Risk

Modelling the risk associated with medical treatments is a very complex process. Using ionising radiation in medicine does increase the risk of doing harm, while at the same time having the potential for aiding diagnosis, providing treatment and promoting recovery. Researchers use the medical records of patients undergoing treatments with radiation to analyse the level of risk for those treatments. The models that they arrive at are mainly statistical as they do not have the benefit of repeating experiments to prove or disprove a theory.

Common cause of death	Average annual risk
Smoking 10 cigarettes a day	1 in 200
Natural causes at or before age 40	1 in 850
Road accident	1 in 9500
Accident at work	1 in 26,000
From radiation exposure of 2 mSv/year	1 in 80,000

Source: NRPB, Living with radiation 1986

Table 9.9.3

Table 9.9.3 shows some estimates for the risk of death from various activities. While there are many grounds on which the validity of these statistics can be questioned the fact remains that there is no clear evidence that low level radiation is a major health problem. The risk associated with average levels of radiation is considerably less than the risk from normal patterns of smoking, drinking alcohol or road traffic accidents.

However, this risk analysis fails to take into account the potential for catastrophic damage from a large scale accident. For many opposed to the proliferation of nuclear power it is this unquantifiable risk which is their major objection.

9.10 Shielding and protection

Penetration

The properties of the different ionising radiations are summarised in figure 9.10.1.

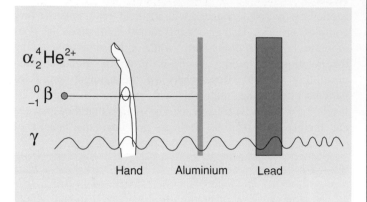

Figure 9.10.1

As alpha and beta particles attempt to pass through a material they lose their kinetic energy during collisions which cause ionisation of the atoms in the material. **After travelling a certain distance they have insufficient energy to cause more ionisation and are then considered to have been absorbed by the material.**

Alpha particles are the most highly ionising and consequently penetrate least. They can pass through only a few centimetres of air and are stopped by thin skin or by a sheet of paper. Beta particles with average kinetic energies can travel approximately 1 m through air and will be stopped by an aluminium plate about 3 mm thick.

Gamma rays are not particles like alpha and beta radiations. Photons of gamma radiation interact with matter in a much more complex manner than alpha and beta particles. Gamma rays, like other types of electromagnetic radiation, will pass almost unaffected through air. They are most strongly absorbed by elements such as lead.

If we assume that the gamma source shown in figure 9.10.2 is emitting equally in all directions, the GM tube will be collecting only a small part of the total energy radiated by the source.

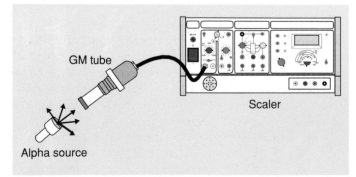

Figure 9.10.2

The gamma source is behaving as a **point source** of electromagnetic radiation radiating energy equally in all directions, so we would expect the intensity, I, at any point a distance, d, from the source to vary as;

$$I \propto \frac{1}{d^2}$$

By measuring the intensities at various distances from the point source this relationship can be confirmed experimentally. It should be noted that the reduced intensity at greater distances from the gamma source is a feature of the electromagnetic nature of the gamma rays and is *not* because some of the gamma radiation is absorbed by the air.

How far can Gamma rays travel through lead?

EXPERIMENT The intensity of a beam of gamma rays can be reduced by putting a piece of lead between the source and the detector. Some of the gamma rays are absorbed by the lead shield.

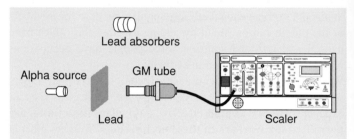

Figure 9.10.3

The thicker the piece of lead between the source and the detector, the lower the intensity reaching the detector. With the apparatus shown in figure 9.10.3 the gamma source is initially aligned with the detector.

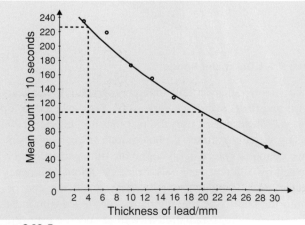

Figure 9.10.4

With no lead between the source and detector, the counts during a few consecutive 10 second intervals are measured and the average count rate found. Placing different thicknesses of lead between the source and detector reduces the intensity of the radiation arriving at the source as shown in figure 9.10.5.

Figure 9.10.5

The thickness of lead needed to reduce the count rate to half its initial value is called the **half-value thickness** for lead. From the graph we can see that the count rate when the radiation is passing through 4 mm of lead is 230 counts in 10 seconds. The count rate is halved to 115 counts in 10 seconds by using 20 mm of lead. Therefore we can say that an additional 16 mm of lead are needed to halve the count rate measured at the detector. The half-value thickness of lead for radiation from the source is 16 mm. If we use a starting count rate of 180 counts in 10 seconds, we can also show that to halve the count to 90 again requires an additional 16 mm of lead.

Half-value thicknesses

Placing one half-value thickness of material in front of a gamma source will halve the intensity of the radiation arriving at the detector. The absorbed dose received by any matter exposed to the shielded source will also be halved as will the absorbed dose equivalent. Shielding with 2 half-thicknesses of absorber quarters the intensity, the absorbed dose and the absorbed dose equivalent.

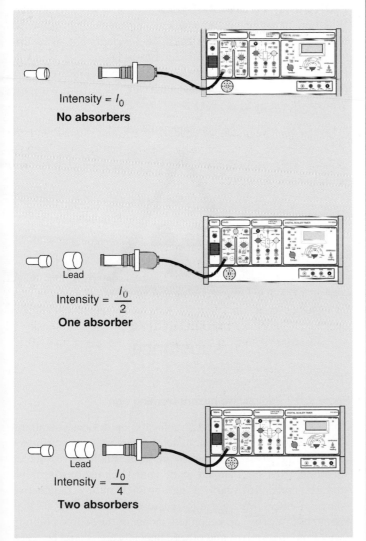

Figure 9.10.6 Intensity and thickness

Q In a research laboratory a gamma source is stored in a 'lead castle'. This places the source at the centre of a lead container which has 2.5 cm thick walls. If the count rate outside the lead container is 60 counts per minute above background what count rate would be expected if the source is removed from its container? Assume that lead has a half-value thickness of 12.5 mm and that the source is kept at a constant distance from the detector.

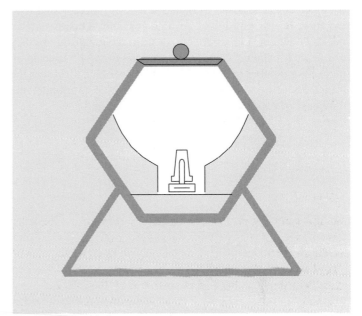

Figure 9.10.7 Lead castle

A The walls of the lead container are 2.5 cm thick and since the half-value thickness for the lead is 12.5 mm we can say that the walls are two half-thicknesses thick. The first half-thickness halves the count rate. The second half-thickness halves the count rate again.

Therefore the count rate due to the emerging radiation is ¼ of what is being created within the container.

The count rate when the source is inside the container will be ¼ of the count rate without the container when the distance between the source and the detector is kept constant.

Count rate caused by the unshielded source – 4 × 60
= 240 counts per minute above background radiation

Shielding

Storing radioactive sources in lead containers is a very effective way of protecting the close environment from the potentially harmful effects of ionising radiation. Radiation from alpha and beta sources are stopped entirely by even the simplest shields while sheets of lead a few centimetres thick are required to reduce the intensity of gamma radiation.

The core of nuclear reactors are surrounded by shields made of lead and concrete. As well as having to contain the coolant gas as it is pumped at high pressure through the core, the containment vessel around the reactor core must be able to minimise the doses experienced by those working in the vicinity of the reactor core.

9.11 Using radiation safely

Using radioisotopes

All of us are exposed to low doses of ionising background radiation from different sources. These levels of exposure are not currently considered to pose a serious health risk. However, there are a small number of people who, as a result of their occupation, are exposed to higher levels of radiation and risk. Doctors, health physicists, technicians and support staff in nuclear medicine clinics are at risk from the radioisotopes they use to treat patients. In industry, radioisotopes are used in an increasing number of applications. Figure 9.11.1 illustrates the principles of how a radioisotope can be used to monitor the thickness of paper as it is made in a rolling mill.

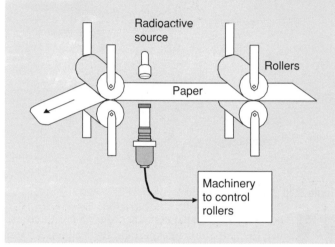

Figure 9.11.1 Making paper

The intensity of radiation arriving at the detector depends on the thickness of the paper. A change in the intensity indicates that the thickness has altered and that the manufacturing process needs to be modified. If the measured count rate increases beyond a specified limit the paper needs to be made thicker but if the intensity reduces the paper is too thick and the manufacturing process needs to make it thinner.

If the information from the radiation detector is fed back to the machinery controlling the rollers the thickness of the paper can be continuously checked and reset. This is an example of negative feedback.

Beta emitters are ideal for thickness control systems in rolling mills producing paper, plastic or aluminium foil. In mills rolling sheets of steel gamma emitters are required.

The late 1990s has seen the widespread introduction of smoke detectors in homes, factories and offices. In the smoke alarm shown in figure 9.11.2 a weak source of the radioisotope americium-241 emits alpha particles inside an open unit. The ionisation of air by the alpha particles causes a small current in the circuit which means that there is a p.d. across the resistor. If smoke particles fill the space surrounding the source, the alpha radiation is absorbed more rapidly and the level of ionisation falls. The circuit current falls and a buzzer or sounder gives an audible warning when the p.d. across the resistor falls below a preset level.

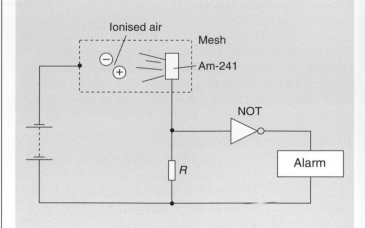

Figure 9.11.2 Smoke alarm

Handling radioisotopes

Whether being used in medical, industrial or domestic environments radioisotopes must be handled in a way that minimises the risk of exposure or contamination.

Places where there is even the slightest possibility of harm from sources of ionising radiation are identified with the symbol shown in figure 9.11.3.

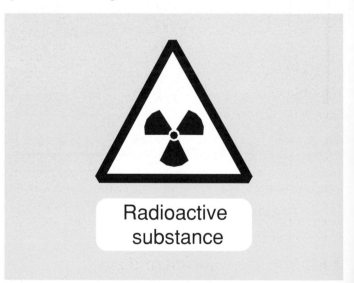

Figure 9.11.3 Radioactive hazard warning sign

This sign must be displayed prominently on devices containing sources of ionising radiation, in areas where radioactive sources are stored or where the sources are used. Radioactive sources should always be handled with care. Regulations prohibit students under 16 years of age from handling radioactive sources. Others who are allowed to use radioactive sources should use forceps and keep the window of the source pointing away from themselves and others. Special care should be taken to avoid pointing the source at particularly sensitive parts of the body, especially the eye.

Figure 9.11.4 Lifting sources with forceps

Very intense sources are manipulated remotely with specialised handling equipment such as that shown in figure 9.11.5.

Figure 9.11.5 Remote handling

Sources should be kept out of their shielded containers for the minimum possible time and those using radioactive sources are encouraged to wash their hands immediately after replacing the source and its container in the locked, labelled store cupboard. These precautions are designed to minimise the dose equivalent received by the user and others in the area where the source is being used.

Reducing the equivalent dose

Equivalent dose values can be used to indicate the susceptibility to harm from exposure to radioactive sources. Apart from handling radioactive sources according to a 'code of best practice', there are a number of practical measures which can be taken to minimise the dose equivalent.

1 Limiting the time for which the source is being used reduces the equivalent dose because the total energy absorbed by the irradiated body is smaller for shorter exposures.
2 Shielding the source reduces the equivalent dose because this decreases the intensity of radiation on the irradiated body.

3 The exposure from gamma sources can be reduced by working at greater distances from the source. The intensity of radiation from a point source of gamma rays decreases rapidly as the distance from the source increases.
4 Using the minimum amount of the radioisotope required for the experiment or investigation is another way of reducing the equivalent dose. Smaller quantities have lower activities so the energy transferred each second to the irradiated object is less.

Figure 9.11.6 shows a technician carrying a radioactive source in its shielded container. The gamma radiation emitted from the source radiates in all directions, so because of the way he is carrying the source a large part of his upper body is being irradiated.

Figure 9.11.6

In figure 9.11.7 the source is being carried by two technicians. The exposure of each technician is now much less. Not only is each technician further from the source but also a much smaller part of their bodies is being irradiated.

Figure 9.11.7

Equivalent dose rate

Each of the measures already outlined has been designed to reduce the equivalent dose. Since the equivalent dose rate, $\dot{H}$, is defined as;

$$\dot{H} = \frac{H}{t}$$

we can see that shielding a source or increasing the distance from the source will also reduce the equivalent dose rate. Similarly using as small a mass as possible of the radioisotope will help keep equivalent dose rate values low.

9.12 The nuclear debate

Introduction
Throughout this chapter on radioactivity we have sought to express the *facts* concerning the potential benefits and risks associated with the use of radioisotopes.

There can be no doubt that many modern medical treatments have only become possible following Henri Becquerel's discovery of radiation in 1896. However, Becquerel's successor, Marie Curie and others, have paid the ultimate price for their research with such dangerous materials.

Two sides of the debate
Any attempt to balance ionising radiation's potential for good with the risk of harm must be guided by scientific facts, but in the nuclear debate, facts and opinions are closely intertwined. On one side of the argument the environmental campaigners are prepared to tolerate radioisotopes used in medicine but see commercial or military uses for nuclear power as unnecessary. They will point to studies highlighting the harm that even low levels of exposure can cause.

On the other side of the debate the supporters of nuclear power will argue that the environmentalists' studies are flawed or at best inconclusive. They will produce their own assessment of risk highlighting the safety record and good health of the vast majority of workers in the nuclear industry.

Figure 9.12.1 This campaigner and a 'mutant lobster' are protesting against the import of foreign spent nuclear fuel at Sellafield.

Most people today welcome the demise of nuclear weapons, although worry still remains about the storage and safe disposal of nuclear arsenals.

Getting at the truth
There are a number of factors which make it difficult for even independent observers to establish the levels of harm that ionising radiation can cause. The scientist's natural instinct is to form conclusions based on the results of experiments. Clearly when assessing the impact of ionising radiations on life expectancy it is not possible to base conclusions on results from a series of carefully controlled tests! Details of the consequences of accidental exposure to high levels of radiation are available but thankfully only a small number of such accidents have occurred and those involved do not constitute a statistically representative sample of the population.

Researchers investigating the effects of exposure to high levels of radiation can look at the aftermath of the 1945 nuclear explosions in Hiroshima and Nagasaki. It is clear that single acute exposures to radiation doses in excess of approximately 10 Sv caused fatal damage to the central nervous system. The supporters of the nuclear industry would not argue over the harm from high doses but would say that in their industry such doses are impossible.

Risk from low level exposures
Exposure to high levels of radiation is much less likely than exposure to low doses over a long period of time. It is in this area that the nuclear debate is most polarised.

Figure 9.12.2 Nuclear explosion with mushroom cloud

The survivors of the explosions at Hiroshima and Nagasaki have shown higher than normal incidences of leukaemia and cancers of the lung, thyroid and breast. These problems have only developed years after the initial exposure, prompting others in different parts of the world to worry about their own exposure to low levels of radiation. One controversial view is that low level exposures may lead to genetic defects that may not be revealed for a number of generations. The supporters of nuclear power question the statistical validity of such theories. They base their objections on the lack of proof between cause and effect and point to the fact that cancer occurs in parts of the world so far unaffected by the nuclear industry.

A recent worry has been the identification of a number of small areas containing clusters of cases of cancer and leukaemia. Some of these areas are in the immediate proximity of nuclear installations but others are not. Again there is debate over the cause of these clusters.

Chernobyl

In the early hours of 26th April 1986 an explosion occurred in the core of reactor number four at the former Soviet Union's nuclear power station at Chernobyl in the Ukraine. For the nuclear industry at large this accident was a major blow to its credibility, but for the immediate vicinity the consequences were catastrophic.

Figure 9.12.3 Chernobyl

An investigation showed that the explosion occurred when most of the reactor's normal safety features were switched off during an experiment on the reactor. Around 300 people in the immediate vicinity of the explosion suffered serious radiation sickness and 31 have died. These included fire fighters and reactor operators who had stayed on site to minimise the release of radioactive particles into the environment.

The initial explosion released a cloud of radioactive gases and fragments into the atmosphere. Small explosions and further releases continued for the next 10 days during which residents from the nearby city and others within a 30 km radius were evacuated.

The cloud of gases released into the atmosphere was carried by winds across Europe, roughly following the path shown in figure 9.12.4.

The contaminated air passed into southern Scotland on the morning of Saturday 3rd May. Heavy rain meant that radioactive iodine-131, caesium-134 and caesium-137 were deposited on the hills of Cumbria and south west Scotland.

The fallout from Chernobyl deposited on southern Scotland posed particular problems for the farming community. Radioactive iodine-131 deposited on the grass found its way into milk although levels of activity never rose above safety limits set by the government. Iodine-131 has a half-life of eight days so the problems it caused were relatively short lived. On the other hand caesium-137 has a half-life of 30 years and posed particular problems for sheep farmers. The levels of caesium-137 in lambs reared on open uplands rose above the permitted limits and stayed high for some considerable time. This led to lengthy restrictions on the movement and sale of sheep and lambs.

Scientific analysis of the after effects of the Chernobyl accident

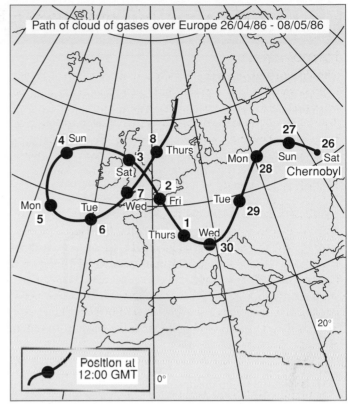

Figure 9.12.4 Radioactive gases from Chernobyl

has added significantly to our understanding of how the food chain is affected by low levels of radioisotopes deposited over a large area.

It can only be hoped that the lessons learnt from the Chernobyl accident will prevent similar accidents. Subsequent studies into the long term effects of the Chernobyl fallout may help those using low levels of radiation in medicine to understand how human tissue reacts to prolonged low level exposure.

Nuclear power

The splitting of the atom was hailed as the dawn of a new era in electricity generation. The small quantities of raw materials needed are available in abundance so nuclear energy was seen as the way of satisfying the needs of a society in which living standards were rising.

Extracting the raw uranium has less environmental impact than mining coal and no ash is released into the atmosphere from plants generating electricity from nuclear reactions. It was felt that nuclear power stations would cause less atmospheric pollution than the coal fired alternative. With all of these factors in its favour it is difficult for some to understand why it has not been accepted and adopted more widely. The storage and disposal of the by-products of nuclear power production is an emotive issue with many feeling that it is morally wrong to pollute the world for our children's children.

Conclusion

In this chapter we have tried to outline some of the key issues and arguments surrounding the use of radioactive materials. However, it is left up to individuals in each generation to make up their own minds on the merits of the arguments put forward by both sides in the nuclear debate.

Worked example 1

A research scientist works with a radioactive beta source which has a very long half-life. The health physicist at the research establishment analyses the worker's dosimeter badge and estimates that the research worker is exposed to an average equivalent dose rate of 20 µSv h^{-1}.

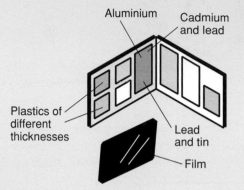

fig 9.13.1

a) Explain how the dosimeter badge of the type shown in figure 9.13.1 is used to monitor the level of exposure to radiation.

b) Calculate the total equivalent dose collected in 1 week if the scientist works with this source for 5 hours per day, 5 days per week.

c) For how many weeks per year is the research worker allowed to use this source before she exceeds the establishment's annual limit of 20 mSv?

d) What is the weekly absorbed dose of this radiation?

Answers and comments

a) The dosimeter badge uses the principle that radiation affects photographic film. When film exposed to radiation is developed it appears 'fogged'. The higher the level of exposure the more fogging occurs. At work, the scientist wears the badge which will be sent away for analysis at regular intervals. In this way a cumulative total for the exposure to radiation can be kept.

b) The equivalent dose rate is defined by the equation

$$\dot{H} = \frac{H}{t}$$

Total equivalent dose = Equivalent dose rate × Time
$$H = \dot{H}t = 20 \times 5 \times 5$$
$$H = 500 \ \mu Sv \ per \ week$$

c) The absorbed dose per week is 500 µSv and the permitted annual limit is 20 mSv. We must calculate the time taken to accumulate 20 mSv at a rate of 500 µSv per week.

$$Permitted \ time = \frac{20 \times 10^{-3}}{500 \times 10^{-6}} = 40 \ weeks$$

The scientist will only be allowed to use this source for 40 weeks per year before exceeding the establishment's annual exposure limit.

d) Since the exposure is from a beta source, we know that the radiation weighting factor, W_R, is 1. Therefore;

$$H = W_R D$$
$$500 \ \mu Sv = 1 \times D$$
$$D = 500 \ \mu Gy$$

The absorbed dose is 500 µGy per week.

Worked example 2

A cobalt-60 source is used by a student to investigate the effect of distance on the count rate from a source emitting gamma radiation.

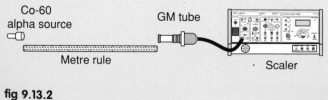

fig 9.13.2

Corrected count rate/counts s^{-1}	1330	680	270	160	108
Distance/cm		13	19.5	31	40	49

Table 9.13.1

a) Use these results to show that the count rate is inversely proportional to the square of the distance from the source.

b) Explain why the count rate is described as the *corrected count rate*.

Answers and comments

a) To test for an inverse proportion between the count rate and the distance squared we must plot a graph of count rate versus $1/d^2$ as shown in figure 9.13.3. Since the graph of corrected count rate versus

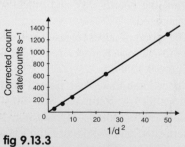

fig 9.13.3

$1/d^2$ is a straight line passing through the (0,0) origin we can say that the count rate is inversely proportional to the square of the distance from the source of gamma rays.

Corrected count rate/counts s^{-1}	1330	680	270	160	108
$1/d^2$ (m^{-2})		51	26.3	10.4	6.25	4.2

Table 9.13.2

b) Some of the radiation detected by the counter is not due to the source but is due to background radiation. Subtracting the background count from the total counts gives the value for the counts due to the source.

Questions

1 A student making notes on radiation writes down the following statements;
I   Effective equivalent dose values take account of the different susceptibilities to harm of irradiated tissue.
II The average annual effective equivalent dose for a person in the UK due to background radiation is 2 mSv.
III Equivalent dose limits are higher than 2 mSv for people working in certain occupations.

Which of these statements is/are true?

A I only
B II only
C I and II only
D II and III only
E I, II and III.

2 A tumour in a patient's liver has a mass of 25 g and is irradiated with 0.05 J of gamma radiation.
   a) What is the absorbed dose?
   b) What is the equivalent dose?
   c) Explain why gamma radiation is the most suitable form of radiation for treating this type of tumour.

3 A sample of tissue with a mass of 10 g is irradiated using a radioactive source emitting alpha and beta particles as well as gamma rays. During an exposure the tissue absorbs 30 μJ of gamma radiation, 20 μJ of alpha radiation and 25 μJ of beta radiation.
   a) Which type of radiation gives most energy to the tissue?
   b) Calculate the absorbed dose of alpha radiation.
   c) Calculate the total equivalent dose.

4 The equivalent dose rate for a certain sample placed 1 m from a radioactive beta source is 20 μSv h^{-1}.
   a) Calculate the dose absorbed by the sample of mass 10 kg irradiated 1 m from this source for 5 hours.
   b) How much energy has been given to the irradiated mass by the radioactive source during the 5 hours?

5 A certain material, used to shield a radioactive source, reduces the count rate of the source from 2×10^6 counts per second to 5×10^5 counts per second. The thickness of the shielding material is 8 cm.
   a) What is the half-value thickness for this material?
   b) What thickness of material would reduce the count rate of the source to 1.25×10^5 counts per second?

6 A radioactive source is stored in a lead lined box and kept in a secure location.

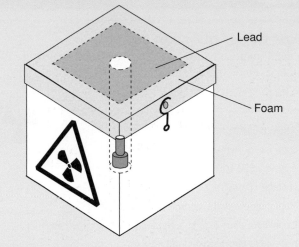

fig 9.13.4

A health physicist monitors the count rate by placing a GM tube against the side of the box and measuring the activity over a period of time.
   a) She states in her report that when the GM tube is placed as shown 'the count rate varies but is approximately constant during the entire test which lasted 30 minutes'.
      (i) Explain why the count rate varies during the test.
      (ii) If the count rate is approximately constant during the test what can be deduced about the half-life of the source?
   b) Explain why the measured count rate is lower if the GM tube is placed against one of the corners of the box.
   c) Shielding a radioactive sample is one way of reducing the equivalent dose. State two other ways in which the equivalent dose can be reduced.

7 A manufacturer of sheets of plastic approximately 1 cm thick wants to automatically test the thickness of the sheets during their production.
   a) Describe, with the aid of a diagram, how a radioactive source and detector can be used to indicate the thickness of the plastic sheets.
   b) What type of radiation should the source emit? Explain your answer.
   c) Sketch a graph to show how the intensity of gamma radiation varies with the thickness of the absorbing material through which it passes.

Past paper question 1

A small radioactive source in a sealed container is used in a nuclear medicine laboratory. The information shown in figure 9.14.1 is displayed on the label.

Radionuclide : $^{131}_{53}$I

Date : 23rd February 1993 (12 noon)

Activity : 300 MBq

Half-life : 8 days

Radiation emitted : gamma (quality factor 1)

Equivalent dose rate at a distance of 1 mm : 16^{-1} Sv h^{-1}

Half-value thickness of lead : 3.3 mm

fig 9.14.1

a) When the source has the activity shown on the label, how many nuclei decay in one minute?

b) A technician needs to work at a distance of 1 m from a freshly prepared source. For what period of time can the technician work at this distance so that the absorbed dose does not exceed 50 μGy?

c) Lead shielding is used around the source to reduce the equivalent dose rate at a distance of 1 m to 2.5 μSv h^{-1}.
   (i) Draw a graph to show how the equivalent dose rate at a distance of 1 m varies with the thickness of lead shielding.
   (ii) Use your graph to estimate the thickness of lead needed to provide the required level of shielding.

SQA 1994

Answers and comments

a) The activity shown on the label is 300 MBq. This means that there are 300×10^6 decays per second. Therefore the decays in one minute are found from;

Decays per minute = Activity × Time (s) = $300 \times 10^6 \times 60$
Decays per minute = 1.8×10^{10}

b) The radiation is gamma radiation with a weighting factor of 1. Therefore the labelled equivalent dose rate of 16 μSv h^{-1} is equal to an absorbed dose rate of 16 μGy h^{-1}. The time for a total absorbed dose of 50 μGy is found from;

$$Time = \frac{Absorbed\ dose}{Absorbed\ dose\ rate} = \frac{50\ \mu Gy}{16\ \mu Gy\ h^{-1}}$$

Time = 3.125 hours

c) The half-thickness of lead is quoted as 3.3 mm. This allows the type of graph shown in figure 9.14.2 to be drawn. The thickness of lead needed to reduce the equivalent dose rate to 2.5 μSv h^{-1} is found from the graph to be 9 mm.

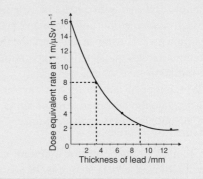

fig 9.14.2

Past paper question 2

While investigating the effect of different types of radiation on the human body, the data in table 9.14.1 is obtained for one particular type of body tissue.

Type of radiation	Absorbed dose rate	Weighting factor
γ rays	100 μ Gy h^{-1}	1
Fast neutrons	400 μ Gy h^{-1}	10
α particles	50 μ Gy h^{-1}	20

Table 9.14.1

a) Show, using the data in the table, which radiation is likely to be the most harmful to this type of tissue.

b) The effect of radiation can be reduced by putting shielding material between the source and the tissue. The effectiveness of this shielding material can be described by the half-value thickness of the material.
   (i) Explain the meaning of *half-value thickness*.
   (ii) The half-value thickness for a particular material is 7 mm. A block of this material of thickness 3.5 cm is inserted between the source and the tissue. What fraction of the radiation, which is directed at the tissue, is received by the tissue? **SQA 1996**

Answers and comments

a) We must calculate the equivalent dose rate for each type of radiation using the equation;

Equivalent dose rate = Absorbed dose rate × Weighting factor
For the gamma rays equivalent dose rate = $100 \times 1 = 100$ μSv h^{-1}
For fast neutrons equivalent dose rate = $400 \times 10 = 4000$ μSv h^{-1}
For the alpha particles equivalent dose rate = $50 \times 20 = 1000$ μSv h^{-1}

The fast neutrons are likely to be the most harmful type of radiation because it is the one with the highest equivalent dose rate.

b) (i) The thickness of a material which must be placed between a radioactive source and detector to halve the intensity of the radiation from the source is called the half-value thickness.
   (ii) If 7 mm of this material is placed between the source and detector the intensity of the radiation is halved; a further 7 mm will half the intensity again. 3.5 cm represents 5 half-thicknesses so the intensity arriving at the detector is $1/2^5$ times the initial intensity. 1/32 of the initial radiation will be received by the detector.

Questions

1 Plutonium-239 is a radioactive source emitting only one type of radiation. The decay equation is as shown in figure 9.14.3.

$$^{239}_{94}\text{Pu} \longrightarrow {}^{235}_{92}\text{X} + \text{Y}$$

fig 9.14.3

a) How many neutrons are there in nuclide X?
b) Element X also exists as an isotope having a mass number of 238. Explain how this differs from the isotope produced from the above decay.
c) How many protons and neutrons are in particle Y?
d) What type of radiation is emitted by plutonium-239?

2 Write down the numerical values of the letters a to l in the following decay equations;

$$^{224}_{88}\text{Ra} \longrightarrow {}^{a}_{b}\text{Rn} + {}^{c}_{d}\alpha$$

$$^{e}_{83}\text{Bi} \longrightarrow {}^{210}_{f}\text{Po} + {}^{g}_{h}\beta$$

$$^{i}_{38}\text{Sr} \longrightarrow {}^{90}_{j}\text{Y} + {}^{g}_{h}\beta$$

$$^{k}_{l}\text{Ra} \longrightarrow {}^{218}_{84}\text{Po} + {}^{c}_{d}\alpha$$

fig 9.14.4

3 Uranium-238 captures a medium speed neutron which initiates the sequence of decays shown in figure 9.14.5.

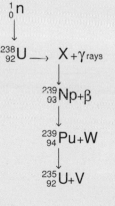

$${}^{1}_{0}\text{n}$$
$$\downarrow$$
$$^{238}_{92}\text{U} \longrightarrow \text{X} + \gamma\text{rays}$$
$$\downarrow$$
$$^{239}_{03}\text{Np} + \beta$$
$$\downarrow$$
$$^{239}_{94}\text{Pu} + \text{W}$$
$$\downarrow$$
$$^{235}_{92}\text{U} + \text{V}$$

fig 9.14.5

a) What are the mass number and atomic number of radionuclide X?
b) What are the products W and V? Explain your answer.
c) Which nuclides are isotopes of each other?

4 Figure 9.14.6 is a diagram summarising the number of protons and neutrons in a nuclide. The position of the radioisotope, Bi, (bismuth-214) which is formed from the beta decay of lead-214, is marked.
a) Copy figure 9.14.6 and mark on the position of the parent lead-214 nuclide.

b) The bismuth can decay either by alpha emission into nuclide P or by beta emission into nuclide Q. Mark the positions of P and Q on your copy of the figure.

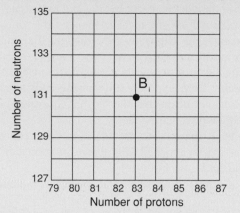

fig 9.14.6

5 Two possible reactions involving uranium are represented by the following equations and the masses of the particles and nuclides involved are as shown.

Equation A $\quad {}^{235}_{92}\text{U} + {}^{1}_{0}\text{n} \longrightarrow {}^{134}_{52}\text{Te} + {}^{98}_{40}\text{Zr} + 4{}^{1}_{0}\text{n}$

Equation B $\quad {}^{235}_{92}\text{U} + {}^{1}_{0}\text{n} \longrightarrow {}^{144}_{56}\text{Br} + {}^{90}_{36}\text{Kr} + 2{}^{1}_{0}\text{n}$

fig 9.14.7

Nuclide	$^{235}_{92}$U	$^{134}_{52}$Te	$^{98}_{40}$Zr	$^{144}_{56}$Ba	$^{90}_{36}$Kr	$^{1}_{0}$n
Mass/3 10^{-25} kg	3.901	2.221	1.626	2.388	1.492	0.017

Table 9.14.2

a) What name is given to the reactions represented by equations A and B?
b) What property of the incoming neutron might determine how a particular uranium-235 nucleus splits?
c) Show that the mass defect for equation A is 3×10^{-28} kg.
d) How much energy is released by a single nucleus splitting as shown in equation A.
e) Calculate the energy released by a single nucleus splitting as shown in equation B.

6 A student making notes on the effects of radiation on living tissue writes down the following statements;
I The risk of harm depends upon the absorbed dose of radiation, the type of radiation and the part of the body irradiated.
II A weighting factor is given to each type of radiation as a measure of its biological effect.
III The equivalent dose is the product of absorbed dose and weighting factor $H = DW_R$.
Which if these statements is/are true?
A I only
B III only
C I and III only
D II and III only
E I, II and III.

7 The National Radiological Protection Board and other international bodies have set permitted levels for exposure to ionising radiation.

a) Explain what is meant by ionising radiation.

b) Absorbed dose measurements are one way of quantifying exposure to radiation. Explain what is meant by absorbed dose.

c) Equivalent dose values are used to indicate the risk to health from exposure to ionising radiations. Explain what is meant by the terms equivalent dose and radiation weighting factor.

d) What is the approximate annual effective equivalent dose that a person in the UK is allowed to receive from natural sources?

e) State three different natural sources of radiation.

8 A piece of tissue of mass 1.5 g is placed close to a long half-life source which emits alpha particles. The tissue absorbs an average of 30,000 alpha particles per second. The alpha particles have an average energy of 1×10^{-10} J and a quality factor of 20.

a) Explain how this source can damage the tissue irradiated.

b) Calculate the total energy received by the tissue in 1 second.

c) What dose of radiation is absorbed in 1 minute?

d) What is the equivalent dose if this source irradiates the tissue for 1 hour?

9 A region of high radioactivity is monitored remotely. The levels of radiation recorded and their weighting factors are as shown in table 9.14.3.

Type of radiation	Absorbed dose rate	Radiation Weighting factor
γ and X-rays	2 mGy h^{-1}	1
Fast neutrons	0.5 mGy h^{-1}	10
α particles	15 mGy h^{-1}	20
β particles	15 mGy h^{-1}	1

Table 9.14.3

a) Which type of radiation contributes most to the overall equivalent dose for radiation emitted into the monitored region?

b) Calculate the total equivalent dose for radiation emitted into the monitored region in one day.

10 A GM tube and counter are placed 20 cm away from a point source of gamma radiation. The count rate at this point is measured as 6000 counts per second.

a) Sketch a graph to show how the count rate for the radiation varies with distance from this point source.

b) What is the count rate if the GM tube is placed 1 m away from the point source?

11 The equivalent dose rate of radiation from a spent fuel rod is found to be 400 Sv h^{-1}. This fuel rod is kept in a storage tank under water which has a half-value thickness of 75 cm.

a) What is the equivalent dose rate at the surface of the tank if the water is 1.5 m deep?

b) What is the equivalent dose rate at the surface of the tank if the water is 3 m deep?

12 Materials X and Y have half-value thicknesses of 12 mm and 18 mm respectively. A gamma source with a long half-life is to be stored in a box and shielded by materials X and Y. The gamma source produces a count rate of 3×10^5 counts per second at a detector placed at a fixed distance from it.

a) The energy of the gamma radiation is absorbed by the materials through which it passes. Which material, X or Y, is the best absorber of gamma radiation? Explain your answer.

b) What is the count rate if the source is now shielded by 36 mm of material Y?

c) What is the count rate when the source is shielded only by 36 mm of material X?

d) What is the count rate if the source is shielded by 36 mm of X as well as 36 mm of Y?

e) Both the container designs shown in figure 9.14.8 are 6 cm thick. Which of the designs will provide the greater shielding? Explain your answer.

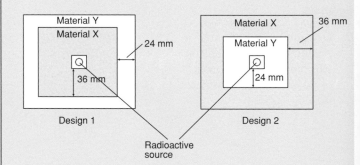

fig 9.14.8

Revision

In sections 10.1 to 10.9 the statements in normal type are from the SQA's arrangements for Higher Physics. The statements in italics detail content in each chapter leading up to these statements or taking them a little further.

PART 4

$$a = \frac{dv}{dt}$$

$$v = \text{lift} \, S \leftarrow \text{drag}$$

$$a = \frac{dv}{dt} = \frac{d}{dt}\left(\frac{ds}{dt}\right)$$

$$= \frac{d^2 s}{dt^2}$$

Section	Content covered
1.1: Average speed	***You should now be able to;*** *Explain the term speed.* *Describe how to measure an average speed.* *Carry out calculations involving distance, speed and time.*
1.2: Instantaneous speed	***You should now be able to;*** *Describe how to measure an instantaneous speed.* *Identify situations where instantaneous speed and average speed are different.*
1.3: Distance and displacement	***You should now be able to;*** *Describe what is meant by vector and scalar quantities (also section 2.1).* *Use a motion sensor to display graphs on a computer.* Distinguish between distance and displacement. Use scale diagrams or otherwise to find the resultant of a number of displacements.
1.5: Velocity	***You should now be able to;*** Explain the term velocity. State the difference between speed and velocity. Distinguish between speed and velocity. Use scale diagrams or otherwise to find the resultant of a number of velocities.
1.6: Acceleration	***You should now be able to;*** *Explain the term acceleration.* *Carry out calculations involving the relationship between initial velocity, final velocity, time and uniform acceleration.* State that acceleration is the change in velocity per unit time. Describe the principles of a method for measuring acceleration. Calculate the approximate random uncertainty in the mean value of a set of measurements using the relationship; approximate random uncertainty in mean $= \dfrac{\text{maximum value} - \text{minimum value}}{\text{number of measurements taken}}$
1.7: Displacement-time graphs	***You should now be able to;*** Use the terms constant velocity and constant accelerations to describe motion represented in graphical or tabular form.
1.8: Velocity-time graphs	***You should now be able to;*** *Describe the motions represented by a velocity-time graph.* *Draw velocity-time graphs involving more than one constant acceleration.* *Discuss the equivalence of displacement and distance for the specific case of an object moving in a straight line in one direction.* *Calculate displacement and acceleration from velocity-time graphs involving more than one acceleration.* Draw an acceleration-time graph using information from a velocity-time graph for motion with a constant acceleration.
1.9: Graphs for a bouncing ball	***You should now be able to;*** Explain the displacement-time, velocity-time and acceleration-time graphs for the motion of a bouncing ball.
1.12: The equations of motion	***You should now be able to;*** Show how the following relationships can be derived from basic definitions in kinematics $s = ut + \frac{1}{2} at^2$; $v^2 = u^2 + 2as$. Carry out calculations using the kinematic equations.
1.13: Equations of motion expts	***You should now be able to;*** Describe experiments using a microcomputer to confirm the kinematic equations. Calculate the mean value of a number of measurements of the same physical quantity.

1.14: Projectiles

You should now be able to;

Explain the curved path of a projectile in terms of the force of gravity.

Explain how projectile motion can be treated as two independent motions.

Solve numerical problems using the above method for an object projected horizontally.

Carry out calculations using kinematic equations.

10.2 Revision: Dynamics

Section	Content covered

2.1: Scalars and vectors	*You should now be able to;*
	    Describe what is meant by vector and scalar quantities.
	    State that force is a vector quantity.
	    Define and classify vector and scalar quantities.
	    State what is meant by the resultant of a number of forces.
	    Use a scale drawing or otherwise to find the magnitude and direction of the resultant of two forces.
	    Carry out calculations to find the rectangular components of a vector.

2.2: Forces	*You should now be able to;*
	    Describe the effect of forces in terms of their ability to change shape, speed or direction of travel.
	    Explain the movement of objects in terms of Newton's First law.
	    State that forces which are equal in size and opposite in direction are called balanced forces and are equivalent to no force at all.
	    State that the force of friction opposes the motion of a body.

2.3: Force, mass and acceleration	*You should now be able to;*
	    Describe the qualitative effect of change in mass or force on the acceleration of an object.
	    Define the newton.
	    Describe the use of a Newton balance to measure force.

2.4: Identifying unbalanced forces	*You should now be able to;*
	    Describe and explain situations where attempts are made to increase or decrease the force of friction.
	    Use free body diagrams to analyse the forces on an object.
	    Carry out calculations using the relationship $F = ma$ in situations where resolution of forces is not required.

2.6: Newton's Third Law	*You should now be able to;*
	    State Newton's Third Law.
	    Identify Newton pairs of forces in situations involving several forces.

2.7: Impulse and momentum	*You should now be able to;*
	    State that momentum is the product of mass and velocity.
	    State that momentum is a vector quantity.
	    State that Impulse = Force $\times$ Time.
	    State that Impulse = Change in momentum.
	    Carry out calculations using the relationship Impulse = Change in momentum.

2.8: Conservation of momentum	*You should now be able to;*
	    State that the law of conservation of momentum can be applied to interactions of two objects moving in one direction in the absence of net external forces.
	    Carry out calculations concerned with collisions in which the objects move in only one dimension.
	    Carry out calculations concerned with explosions in one dimension.
	    Apply the law of conservation of momentum to interactions of two objects moving in one dimension to show that:
	    a) Changes in momentum are equal in size and opposite in direction.
	    b) The forces acting on the object are equal in size and opposite in direction.

2.11: Energy, work and power

You should now be able to;

State that work done is a measure of the energy transferred.

Carry out calculations involving the relationship between work done, force and distance.

Carry out calculations involving the relationship between K.E., mass and velocity.

State that an elastic collision is one where momentum and K.E. are conserved.

State that an inelastic collision is one in which momentum is conserved but K.E. is not.

State the relationship between work done, power and time.

Carry out calculations using the above relationship.

Carry out calculations involving the relationship between efficiency, output power (output energy), and input power (input energy).

Carry out calculations involving work done, potential energy, kinetic energy and power.

2.12: Weight and mass

You should now be able to;

Distinguish between mass and weight.

State that weight is a force and indicates the earth's pull on an object.

State that weight per unit mass is called gravitational field strength.

Carry out calculations involving the relationship between weight, mass and gravitational field strength including situations where g is not equal to 10 N/kg.

Carry out calculations involving the relationship between change in gravitational PE, mass, gravitational field strength and change in height.

Explain the equivalence of the acceleration due to gravity and gravitational field strength.

10.3 Revision: Properties of matter

Section	Content covered
3.1: Density	*You should now be able to;* State that density is mass per unit volume. Carry out calculations involving density, mass and volume. Describe the principles of a method for measuring the density of air. State and explain the relative magnitudes of the densities of solids, liquids and gases.
3.2: Pressure	*You should now be able to;* State that pressure is the force per unit area when the force acts normal to the surface. State that one pascal is one newton per square metre. Carry out calculations involving pressure, force and area.
3.3: Pressure in liquids	*You should now be able to;* State that the pressure at a point in a fluid at rest is given by $h\rho g$. Carry out calculations involving pressure, density and depth.
3.4: Buoyancy and upthrust	*You should now be able to;* Explain buoyancy force in terms of the pressure difference between the top and bottom of an object.
3.7: Pressure in gases	*You should now be able to;* State that the pressure of a fixed mass of gas at constant temperature is inversely proportional to its volume.
3.8: Charles' Law	*You should now be able to;* State that the volume of a fixed mass of gas at constant pressure is directly proportional to its temperature measured in kelvin (K).
3.9: Pressure-Temperature variation	*You should now be able to;* State that the pressure of a fixed mass of gas at constant volume is directly proportional to its temperature measured in kelvin (K).
3.10: Absolute temperature and kinetic theory	*You should now be able to;* Describe how the kinetic model accounts for the pressure of a gas. Explain what is meant by absolute zero of temperature. Carry out calculations to convert temperatures in °C to K and vice versa. Explain the pressure-volume, pressure-temperature and volume-temperature laws qualitatively in terms of a kinetic model. Carry out calculations involving pressure, volume and temperature of a fixed mass of gas using the general gas equation.

10.4 Revision: Electricity and electronics

Section	Content covered
4.1: Static and moving charges	*You should now be able to;* State that electrons are free to move in a conductor. Describe the electrical current in terms of the movement of charges around a circuit.
4.2: Electric fields and currents	*You should now be able to;* State that, in an electric field a charge experiences a force. State that an electric field applied to a conductor causes the free electric charges in it to move.
4.3: Current, voltage and p.d.	*You should now be able to;* Carry out calculations involving $Q = It$. State that when there is an electrical current in a wire, there is energy transformation. State that the voltage of a supply is a measure of the energy given to the charges in a circuit. State that, when charge Q is moved in an electric field, work W is done. State that the potential difference between two points is the work done in moving one coulomb of charge between the two points. State that if one joule of work is done moving one coulomb of charge between two points, the potential difference between the two points is one volt. State the relationship $V = W/Q$. Carry out calculations involving the above relationship.
4.4: Resistance	*You should now be able to;* Distinguish between conductors and insulators and give examples of each. Draw and identify the circuit symbols for an ammeter, voltmeter, battery, variable resistor, switch and lamp. Draw circuits to identify the position of an ammeter and a voltmeter in a circuit. State that an increase in the resistance of a circuit leads to a decrease in the current in that circuit. State that V/I for a resistor remains approximately constant for different currents. Carry out calculations involving the relationship $V = IR$.
4.6: Series and parallel circuits	*You should now be able to;* State that in a series circuit the current is the same in all positions. State that the sum of the potential differences across the components in series is equal to the voltage of the supply. Derive the expression for the total resistance of any number of resistors in series, by consideration of the conservation of energy. State that the sum of the currents in parallel branches is equal to the current drawn from the supply. State that the potential differences across components in a parallel circuit is the same for each component. Derive the expression for the total resistance of any number of resistors in parallel by consideration of the conservation of charge. Carry out calculations involving the relationships; $R_T = R_1 + R_2 + R_3$ and $1/R_T = 1/R_1 + 1/R_2 + 1/R_3$.
4.7: Voltage dividers	*You should now be able to;* State that a potential divider circuit consists of a number of resistors or variable resistors connected across a supply. Carry out calculations involving voltages and resistances in a voltage divider.
4.8: Resistors and voltage dividers	*You should now be able to;* State that the resistance of a thermistor usually decreases with increasing temperature and the resistance of an LDR decreases with increasing light intensity. Carry out calculations for using $V=IR$ for the thermistor and LDR.

4.9: Wheatstone bridges

You should now be able to;
State the relationship among the resistors in a balanced Wheatstone bridge.
Carry out calculations involving the resistances in a balanced Wheatstone bridge.
State that for an initially balanced Wheatstone bridge, as the value of one resistor is changed by a small amount, the out-of-balance p.d. is proportional to the change in resistance.
Use the following term correctly in context: bridge circuit.

4.12: Electrical energy and power

You should now be able to;
State that in a lamp electrical energy is transformed into heat and light.
State that the energy transformation in an electrical heater occurs in the resistance wire.
State that the electrical energy transformed each second = VI.
State the relationship between energy and power.
Carry out calculations using $P = IV$ and $E = Pt$.
Explain the equivalence between VI, I^2R and V^2/R.
Carry out calculations involving the relationships between power, current, voltage and resistance.
Draw and identify the circuit symbol for a fuse.

4.13: Alternating currents

You should now be able to;
Explain in terms of current the terms d.c. and a.c.
State that the frequency of the mains supply is 50 Hz.
Describe how to measure frequency using an oscilloscope.
State that a d.c. supply and an a.c. supply of the same quoted value will supply the same power to a given resistor.
State the relationship between peak and r.m.s. values for a sinusoidally varying voltage and current.
Carry out calculations involving peak and r.m.s. values of voltage and current.

4.15: E.m.f. and internal resistance

You should now be able to;
State that the e.m.f. of a source is the electrical potential energy supplied to each coulomb of charge which passes through the source.
Use the following terms correctly in context; terminal p.d., load resistor, lost volts, short circuit current.
Explain how the conservation of energy leads to the sum of the e.m.f.s round a closed circuit being equal to the sum of the p.d.s round the circuit.
State that an electrical source is equivalent to a source of e.m.f. with a resistor in series (internal resistance).

4.16: Measuring internal resistance

You should now be able to;
Describe the principles of a method for measuring the e.m.f. and internal resistance of a source.
Explain why the e.m.f. of a source is equal to the open circuit p.d. across the terminals of the source.

10.5 Revision: Electronics

Section	Content covered
5.1: Preparing for Analogue Electronics	*You should now be able to;* Give examples of output devices and the energy conversions involved. Describe the energy transformations involved in the following devices; microphone, thermocouple, solar cell. Draw and identify the symbol for a diode and an LED. State that an LED will only light if connected one way round. Describe by means of a diagram a circuit which will allow an LED to light. Calculate the value of the series resistor for an LED and explain the need for this resistor.
5.2: Transistors and Feedback	*You should now be able to;* Draw and identify the circuit symbol for an NPN transistor. Draw and identify the circuit symbol for an n-channel MOSFET. State that a transistor can be used as a switch. Explain the operation of a simple transistor switching circuit.
5.3: Amplifiers	*You should now be able to;* Identify, from a list, devices in which amplifiers play an important part. State that the output signal of an audio amplifier has the same frequency as, but a larger amplitude than, the input signal. Carry out calculations involving input voltage, output voltage and voltage gain of an amplifier.
5.4: Operational amplifiers	*You should now be able to;* State than an op-amp can be used to increase the voltage of a signal. State that for the ideal op-amp in the inverting mode: a) Input current is zero, i.e. it has infinite input resistance. b) There is no potential difference between the inverting and non-inverting inputs, i.e. both input pins are at the same potential.
5.5: Inverting amplifier	*You should now be able to;* Identify circuits where the op-amp is being used in the inverting mode. State that an op-amp connected in the inverting mode will invert the input signal. State the inverting mode gain expression $V_o/V_i = -R_f/R_i$. Carry out calculations using the above gain expression.
5.6: Using inverting amplifiers	*You should now be able to;* State that an op-amp cannot produce an output voltage greater than the positive supply voltage or less than the negative supply voltage. Multiplication, division, addition and D to A converter, square wave generation.
5.9: Difference amplifiers	*You should now be able to;* Identify circuits where the op-amp is being used in the differential mode. State that a differential amplifier amplifies the potential difference between its two inputs. State the differential mode gain expression $V = (V_2-V_1)R_f/R_i$. Carry out calculations using the above gain expression. Describe how to use the differential amplifier with resistive sensors connected in a Wheatstone bridge arrangement.
5.10: Op-amp drive circuits	*You should now be able to;* Describe how an op-amp can be used to control external devices via a transistor.

10.6 Revision: Storing charge and delaying signals

Section	Content covered
6.1: Storing charge	*You should now be able to;* State that the charge Q on two parallel conducting plates and the p.d. V between them are proportional to each other. Describe the principles of a method to show that the p.d. across a capacitor is proportional to the charge on the plates. State that capacitance C is the ratio of charge to p.d. State that the unit of capacitance is the farad and that one farad is one coulomb per volt. Carry out calculations using $C = Q/V$. Describe and explain the possible functions of a capacitor – storing charge.
6.2: Capacitors storing energy	*You should now be able to;* Explain why work must be done to charge a capacitor. State that the work done to charge a capacitor is given by the area under the graph of charge against p.d. State that the energy stored in a capacitor is given by ½ QV (and equivalent expressions). Carry out calculations using ½ QV or equivalent expressions. Describe and explain the possible functions of a capacitor – storing energy
6.4: Charging capacitors	*You should now be able to;* Draw qualitative graphs of current against time and of voltage against time for the charging of a capacitor in a d.c. circuit containing a resistor and capacitor in series. Carry out calculations involving voltage and current in CR circuits (calculus methods are not required).
6.5: Discharging capacitors	*You should now be able to;* Draw qualitative graphs of current against time and of voltage against time for the discharging of a capacitor in a d.c. circuit containing a resistor and capacitor in series. Carry out calculations involving voltage and current in CR circuits (calculus methods are not required).
6.6: Capacitors and a.c.	*You should now be able to;* State the relationship between current and frequency in a capacitive circuit. State the relationship between current and frequency in a resistive circuit. Describe the principles of a method to show how the current varies with frequency in a capacitive circuit. Describe and explain the possible functions of a capacitor – blocking d.c. while passing a.c.

10.7 Revision: Waves and light

Section	Content covered
7.1: Waves revisited	*You should now be able to;* State that a wave transfers energy. Describe a method of measuring the speed of sound in air, using the relationship between distance, time and speed. Use the following terms correctly in context; wave, frequency, wavelength, speed, amplitude, period. State the difference between a transverse and longitudinal wave and give examples of each. State that the frequency of a wave is the same as the frequency of the source producing it. State that period equals l/frequency. State that the energy of a wave depends on its amplitude.
7.2: Electromagnetic spectrum	*You should now be able to;* State that radio and television signals are transmitted through air at 300 million $m\,s^{-1}$ and that light is also transmitted at this speed. State in order of wavelength the members of the electromagnetic spectrum: gamma rays, X-rays, ultraviolet, visible light, infrared, microwaves, TV and radio. State approximate values for the wavelengths of red, green and blue light
7.3: Handling large and small numbers	*You should now be able to;* Use the prefixes p, n, μ, m, k, M, G. Use scientific notation.
7.5: Reflection	*You should now be able to;* State that light can be reflected. Use correctly in context the terms: angle of incidence, angle of reflection and normal when a ray of light is reflected from a plane mirror. State the principle of reversibility of a ray path. Explain the action of curved reflectors on certain received signals. Explain the action of curved reflectors on certain transmitted signals. Describe an application of curved reflectors used in telecommunication.
7.6: Refraction	*You should now be able to;* State what is meant by the refraction of light. Draw diagrams to show the change in direction as light passes from air to glass and glass to air. Use the following terms correctly in context; angle of incidence, angle of reflection and normal. Describe the shapes of converging (convex) and diverging (concave) lenses.
7.7: Refraction equations	*You should now be able to;* State the ratio $\sin\theta_1/\sin\theta_2$ is a constant when light passes obliquely from medium 1 to medium 2. State that the absolute refractive index, n, of a medium is the ratio $\sin\theta_1/\sin\theta_2$ where θ_1 is in a vacuum (or air as an approximation) and θ_2 is in the medium. Describe the principles of a method for measuring the absolute refractive index of glass for monochromatic light. State that the refractive index depends on the frequency of the incident light. State that the frequency of a wave is unaltered by a change in medium. State the relationships $$\sin\theta_1/\sin\theta_2 = \lambda_1/\lambda_2 = v_1/v_2$$ for refraction of a wave from medium 1 to medium 2.

7.10: Total internal reflection	*You should now be able to;* Explain what is meant by total internal reflection. Explain what is meant by critical angle θ_c. Describe the principles of a method for measuring a critical angle. Derive the relationship $\sin \theta_c = 1/n$ where θ_c is the critical angle for a medium of absolute refractive index n.
7.11: Fibre optics	*You should now be able to;* *Describe the principle of operation of an optical fibre transmission system.*
7.12: Interference	*You should now be able to;* Use correctly in context the terms; in phase, out of phase and coherent, when applied to waves. Explain the meaning of; constructive interference and destructive interference, in terms of superposition of waves.
7.13: Interference equations	*You should now be able to;* State that reflection, refraction, diffraction and interference are characteristic behaviours of all types of waves. State that interference is the test for a wave. State the conditions for maxima and minima in an interference pattern formed by two coherent sources in the form: *path difference = nλ (for maxima)* *path difference = (n+1/2)λ (for minima)* where *n* is an integer Carry out calculations using these relationships.
7.14: Interference experiments	*You should now be able to;* Demonstrate the interference pattern produced by two coherent sources. Investigate the effect of path difference on the intensity of radiation in the interference pattern produced by two coherent sources of microwaves.
7.17: Diffraction	*You should now be able to;* Describe the effect of a grating on a monochromatic light beam. Carry out calculations using the grating equation $d \sin \theta = n\lambda$
7.18: The wavelength of laser light	*You should now be able to;* Describe the principles of a method for measuring the wavelength of a monochromatic light source, using a grating. State that measurement of any physical quantity is liable to uncertainty. State that the scale reading uncertainty is a measurement of how well an instrument scale can be read. Estimate the scale reading uncertainty when using an analogue display or a digital display. Express uncertainty in absolute or percentage form. Identify in an experiment where more than one physical quantity has been measured, the quantity with largest percentage uncertainty. State that this percentage uncertainty is often a good estimate of the percentage uncertainty in the final numerical result of the experiment. Express the numerical results of an experiment in the form: final value ± uncertainty.
7.19: Gratings, prisms and white light	*You should now be able to;* Describe and compare the white light spectra produced by a grating and a prism. State approximate values for the wavelengths of red, green and blue light.

10.8 Revision: Opto-electronics and semiconductors

Section	Content covered
8.1: Irradiance	*You should now be able to;* State that the irradiance I at a surface on which radiation is incident is the power per unit surface area. Describe the principles of a method for showing that the irradiance I is inversely proportional to the square of the distance d from a point source. Carry out calculations involving the relationship $I = k / d^2$ Distinguish between random uncertainties and systematic effects. Explain why repeated readings of a physical quantity are desirable. State that the mean is the best estimate of a 'true' value for the quantity being measured. State that where a systematic effect is present the mean value of the measurements is offset from a 'true' value of the quantity being measured.
8.2: Photoelectric effect	*You should now be able to;* State that photoelectric emission from a surface occurs only if the frequency of the incident radiation is greater than some threshold frequency f_o which depends on the nature of the surface. State that for frequencies smaller than the threshold value, an increase in the irradiance of the radiation at the surface will not cause photoelectric emission. State that for frequencies greater than the threshold value, the photoelectric current produced by monochromatic radiation is proportional to the irradiance of the radiation at the surface.
8.3: Planck and Einstein	*You should now be able to;* State that a beam of radiation can be regarded as a stream of individual energy bundles called photons each having an energy $E = hf$, where h is Plank's constant and f is the frequency of the radiation. Carry out calculations involving the relationship $E = hf$. Explain that if N photons per second are incident per unit area on a surface the irradiance at the surface is $I = Nhf$. State that photoelectrons are ejected with a maximum kinetic energy (E_k) which is given by the difference between the energy of the incident photon (hf) and the work function (hf_o) of the surface: $\quad E_k = hf - hf_o$.
8.5: Electrons and emission spectra	*You should now be able to;* State that electrons in a free atom occupy discrete energy levels. Draw a diagram which represents qualitatively the energy levels of a hydrogen atom. Use the following terms correctly in context: ground state, excited state, ionisation level. State that an emission line in a spectrum occurs when an electron makes a transition between an excited energy level W_2 and a lower level W_1, where $W_2 - W_1 = hf$.
8.6: Absorption spectra	*You should now be able to;* State that an absorption line in a spectrum occurs when an electron in energy level W_1 absorbs radiation of energy hf and is excited to energy level W_2, where $W_2 = W_1 + hf$. Explain the occurrence of absorption lines in the spectrum of sunlight.

8.9: Lasers

You should now be able to;

State that spontaneous emission of radiation is a random process analogous to the radioactive decay of a nucleus.

State that when radiation of energy hf is incident on an excited atom the atom may be stimulated to emit its excess energy hf.

State that in stimulated emission the incident radiation and the emitted radiation are in phase and travel in the same direction.

State that the conditions in a laser are such that a light beam gains more energy by stimulated emission than it loses by absorption – Light Amplification by the Stimulated Emission of Radiation.

Explain the function of the mirrors in a laser.

Explain why a beam of laser light having a power even as low as 0.1 mW may cause eye damage.

8.10: Conductors and semiconductors

You should now be able to;

State that materials can be divided into three broad categories according to their electrical properties – conductors, insulators and semiconductors.

Give examples of conductors, insulators and semiconductors.

State that the addition of impurity atoms to a pure semiconductor (a process called doping) decreases its resistance.

Explain how doping can form an n-type semiconductor in which the majority of the charge carriers are negative, or a p-type semiconductor in which the majority of the charge carriers are positive.

8.11: p-n junction

You should now be able to;

Describe the movement of the charge carriers in a forward/reverse biased p-n junction diode.

State that in the junction region of a forward biased p-n junction diode, positive and negative charge carriers may recombine to give quanta of radiation.

8.12: Light on p-n junctions

You should now be able to;

State that a photodiode is a solid-state device in which positive and negative charges are produced by the action of light on a p-n junction.

State that in the photovoltaic mode, a photodiode may be used to supply power to a load.

State that in the photoconductive mode, a photodiode may be used as a light sensor.

State that the leakage current of a reverse biased photodiode is directly proportional to the irradiance of the light and fairly independent of the reverse biasing voltage (below the breakdown voltage).

State that the switching action of a reverse biased photodiode is extremely fast.

8.13: MOSFETs

You should now be able to;

Describe the structure of an n-channel MOSFET using the terms; gate, source, drain substrate, channel implant and oxide layer.

Explain the electrical ON and OFF states of an n-channel MOSFET.

State that an n-channel MOSFET can be used as an amplifier.

10.9 Revision: Nuclear physics and its effects

Section	Content covered
9.1: Atomic theory	*You should now be able to;* Describe a simple model of the atom that includes protons, neutrons and electrons. Describe how Rutherford showed that; a) the nucleus has a relatively small diameter compared with that of the atom. b) most of the mass of the atom is concentrated in the nucleus.
9.2: The nuclear model	*You should now be able to;* *Describe a simple model of the atom which includes protons, neutrons and electrons.* Explain what is meant by alpha, beta and gamma decay of radionuclides. Identify the processes occurring in nuclear reactions written in symbolic form.
9.3: Nuclear reactions	*You should now be able to;* *Describe in simple terms the process of fission.* State that in fission a nucleus of large mass number splits into two nuclei of smaller mass numbers, usually along with several neutrons. State that fission may be spontaneous or induced by neutron bombardment. State that in fusion two nuclei combine to form a nucleus of larger mass number. Explain, using $E = mc^2$, how the products of fission and fusion acquire large amounts of kinetic energy. Carry out calculations using $E = mc^2$ for fission and fusion reactions.
9.4: Harnessing nuclear power	*You should now be able to;* *Explain in simple terms a chain reaction.* *Describe the principles of the operation of a nuclear reactor in the following terms; fuel rods, moderator, control rods, coolant, containment vessel.*
9.6: Activity and half-life	*You should now be able to;* State that the average activity A of a quantity of radioactive substance is N/t where N is the number of nuclei decaying in the time t. *State that the activity of a radioactive source is measured in becquerels, where one becquerel is one decay per second.* State that one becquerel is one decay per second. Carry out calculations involving the relationship $A = N/t$. *State that the activity of a radioactive source decreases with time.* *State the meaning of the term half-life.* *Describe the principles of a method for measuring the half-life of a radioactive source.* *Carry out calculations to find the half life of a radioactive element from appropriate data.*
9.7: The effect of radiation	*You should now be able to;* *Explain the term ionisation.* *State that alpha particles produce much greater ionisation density than beta particles or gamma rays.* *Describe how one of the effects of radiation is used in a detector for radiation.* *State that radiation can kill living cells or change the nature of living cells.* *Describe one medical use of radiation based on the fact that radiation can destroy cells.* *Describe one use of radiation based on the fact that radiation is easy to detect.*

9.8: Radiation and dose

You should be able to;

State that the absorbed dose D is the energy absorbed per unit mass of the absorbing material.

State that the gray (Gy) is the unit of absorbed dose and that one gray is one joule per kilogram.

State that the risk of biological harm from an exposure to radiation depends on:
a) the absorbed dose,
b) the kind of radiations
c) the body organs or tissues exposed.

State that a radiation weighting factor Q is given to each kind of radiation as a measure of its biological effect.

State that the equivalent dose H is the product of D and W_R and is measured in sieverts (Sv).

Carry out the calculations involving the relationship $H = DW_R$.

State that equivalent dose rate $= H/t$.

State that the effective equivalent dose takes account of the different susceptibilities to harm of the tissues being irradiated and is used to indicate the risk to health from exposure to ionising radiation.

9.9: Background radiation and dose limits

You should now be able to;

Describe the factors affecting the background radiation level.

State that the average annual effective equivalent dose which a person in the UK receives due to natural sources (cosmic, terrestrial and internal radiation) is approximately 2 mSv.

State that the annual effective equivalent dose limits have been set for exposure to radiation for the general public, and higher limits for workers in certain occupations.

9.10: Shielding and protection

You should now be able to;

State that radiation energy may be absorbed in the medium through which it passes.

State the range through air and absorption of alpha, beta and gamma radiation.

Sketch a graph to show how the intensity of a beam of gamma radiation varies with the thickness of an absorber.

Describe the principles of a method for measuring the half-value thickness of an absorber.

Carry out calculations involving half-value thickness.

9.11: Using radiation safely

You should now be able to;

Describe the safety procedures necessary when handling radioactive substances.

State that the equivalent dose is reduced by shielding, limiting the time of exposure or by increasing the distance from a source.

State that the equivalent dose rate is reduced by shielding or by increasing the distance from a source.

Identify the radioactive hazard sign and state where it should be displayed.

9.12: The nuclear debate

You should now be able to;

State the advantages and disadvantages of using nuclear power for the generation of electricity.

Describe the problems associated with the disposal and storage of radioactive waste.

10.10 Avoiding common mistakes

Introduction

Higher Physics is a very precise subject. Not only do you have to be careful with the way that you manipulate numbers but you also have to use a specialised vocabulary. This is not always easy since words such as 'weight' have a different meaning in physics from the way that they are used in day to day speech. Throughout this book we have sought to explain the scientific meaning of specialist words. If you meet a scientific term that you don't recognise you should use the index to find the place in the text where it is explained.

Mathematical manipulations

In section 7.3 we looked at some of the difficulties in handling large and small numbers. Using powers of 10 correctly is very important. We can find the frequency of an electromagnetic wave whose wavelength is 500 nm using;

$$f = \frac{c}{\lambda} = \frac{3 \times 10^8}{500 \times 10^{-9}}$$

Entering these numbers into a calculator properly will show that the waves have a frequency of 6×10^{14} Hz. In the rush of the exam some candidates write this down as 6×10^{-14} Hz or even 6^{14} Hz. Both of these answers are incorrect even though the number '6' shows that the student used the formula correctly. In this answer we quoted the frequency as 6×10^{14} Hz. In doing so we actually mean that it is 6.000 (etc) $\times 10^{14}$ Hz. We are able to shorten this to 6×10^{14} Hz because the numbers in the equation give a whole number as the answer.

If we were asked to calculate the current in a 470 ohm resistor when it is connected to a 6 V battery we would need to use Ohm's law. In this case the answer is not a simple number. The calculator display will show that the answer is 0.012765957 A. It is bad practice to copy all of the digits from the calculator display. However it is quite acceptable to round this number to a more meaningful 0.0128 A. Some students get confused as to how many numbers to quote in the answer. In this question the quoted resistance has two digits which are not zero, (the 4 and the 7). Consequently you should round your answer so as to give two or at the most three non zero digits. We must remember to round numbers accurately. Quoting the current as 0.012 A would not gain full marks.

Tackling problems

Students often have difficulty in deciding how to start a problem. In many questions we are expected to read through some words and may then be required to find other relevant data in graph or in tables. One particularly good way to start mechanics and electricity questions is to draw a diagram. If asked to find the time taken for an object projected upwards with a velocity of 19.6 m s^{-1} to reach its maximum height we could draw a diagram like figure 10.10.1.

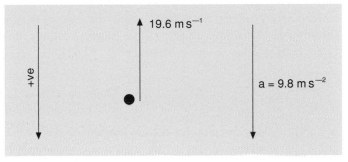

Figure 10.10.1

On this diagram we have added an arrow to indicate the acceleration due to gravity. This has a value of 9.8 m s^{-2} and acts downwards. Since the object is initially projected upwards we need to be careful with signs when substituting numbers into equations. Adding the arrow to the diagram to indicate which direction we have chosen as positive for vector quantities helps us use + and – signs consistently.

It is also important to use signs consistently when considering the forces acting on an object. In figure 10.10.2 a submerged 9.8 N weight experiences an upthrust of 3 N. Adding the vector positive arrow to the diagram shows that the unbalanced force of 6.8 N acts downwards. If released, this weight will accelerate towards the bottom of the water container.

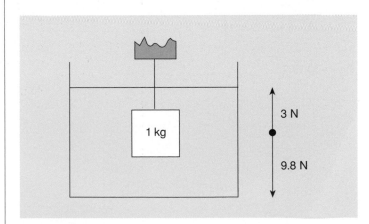

Figure 10.10.2

Wrong physics

There are a number of areas where students fail to score marks in exams because they have not used their knowledge of physics correctly. A student who quotes the second equation of motion as $s = ut + at^2$ is almost but not totally correct. No credit is given for equations remembered incorrectly so you must learn equations thoroughly.

Students calculating the combined value of two 47 kilohm resistors joined in parallel often recall the formula accurately.

$$\frac{1}{R_T} = \frac{1}{R_1} + \frac{1}{R_2}$$

Unfortunately mistakes occur when the numbers are substituted. Frequently students in a hurry continue this example by writing;

$$R_T = \frac{1}{47} + \frac{1}{47}$$

A student making this error has made a mathematical error that gives a meaningless answer. At the end of a calculation it is always good practice to think if the answer is in line with what you would expect. If it seems odd you should recheck for errors in the manipulation of data.

The energy gained by an electron when it moves from plate X to plate Y in figure 10.10.3 is calculated using the formula, $W = QV$, where Q is the charge on the electron and V is the constant potential difference maintained by the power supply connected to the plates.

As the capacitor shown in figure 10.10.4 charges the p.d across its plates increases. The energy, W, needed to transfer Q coulombs of charge to a capacitor so that a p.d. of V volts is created across its plates is calculated from; $W = \frac{1}{2} QV$. This equation and the earlier one connect similar terms but it is important that you consider the context of the question before you decide which equation to use.

Preparing for exams

The best way to be sure of minimising the number of mistakes made in a test or exam is to know the course content well and do lots of practice questions. In this book sections 10.1 – 10.9 provide a checklist of what you are expected to know. The consolidation and exam style questions will give you an idea of what to expect in the exam. The book index should help you find more details of specific content areas where you are uncertain and the formula index is a summary of the important formula.

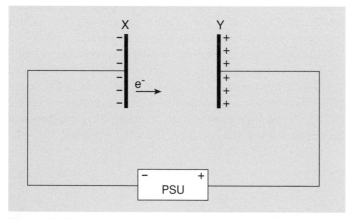

Figure 10.10.3

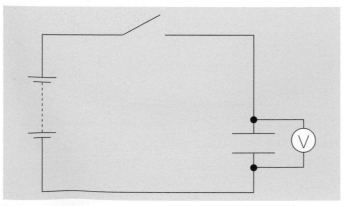

Figure 10.10.4

10.11 Formula index

Formula	Page number
Average speed = $\dfrac{\text{Total distance travelled}}{\text{Total journey time}}$	1
Average speed = $\dfrac{u+v}{2}$	4
Speed at light gate = $\dfrac{\text{Length of Card}}{\text{Time that light gate is interrupted}}$	5
Average Velocity = $\dfrac{\text{Displacement}}{\text{Total time for Journey}}$	10
Acceleration = $\dfrac{v-u}{t}$	12
Uncertainty = $\dfrac{\text{Range of readings}}{\text{Number of readings}}$	13, 207
Gradient = $\dfrac{\text{Change in } y \text{ - axis variable}}{\text{Change in } x \text{ - axis variable}}$	14
$v = u + at$	26
$s = ut + \dfrac{1}{2}at^2$	26
$v^2 = u^2 + 2as$	27
$F_H = F\cos\theta$	39
$F_v = F\sin\theta$	39
$F = ma$	43
Momentum = Mass × Velocity	38, 52
$Ft = mv - mu$	52
$m_1 u_1 + m_2 u_2 = m_1 v_1 + m_2 v_2$	54
K.E. = $\dfrac{1}{2}mv^2$	38, 60
Work done = $F \times s$	60
$F \times s = \dfrac{1}{2}mv^2 - \dfrac{1}{2}mu^2$	60
Power = $\dfrac{\text{Energy transferred}}{\text{Time}} = \dfrac{\text{Work done}}{\text{Time}}$	61

Formula	Page number
% Efficiency = $\dfrac{\text{Useful output energy}}{\text{Total energy input}} \times 100$	61
Weight = mg	62
P.E. = mgh	63
Density = $\dfrac{m}{V}$	66
Pressure = $\dfrac{\text{Force}}{\text{Area}}$	68
$P = h\rho g$	69
Upthrust = $H\rho gA$	70
$P \propto \dfrac{1}{V}$	77
$V \propto T_{ABS}$	78
$P \propto T_{ABS}$	79
$T_K = T_{^\circ C} + 273$	81
$\dfrac{P_1 V_1}{T_1} = \dfrac{P_2 V_2}{T_2}$	81
$I = \dfrac{Q}{t}$	92
$V = \dfrac{W}{Q}$	93
$R = \dfrac{V}{I}$	94
$R_T = R_1 + R_2 \ldots$	99
$\dfrac{1}{R_T} = \dfrac{1}{R_1} + \dfrac{1}{R_2} \ldots \qquad R_T = \dfrac{R_1 R_2}{R_1 + R_2}$	99
$V_X = \dfrac{R_2}{R_1 + R_2} \times V_S$	104
$\dfrac{R_1}{R_2} = \dfrac{R_3}{R_4}$	104
$P = V \times I \quad P = I^2 R \quad P = \dfrac{V^2}{R}$	113

Index

D

E

F

G

Answers

Answers

Chapter 1: Kinetics
Section 1.4 (page 9)

1   E

2   D

3 a)   100 seconds
   b)   2250 m
   c)   22.5 ms^{-1}

4 a)   Speed 1 = 0.50 ms^{-1}
      Speed 2 = 0.59 ms^{-1}
   b)   0.09 ms^{-1}

5 a)   25 ms^{-1}
   b)   12.5 ms^{-1}
   c)   62.5 m
   d)   42.5 ms^{-1}

6 a)   3 m
   b)   3 m

7 a)   4 m
   b)   1.6 m to the right of starting point

8   3 km bearing 045

Section 1.10 (page 20–21)

1   A

2   E

3 a)   74.3 ms^{-1} bearing 017
   b)   22.3 km

4 a)   501 ms^{-1}
   b)   5010 m

5   A

6   car A = 3.5 ms^{-2}
     car B = 3.33 ms^{-2}

7   u = 4.5 ms^{-1}

8 a)   21.56 ms^{-1}
   b)   10.78 ms^{-1}
   c)   23.72 m
   d)   12.69 m

9   E

10 a)   5 m
   b)   6 m
   c)   8 m
   d)   No, because there is a change in direction

11   C

12 a)   Deceleration from 10 ms^{-1} to rest
   b)   −5 ms^{-2}
   c)   10 m
   d)   4 m
   e)   6 m

13 a)   Car A 50 m
      Car B 100 m
   b)   Car B
   c)   Approximately 78 m

14   B

Section 1.11 (pages 23–25)

1 a)   30 ms^{-1}
   b)   20 ms^{-1}
   c)   200 m

2 a)   8 m
   c)   68 m

3 a)   15 kmh^{-1} due North
   b)   0.5 hours
   c)   21 kmh^{-1} south
   d)   0.86 hours

4 a)   A
   b)   1.115, 1.638, 2.20, 2.74, 3.32
   d)   ±0.018, ±0.020, ±0.010, ±0.013, ±0.018

5 a)   Object A (fig 1.11.7)
   b)   36.5 m

6   D

7   B

9 a)   50 ms^{-1}
   b)   25 seconds
   c)   875 m

10   B

11 a)   10 m
   b)   6 m in the initial direction

12   D

13   E

14   C

Section 1.15 (page 33)

1 a)   40 m
   b)   2.86 s
   c)   5.7 s

2 b)   360 m
   c)   396 m

3 a)   −2.5 ms^{-2}
   b)   80 m
   d)   180 m

4 b)   80 m
   c)   4 seconds
   d)   10 ms^{-2}
   e)   40 ms^{-1}

6 a)   2 seconds
   b)   20 ms^{-1}
   c)   10 m

Section 1.16 (pages 35–37)

1 **b)** 28 m
  c) 60 m

2 **a)** 2.4 ms^{-2}
  b) 5 s
  c) 2.5 s
  d) 7.65 s

3 **a)** 12 ms-1 due east
  c)(ii) 78 m

4 **a)** -0.25 ms^{-2}
  b) 4.5 m
  c) 16 s

5 **a)** 12m, 24m, 36m, 48m
  b) 4.3 ms^{-1}, 8.6 ms^{-1}, 12.9 ms^{-1}, 17.2 ms^{-1}
  d) 2.8 s
  e) 34.4 m
  f) B in front by 13.6 m

6 **a)** 100 kmh^{1}

7 **a)** 161.5 ms^{-2}

8 **a)** 121.7 ms^{-1} on a bearing of 010
  b) 30 ms^{-1} from the south

9 **a)** 18ms^{-1}
  b) 1 ms^{-1}
  c) 18.03 ms^{-1} angle 3.2 east of north
  d) 54.09 m

10 **a)** constant acceleration
  b) 2.5 ms^{-2}
  c) 10 m
  d) 20 m

12 **b)** 50 m
  c) 325 m in original direction.

13 **b)** BC and EF
  c) 1 m
  e) 2 ms^{-2}

14 **a)** 0.7 s
  b) 1.92 s
  c) 75 m

15 **a)** 2.75 s
  b) 29 ms^{-1}
  c) 27 ms^{-1}
  d) 39.6 ms^{-1} at 43º to the horizontal

Chapter 2: Dynamics
Section 2.5 (pages 47–49)

1 **e)** Kinetic energy

2 **a)** 27 km North of start
  b) 1.125 kmh^{-1} due north

4   43.5 N on a bearing of 073

5   23.1 N

6   153 N

8 **a)** 3 N
  b) 5 N

9 **a)** approx 10 ms^{-2}

10 **a)** 750 N

11 **a)** 5 ms^{-2}
  b) 2750 N
  c)(i) -5 ms^{-2}
    (ii) -2750 N

12 **a)** 1275 N
  b) 487.5 N

  c) 1762.5 N
  d) 0.74 ms^{-2}

13 **a)** 784 N
  b) 216 N
  c) 2.7 ms^{-2}
  d) 7.7 s

14 **a)** 1080 kN

16 **c)** 800 N
  d) Approx. 0.29 ms^{-2}

17 **a)** 5000 N
  b) -0.4 ms^{-2}
  c) 0 ms^{-1}
  e) 5 m

Section 2.9 (page 57)

1 **a)** 6 kg ms^{-1} and 0 kg ms^{-1}
  b) 6 kg ms^{-1}
  c) 1.71 ms^{-1}

2 **a)** 0.22 ms^{-1} to the right
  b) 20 J

3 **a)** 2.5 kg ms^{-1}
  b) -2 kg ms^{-1}
  c) 4.5 kg ms^{-1}
  d) 4.5 Ns
  e) 45 N

4 **a)** 1.33 ms^{-1} to the right

5 **a)** 300 Ns
  b) 4 ms^{-1}
  c) 60 000 N

Section 2.10 (page 59)

1 **a)** 20 kgms^{-1}
  b) 2000 N

2 **a)** 0.67 ms^{-1}
  b) 1.67 ms^{-1}

3 **b)** 3 ms^{-1} to the right

4 **a)** 9 kg ms^{-1}
  b) -11.25 kg ms^{-1}
  c) 20.25 kg ms^{-1}
  d) 135 N

5 **a)** 6.86 ms^{-1}
  b) 1.715 kg ms^{-1}
  c) 4.85 ms^{-1} upwards
  d) 2.93 kg ms^{-1}

6 **a)** 0.4 kg ms^{-1}
  c) 0.125 ms^{-1} to the right

Section 2.13 (pages 64–65)

1 **b)** 1.84 m
  c) 1.28 m

2 **a)** 340 kg
  b) 3332 N
  c) 16 660 J
  d) 340 W

3 **e)** 1.54 kgms^{-1} up the slope

4 **a)** 120 kJ
  b) 120 kJ
  d) 2222 N

5 **b)** 490 N
  c)(i) 50 N
    (ii) 540 N

6   18 ms^{-1}

7 **a)** 550 N

8   C

9   A

Chapter 3: Properties of matter
Section 3.5 (page 73)
1 a) 0.16 m^2
  b) 0.32 m^3
  c) 800 kg
  d) 7840 N
  e) 49 kPa

2 a) 2.85 kg
  b) 5586 Pa
  c) 18620 Pa

3 a) 0.0619 m^3
  b) 4851 N
  c) 606 N

4 b) 338100 N
  c) 439100 Pa
  d) 8.78 m

5 a) $2.4 \times 10^{-4} \text{ m}^3$
  b) 4.56 kg
  c) 44.7 N
  d) 2.4 N

Section 3.6 (page 75)
1 a) $5 \times 10^{-3} \text{ m}^3$
  b) 1050 kgm^{-3}
  c) $9.9 \times 10^{-5} \text{ m}^3$

2 b) 4900 Pa
  c) 49 N

3 b) 72 kg
  c) $1.2 \times 10^{-3} \text{ kg}$
  d) 1784 m

4 a) $5 \times 10^{-4} \text{ m}^2$
  b) 19.6 N
  c) 39 200 Pa
  d) 139 200 Pa
  e) 154 880 Pa

5 a) 0.49 N
  b) $1 \times 10^{-6} \text{ m}^2$
  c) $4.9 \times 10^5 \text{ Pa}$

6 a) 63 kg
  b) 617.4 N
  c) 8232 N
  d) 777 kg

Section 3.11 (page 83)
1 a) 78 mm

2 c) 600 K

3 b) 364 K

4   302 K

5 d) 6.6 cm^3

6 a) 519.8 cm^3

7   41.5 m^3

Section 3.12 (page 85)
1   B

2   C

3 a) 300 K
  b) 650 cm^3
  c) 150 cm^3

4 a)(i) 1.08 m^3
  (ii) 0.91 m^3
  (iii) 910 mins

---

b) 295 K

5 a) 1500, 1480, 1500, 1500
  b) 125 kPa
  c)(i) $7.85 \times 10^{-5} \text{ m}^2$
  (ii) 23.2 N

Chapter 4: Basics Electricity
Section 4.5 (page 97)
2 b) 1500 V
  c) 4.5 mJ

3 a) 96 000 C
  b)(i) 360 000 C
  (ii) 3.75

4   36 %

5 a) I only
  c) $2.45 \times 10^{-5} \text{ N upward}$

6 b) $1.6 \times 10^{-14} \text{ J}$
  c) $2.22 \times 10^6 \text{ ms}^{-1}$

Section 4.10 (pages 108–109)
1 a) 0.2 A
  b) 5.5 V
  c)(i) $17.5 \ \Omega$
  (ii) $R = 45 \ \Omega, \ R_2 = 27.5 \ \Omega$

2 a) 5 V
  b) 2.27 mA
  c) 1.07 mA
  d) $1.50 \text{ k}\Omega$

3 a) $7.96 \text{ k}\Omega, \ 15.2 \text{ k}\Omega, \ 24.1 \text{ k}\Omega$
  b) $5.84 \text{ k}\Omega$

4 b) 2.7 V

5 a) 3.75 V
  b) $15 \ \Omega$

6 a) 3.5 V
  b) 5 mA
  c) 8.5 mA

7 a) 2.5 V
  b) 25 mA
  c) $140 \ \Omega$
  d)(ii) 1.31 V

8 c) $21.36 \text{ k}\Omega$
  d) $0\mu\text{A}$

Section 4.11 (page 111)
1 a) $6 \ \Omega$
  b) 2 A
  c) 1.71 A
  d) 1.33 A

2 a) 0.4 A
  b) 0.53 A, 2.0 V
  c) 0 A, 2V

4 a)(i) 50 mA
  (ii) 20 mA
  (iii) A
  b)(i) 70 mA
  (ii) $42.9 \ \Omega$
  (iii) 70 mA
  c) 6 V

Section 4.14 (page 115)
1 a) 49.5 mA
  b) 245 mW
  c) 99 mA
  d) 4.95 V

2 a) 7.1 V

b) 7.1 V

3 b) 3.4 mA

Section 4.17 (page 121)

1 E

2 a) 4 A

b) 8 V

c) 32 W

d) 12 A

4 a) 1.1 A

b) 11.9 W

c) 2 A

d) 2.0 V

e) 10 V

f) 20 W

5 a) 24Ω

b) 15 J

c)(i) 5 C

(ii) 30 J

Section 4.18 (page 123)

1 a) 2.2 V

b) 550 J

c)(i) 2 mins

(ii) 5.5 W

2 a) 1.2 V

b) 200 A

c) 0.006 Ω

3 a) 0.625 A

b) 54.7 W

c) 87.5 V

4 b) 0.44 A

5 a) 120 Ω

b) 26.7 W

Chapter 5: Electronics

Section 5.7 (page 137)

1 b) 10 V

c) 25 mV

2 Top left   Gain -0.1

Top Right   Gain -0.1

Bottom Left   Gain -2.2

3 b) -3 V to $+3$ V

4 a) -5

b) -2 V

5 a) -14

b) -175 mV

c)(i) 12.5 mV

(ii) 280

(iv) 53.6 mV

Section 5.8 (page 139)

1 a) 14.68

b) -3

2 a) -8.4

b) 7.55 V

c) 0.16 mA

d) 6.05 V

e) 37.6 kΩ

3 a) -1

4 a) 0 V

b)(i) 1.5 V and 3 V

(ii) 1.5 mA and 3 mA

c) 0.45 mA

d) 4.5 V

e) -4.5 V

f) $V_{out} = -(V_1 + V_2)$

g) $V_{out} = -2(V_1 + V_2)$

Section 5.11 (page 143)

1 a) 1.5 V

c) 3 V

d) 4.5 V

e) 4 V

3 a) 3.26 V

b)(i) 3.26 V

(ii) 8.39 kΩ

c) 0 V

Section 5.12 (page 145)

1 b) -6.7 V, -4V, -20 V, -12 V.

c) S_1 closed, S_2 open

3 b) 0.01 A

Chapter 6: Storing Charge and delaying signals

Section 6.3 (page 151)

1 a) 0.005 F

b) 6 V

c) 1300 V

2 a) 2 A

b) 0.004 F

c) 20 J

d) 100 W

e) 50 V

3 a) 3.76×10^{-4} C

b) 1.5 mJ

d) 19 %

4 b) 109 μA

c) 0.88 mC, 1.65 mC, 2.59 mC, 3.36 mC

5 a) 8 V

b) 5 mJ, 1.6 mJ

c) 5.66 V

Section 6.7 (page 159)

1 a) 3 mA

c) 1.32 mC

2 a) 6 V

4 a) 1 mA

b)(i) 4 V

(ii) 6 V

c) 0.75 mC

Section 6.8 (page 161)

1 a) 0.01 C

b) 0.025 J

c) 2 mF

e) 10 V

2 a) 2 V

4 a) 20 mA

b) 7 V

c) 1.21 mC

d) 11 mJ

e) 4.4 mJ

Chapter 7: Waves and light

Section 7.4 (page 169)

1 a) 22 cms^{-1}

b) 5.5 Hz

2 a) 1.65 s

b) 1×10^{-6} s

3 a) 15 mm
b) 1.25 Hz
c) 4 cm
d) 37.5 cms^{-1}

4 C

5 E

6 D

7 C

Section 7.8 (page 177)
3 b) 5×10^{14} Hz
c) 19.5°
d) 4×10^{-7} m

4 a) 18.4°
b) 18.3°
c) 0.10

5 a) 1.24×10^8 ms^{-1}
b) 5.5×10^{14} Hz
c) 2.25×10^{-7} m

6 C

Section 7.9 (page 179)
1 E

2 a) 21.3°
b) 5.08×10^{14} Hz
c) 5.08×10^{14} Hz
d) 3.7×10^{-7} m

3 b) 27.4°
c) 1.5

4 a) 19.3°
b) 90°

5 a) 36.1°
b) 47°
c) 2.04×10^8 ms^{-1}

6 D

Section 7.15 (page 189)
2 D

3 a) 15 cm
b) 330 ms^{-1}

4 a) 1400 nm

5 B

Section 7.16 (pages 191–193)
1 a) 19.3°
b) 1.33

2 a) 24.4°
b) 1.24×10^8 ms^{-1}
c) 243 nm

3 a) 1.46
b) 43°

4 a) 1.33
b) 48.6°

5 a) 26.6°
b) 48.7°

6 C

7 D

9 C

10 a) 50 cm
b) 5

11 D

Section 7.20 (page 201)
1 E

2 D

3 a) 3
b) 45°

4 E

5 a) 39.8°
b)(i) 512 nm
(ii) Green

Section 7.21 (pages 204–205)
1 a) 500 nm
b) 3×10^8 ms^{-1}
c) 6×10^{14} Hz
d) 333 nm

3 b) 1.46
c) 29°

4 a) 1×10^{10} Hz
b) 2.12×10^8 ms^{-1}
c) 2.12 cm

5 a) 16.7°
b) 35°

6 a) 19.24° for red, 18.97 for blue
b) 0.27°
d) 1.98×10^8 ms^{-1}
e) 3×10^6 ms^{-1}

8 a) 2.75

9 a) 508 nm
b) Green
c) 61°

10 b) 1.5
c) 18.2 cm

11 a) 41.3°
c) 7

13 a) 16.7°
b) 639 nm

Chapter 8: Opto-electronics and semiconductors
Section 8.4 (page 213)
1 c) 1 m^2

3 b) 2.6×10^{23} m^2
c) 3.54×10^{26} W

5 a) 4.64×10^{-19} J
b) 6.63×10^{-19} J
c) Y
d) 1.13×10^{-19} J

6 a) 7.5×10^{14} Hz
b) 3×10^{-19} J
c) 4.52×10^{14} Hz

7 a) 4.49×10^{14} Hz
b) 1.27×10^{-19} J
c)(i) 1.38×10^{-5} J
(ii) 0.138 Wm^{-2}
(iii) 5.12 μA

Section 8.7 (page 219)
2 b) 1×10^{-19} J, 2×10^{-19} J, 3×10^{-19} J, 5×10^{-19} J
c) 4.0×10^{-19} J
d) 7.54×10^{14} Hz

3 c) 3.38×10^{-19} J

4 b) 3.38×10^{-19} J

c) 3.38×10^{-19} J

5 b)(i) P 4.97×10^{-19} J
      Q 2.13×10^{-19} J
      R 2.84×10^{-19} J

Section 8.8 (page 221)
2 a) 1.25 Wm^{-2}
b) 4.97×10^{-19} J
c) 1×10^{19} photons at P

3 a) 3.75×10^{18}

5 a) 3.315×10^{-19} J
b) 17.5°

6 c) P $\;\;63$ nm
    Q $\;\;71$ nm

Section 8.14 (page 233)
1   E

2 a) 3.14×10^{-19} J
b) 3.18×10^{15}
c) 127 kWm^{-2}

4 c)(i) $\;633$ nm
   (ii) Red

Section 8.15 (page 235)
4 a)(i) 4 V
b)(i) 3.98×10^{-19} J
   (ii) 5.03×10^{17}

Chapter 9: Nuclear physics and its effects
Section 9.5 (page 245)
1 a) α
      β

3 a) 92
c)(i) α
  (ii) β
  (iii) β
  (iv) α

4 a) Atomic No. = 84
     Mass No. = 216
b) Atomic No. = 82
     Mass No. = 212
e) Q

5 2.7×10^{-13} J

6 c) 7
d) 1.71×10^{-12} J

Section 9.13 (page 261)
1   E

2 a) 2 Gy
b) 2 Sv

3 a) Gamma
b) 7.5 mGy
c) 9.5 µSv

4 a) 100 µGy
b) 1 mJ

5 a) 4 cm
b) 16 cm

Section 9.14 (pages 263–264)
1 a) 143
c) 2 protons and 2 neutrons
d) alpha

2   a   220
       b   86
       c   4
       d   2
       e   210
       f   84
       g   0
       h -1
       i   90
       j   39
       k   222
       l   86

3 a) mass number 239
      atomic number 92

5 d) 2.7×10^{-11} J
e) 3.6×10^{-11} J

6   E

7 d) 2 mSv

8 b) 3×10^{-6} J
c) 0.12 Gy
d) 2.4 Sv

9 a) α
b) 7.73 Sv

10 b) 240 Bq

11 a) 100 Svh^{-1}
b) 25 Svh^{-1}

12 b) 7.5×10^4 Bq
c) 3.75×10^4 Bq
d) 9375 Bq